D1526219

ACCELERATOR INSTRUMENTATION

CONFERENCE PROCEEDINGS NO. **252**

PARTICLES
AND FIELDS
SERIES 46

ACCELERATOR INSTRUMENTATION

THIRD ANNUAL WORKSHOP

NEWPORT NEWS, VA 1991

EDITORS:
WALTER BARRY
& PETER KLOEPPEL
CEBAF

American Institute of Physics **New York**

L.C. Catalog Card No. 92-70356
ISBN 0-88318-934-8
DOE CONF-9110118

Printed in the United States of America.

CONTENTS

PREFACE

The third annual Workshop on Accelerator Instrumentation met at CEBAF in Newport News, Virginia, on 28–31 October 1991. Showing the same high level of interest that was seen in the previous conferences, this year 111 persons from 4 countries registered for the Workshop. Participants from both the new accelerators, such as the Superconducting Super Collider and the Advanced Photon Source, and established machines such as Fermilab and CERN, met to share ideas and experiences.

The program this year was similar to that of the previous year, with three types of formal presentations: invited talks, contributed oral presentations, and poster sessions. In addition, several manufacturers and representatives set up booths to display their wares. Finally, in sessions that are not reflected in these Proceedings, topical discussions addressed several of the common technical problems that instrumentation specialists face; unfortunately, in some cases (such as radiation damage) no solutions were produced by sharing our misery, so perhaps future conferences will not run out of material.

We are grateful for the support provided by CEBAF and its director, Hermann A. Grunder. All technical sessions took place in the auditorium and meeting rooms of CEBAF Center, which proved to be an excellent venue.

Particular thanks are extended to the Organizing Committee, listed below. Deserving special mention are the local organizers, Julia Leverenz and Cela Callaghan; without them, to put it bluntly, the conference would not have taken place. Also helping during the workshop was Tracey Stewart. The editors wish in addition to thank all the participants who submitted papers for publication; cooperation was excellent, so that a potentially unpleasant task was rendered much less burdensome than was expected.

Organizing committee:

W. Barry, CEBAF	G. Bennett, BNL
C. Bovet, CERN	G. Decker, ANL
J. Heefner, CEBAF	J. Hinkson, LBL
H. Koziol, CERN	D. Martin, SSCL
R. Pasquinelli, FNAL	M. Plum, LANL
M. Ross, SLAC	R. Rossmanith, CEBAF
R. Shafer, LANL	G. Stover, LBL
R. Webber, SSCL	R. Witkover, BNL

DEVELOPMENT OF A LOW INTENSITY CURRENT MONITOR SYSTEM*

F. R. Gallegos,
MP Division, MS H812, Los Alamos National Laboratory
Los Alamos, NM 87545

Abstract

This report documents the development of a current transformer system used to measure pulsed ion beam currents with a wide dynamic intensity range (nA to mA, a factor of 10^6). Peak beam currents at the LAMPF accelerator typically range from 100 nA to 40 mA with pulse widths varying from 30 to 1000 µs. Signal conditioning of the peak current output provides an average current readout with a range of 1 nA to 2 mA, noise of approximately ±0.5 nA, and accuracy of ±0.1%. Since the system has proved stable and highly reliable, calibration is performed yearly. The prototype unit was built in 1985 and the final production unit was completed in early 1989.

Introduction

The Los Alamos Meson Physics Facility (LAMPF) presently operates three pulsed beam injectors, generating H^+, H^-, and polarized H^- (P^-) ion beams over the energy ranges of 113 to 800 MeV. Each injector is capable of variable pulse widths (typically 800 µs) and repetition rates (maximum frequency of 120 Hz). Typical peak currents range from 30 µA for the P^- injector to 40 mA for the H^+ injector. Beam duty factors are limited to approximately 10%, but can be varied according to the needs of the experimental programs.

Accelerator operation requires a real time device that (1) can measure low intensity beams, (2) is nonintercepting, (3) has high gain stability, (4) is independent of the beam energy, and (5) has an absolute accuracy near 0.1% of full scale.

Nonintercepting current monitors are used to tune and operate the accelerator as well as to measure the beam currents delivered to

---

* Work supported by US DOE

the various experiments. The current monitors also provide protection against excessive loss by measuring transmission through various sections of the facility. This information is then supplied to the Hardware Transmission Monitor[1] (HWTM), which converts peak to average current and shuts off beam if a preset trip level is exceeded.

The High Sensitivity Current Monitor System (HSCM) is the latest in a sequence of evolving systems based on toroidal current transformers. The original installations were designed only for high-intensity beams.

The installation of the Lamb-Shift Polarized Source in 1977 provided the incentive to develop an accurate, reliable, nonintercepting current monitor that could measure both high and low intensity beams. When performing well, this source would generate 1 μA peak polarized current in the low energy transport with approximately 70% transmission to the experimental areas. In 1989 it was replaced by an Optically Pumped Polarized Ion Source (OPPIS). Peak polarized beam current in the transport is typically 15 to 30 μA. Approximately 20% reaches the experimental areas with the loss due to bunching efficiency, beam chopping and accelerator acceptance.

To use this device as a real time tuning aid, a system that accurately transmits analog signals over the length of the accelerator (0.5 mile) to the operator control room is required.

The desired response specified a current measuring system with a dynamic range of 10^6:1 and an input of 40 mA peak current providing a full scale output of 10 V. This meant that 40 nA of current provide 10 μV output from the initial stage. The major problem was minimizing the noise in the input stage.

Physical Description

There are three models of low intensity current monitors in service that are basically scaled versions of the first production design (4.0 inch bore). They fit three LAMPF standard beam pipe diameters. All further references to dimensions will pertain to the initial production unit.

The toroid is a 2-mil Supermalloy tape-wound core (O.D. = 10 inches, I.D. = 5 inches, length = 4 inches) vacuum potted in epoxy. Two hundred turns of 20-gauge copper magnet wire are wound over the core. The core is epoxy vacuum potted to provide stabilization of the windings and dimensional definition of the toroid. The potting process causes a 40 to 50% loss of permeability. This was considered when determining the correct toroid size to achieve the minimum inductance requirements.

The toroid is encased in a Mu-metal canister to provide an electromagnetic (EM) shield. To optimize shield effectiveness the length of the inner canister was determined to be twice the diameter of the beam-line opening in the Mu-metal plus the toroid height. This 14 inch cylinder is installed in another Mu-metal cylinder 18 inches long. The final shield is a steel cylinder 14 in. long and 18.5 in. in diameter, with 0.25 in. thick walls. This cylinder, which has removable end plates, serves as the outer chassis. Four insulated spring supports provide the ground isolation and mechanical decoupling from beam line support structure and the beam line. The outer chassis is suspended by the springs within a rectangular steel box structure, lined with sound-absorptive material, that functions as an anechoic chamber. See Figure 1 below.

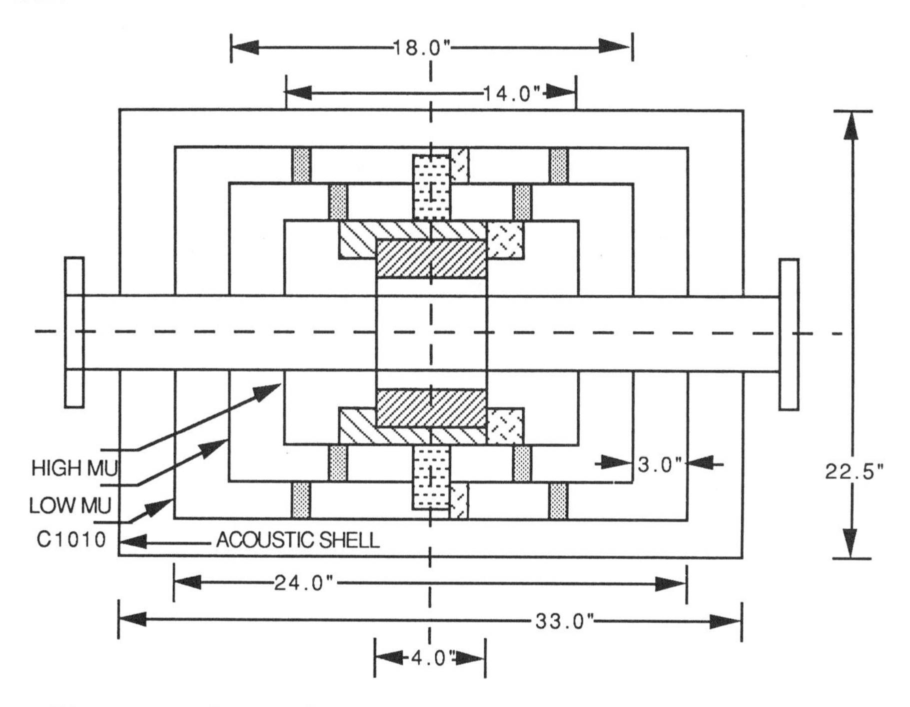

Figure 1 Cross Section of the High Sensitivity Current Monitor

A sheet of insulating material internally attached to the outer chassis provides the support structure for the amplifier card, zeroing card, and the gating circuit. There are penetrations in the cover plates for the power and signal connections. An external NIM-bin chassis houses the final stages of zeroing and signal conditioning. See figures 2 through 4.

Presently there are five current monitor units on line. A spare unit is used for electronic maintenance and development.

Design

Noise Minimization

Minimizing the noise while maximizing the signal-to-noise ratio (SNR) was the prime consideration in the development of this system. This is accomplished by:

1. selecting a toroid core material with a permeability such that a large inductance can be developed with a minimum core size and a minimum number of coil turns.

The inductance was large enough that the L/R time constant is not a factor in the design. If L/R is small in comparison to the beam pulse width, the signal will droop.

2. limiting the bandwidth of the device to approximately 10 kHz.

This frequency range is a good compromise between attenuating the high-frequency noise components and obtaining an analog output that is a true representation of the input.

3. using electronic filtering techniques and gating of the offset voltage to reduce the low-frequency noise.

4. selecting electronic components that have low voltage and current noise densities.

Earlier attempts at designing a low intensity current monitor were unsuccessful. With the development of low noise density electronic components, the project was renewed and successfully completed.

5. laying out signal, power, and chassis grounds to eliminate ground loops and to prevent power supply return currents from flowing through signal return paths.

Figure 2 HSCM 2.0 inch Bore, Acoustic Shell

Figure 3 HSCM 2.0 inch Bore, Outer Chassis

Figure 4 HSCM 6.75 inch Bore, Electronics

Signal and gating cabling is shielded and physically separated to minimize cross coupling of gating noise.

Analog and digital circuitry are placed on separate printed circuit boards with individual power supplies to minimize gating feedthrough. The power supplies are well regulated and located remote to the current monitor to minimize the 60 cycle induced pickup.

6. shielding the toroid from electromagnetic fields with concentric Mu-metal cans and a steel outer chassis.

The prototype design had a single Mu-metal shell, but this was insufficient to shield the toroid from accelerating RF cavity leakage fields (201.25 MHz). The permeability of the shielded material is graded to minimize the effects of magnetic saturation and thus increase its effectiveness to shield external fields. The outer shield is C1010 steel, the center shield is AD-MU-00 and the inner shield is AD-MU-80. The number and size of the penetrations in the shielding were minimized.

7. suppressing microphonic noise.

This is accomplished by decoupling the chassis from the beam line support structure and placing a box lined with sound absorptive material around it.

The toroid windings are also potted to minimize generating noise currents due to interaction with external magnetic fields.

Electronic Design

The HSCM system was designed using a noise voltage and current model. It consists of an ideal (noiseless) device connected to voltage and current noise sources. In the analysis the correlation coefficient between the voltage and current noise sources was assumed to be zero. Figure 5 shows the noise voltage and current model that was used to determine the components in the transimpedance (I to V) amplifier. The primary design objective was to achieve an SNR equal to 1 with peak beam current, IBEAM, equal to 20 nA.

The inductance of the toroid (L) and the current into the initial stage of amplification (I_i) is given by

$$L = k_L \times N^2 \quad \text{and}$$

$$I_i = IBEAM/N$$

where N equals the number of turns and k_L the inductance to turns-squared ratio.

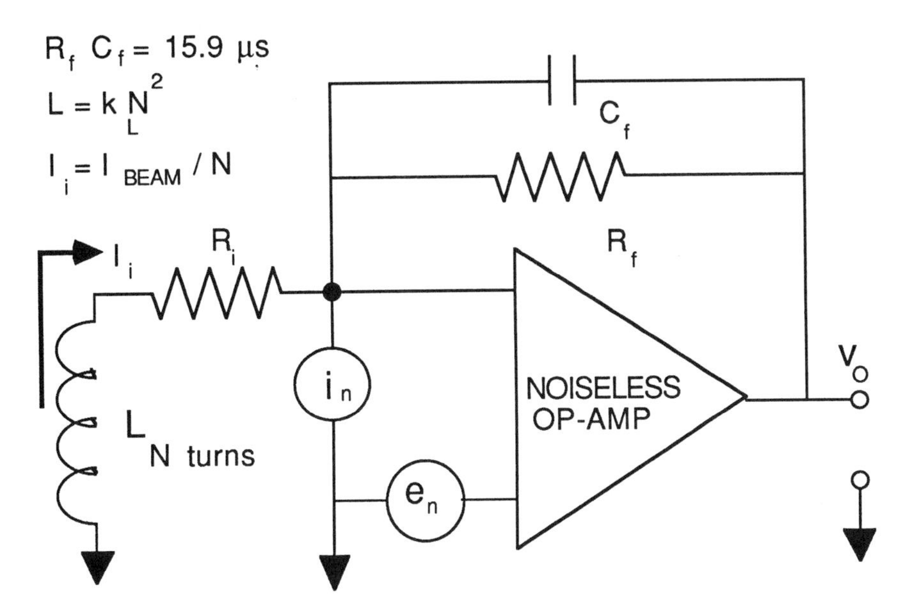

Figure 5 Noise Voltage and Current Model

If we assume all noise sources are random and uncorrelated, then the noise power in the system is additive[2,4].

Total RMS noise of the circuit (V_{nt}) is given by

$$V_{nt} = \sqrt{V_{ne}^2 + V_{ni}^2 + V_{nr}^2}\ , \qquad (1)$$

where V_{ne} is the circuit voltage noise, V_{ni} is the circuit current noise, and V_{nr} is the resistance noise.

For this configuration with a large loop gain ($A\beta \gg 1$),

$$V_{ne} = e_n/\beta \quad ; \quad \beta = \frac{Z_{in}}{Z_{in} + Z_f}, \qquad (2)$$

where A is the open loop gain, β is the feedback factor, e_n is the operational amplifier (op amp) voltage noise, and i_n is the op amp current noise.

Using the equations in (2) and circuit analysis and solving for V_{ne}

$$V_{ne} = \left\{1+\frac{R_f}{[[R_i - \omega^2 R_f C_f L]^2 + [\omega L + \omega R_i R_f C_f]^2]^{1/2}}\right\} e_n \qquad (3)$$

$$V_{ni} = i_n R_f \qquad \text{for } R_f \gg R_i \tag{4}$$

$$V_{nr} = [\ 4KTR_fB\ ]^{1/2} \tag{5}$$

The values for this application are B (bandwidth) = 10 kHz, $K = 1.38 \times 10^{-23}$ J/°K, T = 290°K, R_fC_f = 15.9 μs, ω = 628 rad/s, and R_i = 1.0 Ω.

Equations (4) and (5) illustrate the need to minimize R_f. However, R_f also determines the signal amplitude. Using Eqs. (1) through (5), the signal to noise ratio for this application can be determined.

$$SNR = \frac{\frac{IBEAM}{N} \times R_f}{V_{nt}} \tag{6}$$

Two op-amps were selected for evaluation, one for its low-voltage noise density (e_n = 3 nV/√Hz and i_n = 0.6 pA/√Hz) and the other for its low-current noise density (e_n = 11 nV/√Hz and i_n = 1 fA/√Hz). Using the preceding equations it was found that the op amp with the low-current noise required an excessive number of turns (1340) to maximize the SNR. Also the SNR did not equal 1 until IBEAM was 38 nA. The low-voltage noise op amp, on the other hand, provided a maximum at 184 turns. On this basis the voltage noise density was determined to be the dominant factor in the design of the I to V amplifier. See figure 6. It was decided that 200 turns would be a convenient number. With N = 200, R_f = 50 kΩ , C_f = 330 pF, and L = 8.0 H, an SNR of 1.04 could be achieved at 22 nA. The calculations indicated that, given the initial specifications, a design with approximately 20 nA minimum detectable current was possible.

Given the above constraints, one can see from figure 6 that any low frequency component in the input would swamp the signal at the output. Because the inductive reactance is small for these frequencies, the gain will be large as indicated by the following equation.

$$A = -Z_f/Z_{in} = -50\ K\Omega/1\ \Omega = -50\ K\ \text{(DC gain)} \tag{7}$$

An additional amplifier feedback network reduced this effect. This network, combined with the original transresistance amplifier, serves as a band pass filter with 3 dB roll-off at 0.01 Hz and 11.1 kHz. The stability of this circuit is critical and can be controlled by the gain of the feedback network. Examination of the input stage lead to the feedback control model depicted in figure 7.

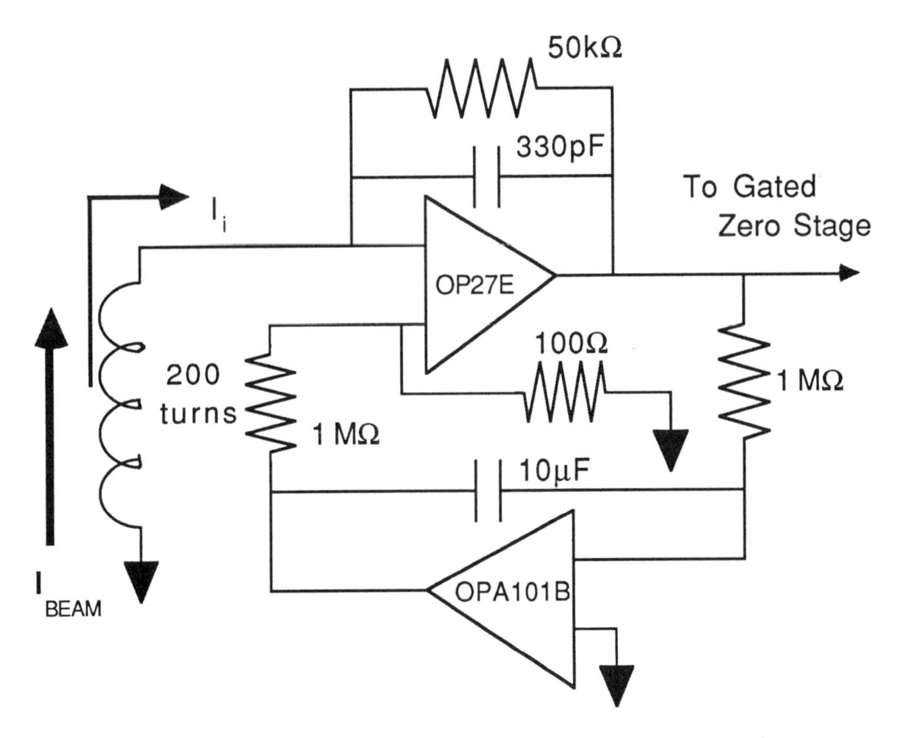

Figure 6 Current to Voltage Stage

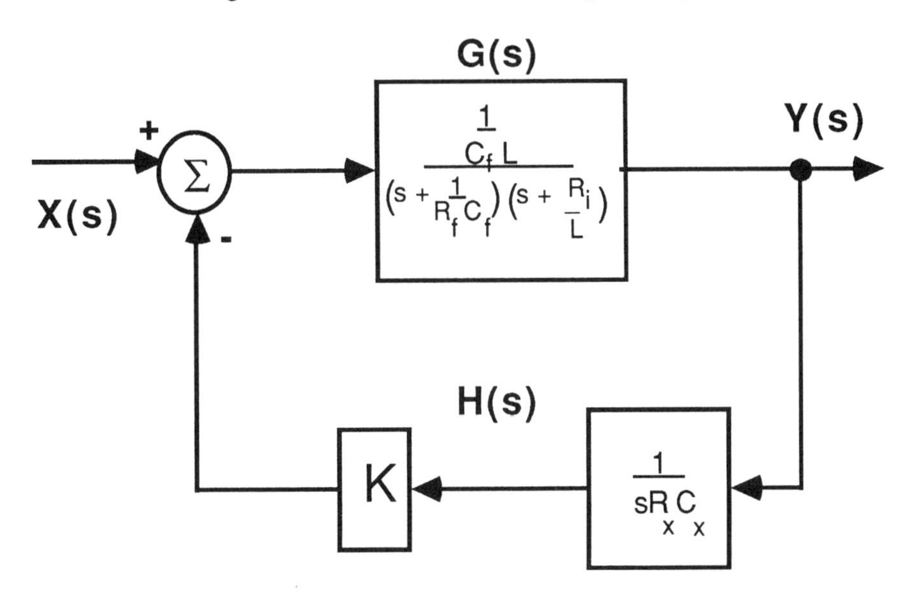

Figure 7 First Stage Feedback Control Model

The open loop poles are at zero, $-1/(R_fC_f)$ and $-R_i/L$. The open loop transfer function is

$$G(s)\,H(s) = \frac{\dfrac{K}{C_f L R_x C_x}}{s(s + 1/(R_fC_f))\,(s + R_i/L)} \quad . \qquad (8)$$

The gain of the circuit is controlled by the voltage divider on the output of the feedback stage. The outer pole ($-1/R_fC_f$) has little effect on the inner poles.

The closed loop transfer function is

$$\frac{Y(s)}{X(s)} = \frac{G(s)}{1 + G(s)H(s)} = \frac{s[1/(C_fL)]}{s[s + 1/(R_fC_f)][s + (R_i/L)][K/(C_fLR_xC_x)]} \quad . \qquad (9)$$

Using equation 9 and the following values a root locus plot (see figure 8) was generated.

R_f = 50 kΩ, C_f = 330 pF, R_x = 1 MΩ, C_x = 10 μF, R_i = 1 Ω, L = 8 H

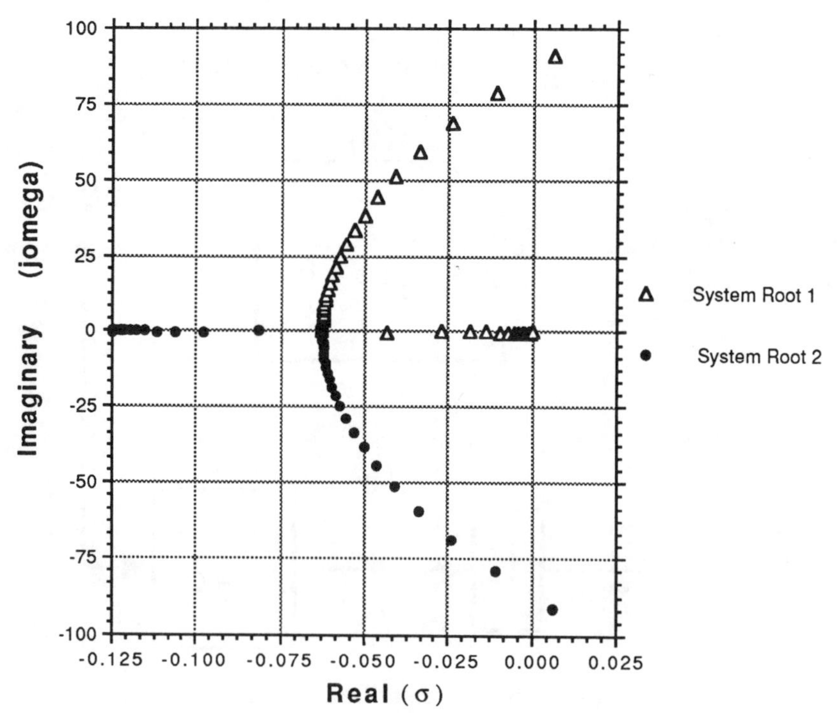

Figure 8 Root Locus Plot

Plotting the values of the poles as the feedback gain increases provides insight into the first stage stability. Using this information, Bode plots and the step input response for specific gains, the desired response and stability was determined.

The gated zero stage employs a different filtering technique. To attenuate frequencies below 200 Hz, the feedback network is gated so that the amplifier is zeroed during beam-off time. In this manner low frequency hum is reduced before the amplification stages. To minimize gating feedthrough, the gating circuitry is separated from the analog circuitry, both through ground isolation and physical location. See figure 9.

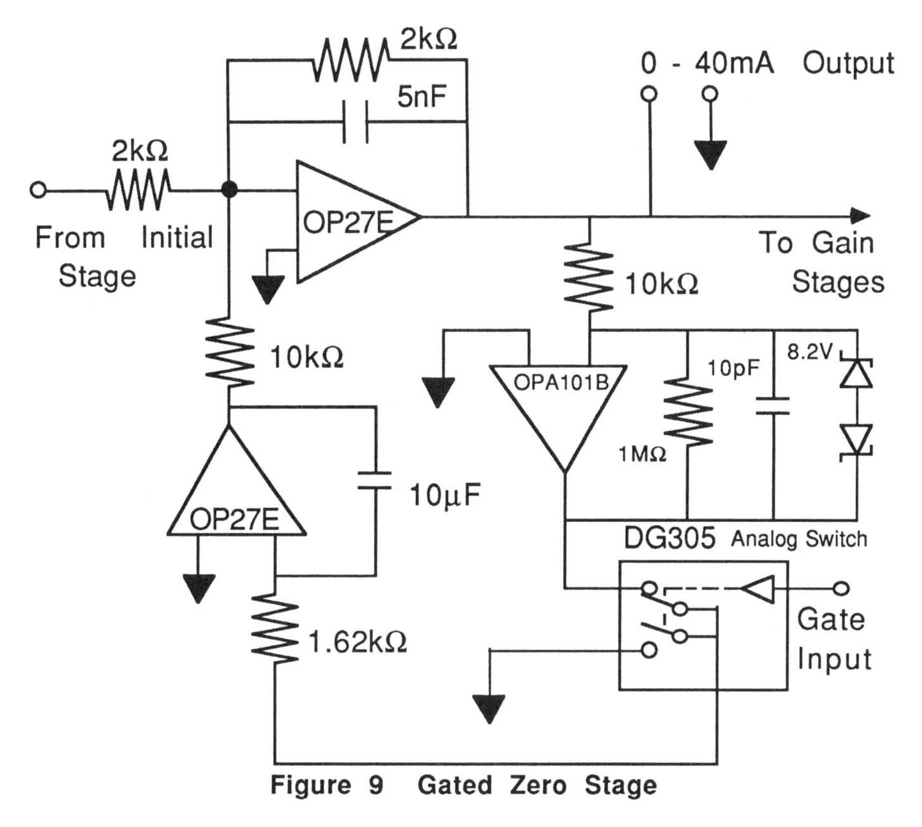

Figure 9 Gated Zero Stage

Following the gated zero stage are two stages of amplification. There are three output ranges, each with a full scale of 10 V. The output scale of these ranges is determined by the gain resistors selected in the amplification stages. Maximum measurable peak current, 40 mA, is determined by the current to voltage stage.

All components mentioned thus far are located within the current monitor chassis. The signals are then fed to an external chassis where there is another stage of auto zeroing and a final stage of signal conditioning can be used to eliminate noise outside the beam gate.

Mechanical Design

Early testing of the core and prototype amplifier showed coupling between the current monitor and local vibration sources. To evaluate the vibration isolation requirements, an analysis of the toroid and amplifier response to a variable-frequency audio signal was performed by attaching an accelerometer to the toroid. The amplifier showed a natural frequency of 180 Hz and the toroid resonance was at 550 Hz. The support structure was also analyzed. Forcing frequencies of 60 Hz were found in the structure. The prototype used pneumatic spring mounts attached to the outer cylinder to decouple vibration sources from the current monitor. Due to air leakage from the bellows, they were replaced by vertically supported springs and a means to adjust the amount of deflection. See figures 4 and 5.

Operation

An HSCM Display System provides real time analog signals and indication of average currents to the accelerator and injector control rooms. It also interfaces the HSCM to the accelerator binary control (RICE) and analog data (ADS) systems. See figure 10.

The desired gain range can be selected through the control computer. Selecting the range will automatically display and scale the computer readouts for that current monitor. The analog current signal is fed to the accelerator control room through a constant-current, line-driver receiver combination. This ensures that low-level signals generated at the current monitor are reproduced within 0.1% at the accelerator control room approximately one-half mile away. The signal serves as an input to an oscilloscope for on-line viewing of peak current and to a HSCM receiver. This receiver was initially developed for use in the Hardware Transmission Monitor (HWTM) system. The receiver integrates the signal and provides an indication of average current. The prototype HSCM required an averaging digital oscilloscope, but improvements in shielding the production unit eliminated this need.

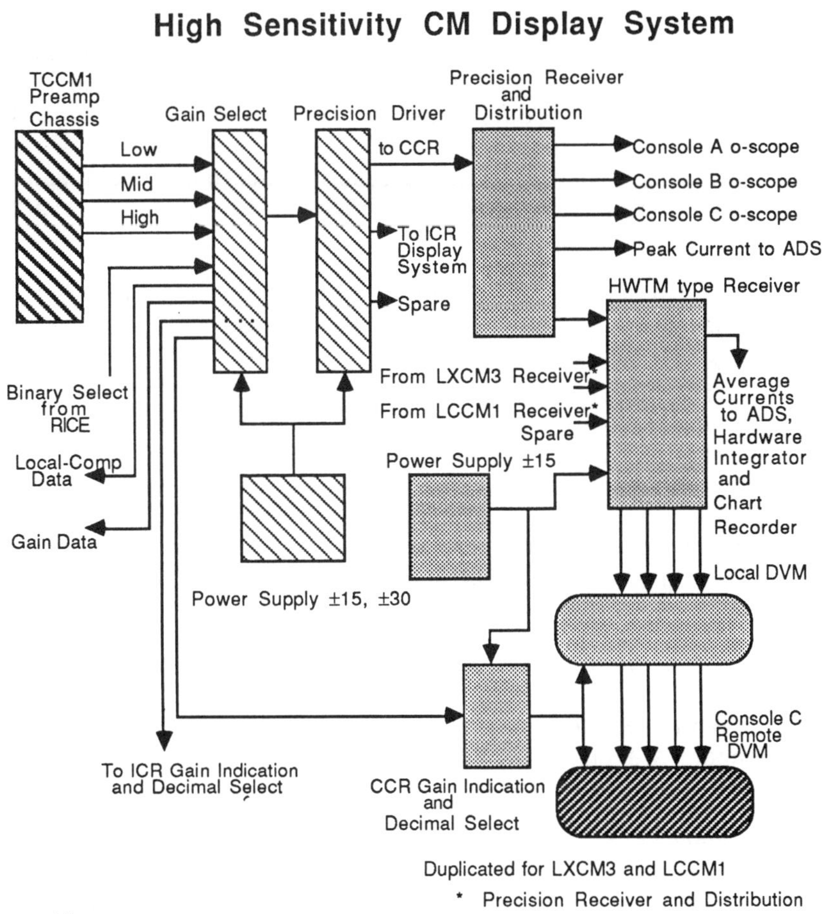

High Sensitivity CM Display System

Notes:

1) Selectable gains
   a) Low 0 - 400μA peak @ 12% D.F. = 48.000μA average
   b) Mid 0 - 40μA peak @ 12% D.F. = 4.8000μA average
   c) High 0 - 4μA peak @ 12% D.F. = 0.4800μA average
2) Gain selection using the control computer or manually.
3) Automatic scaling of peak and average data.
4) Average current indication warning, if peak current analog channel saturates (>2000 counts).

Figure 10 HSCM Display System

Figure 11, showing the output of a chart recorder connected to the HSCM receiver output, illustrates a production P^- average current of 30 nA. A quiescent noise level of approximately 0.4 nA can be seen when beam is off. The differences in current magnitude, due to different states of injector operation, are indicated. Beam quenches, which are a means for measuring the beam polarization with the Lamb-Shift source, are also shown. The minimum reliable average-current measurement is 1 nA.

Figures 12 and 13 are photographs of the prototype HSCM output with a production P^- beam of 240 nA peak current. Circuit noise is approximately 100 mV or 40 nA. This is a factor of 2 greater than the calculated value. This may be due to the assumptions made in the calculations. Averaging the output signal over a number of pulses virtually eliminated all other noise components. Excellent results were obtained using 8 to 16 pulses. See figure 13. The photos illustrate that the averaging technique will allow one to determine peak currents to within the accuracy of the oscilloscope. The minimum reliable peak current measurement is approximately 50 nA.

Results from the production unit can be seen from a computer plot output from the LAMPF Analog Data System. See figure 14. Low energy transport (18 μA peak) and experimental area (3.5 μA peak) current monitors are displayed on the same pulse plot.

The peak and average current outputs of the HSCM are checked and calibrated once per year before accelerator startup. A precision current pulse is fed to a single calibrate winding through the current monitor. This simulates beam current. A single potentiometer in the HSCM receiver is adjusted as necessary. From past experience this adjustment has been minimal or not required. We have found, through experience, that the unit will provide accurate and reliable information.

The current monitor system was designed to minimize the requirement for recalibration when components of the system are replaced. Precision and high stability electronic components and system design specifications well within component limitations help to meet this constraint. The absolute accuracy of the electronics is ±0.1%. Flux loss in the toroid (supermalloy) may contribute another 0.2 to 0.3%.

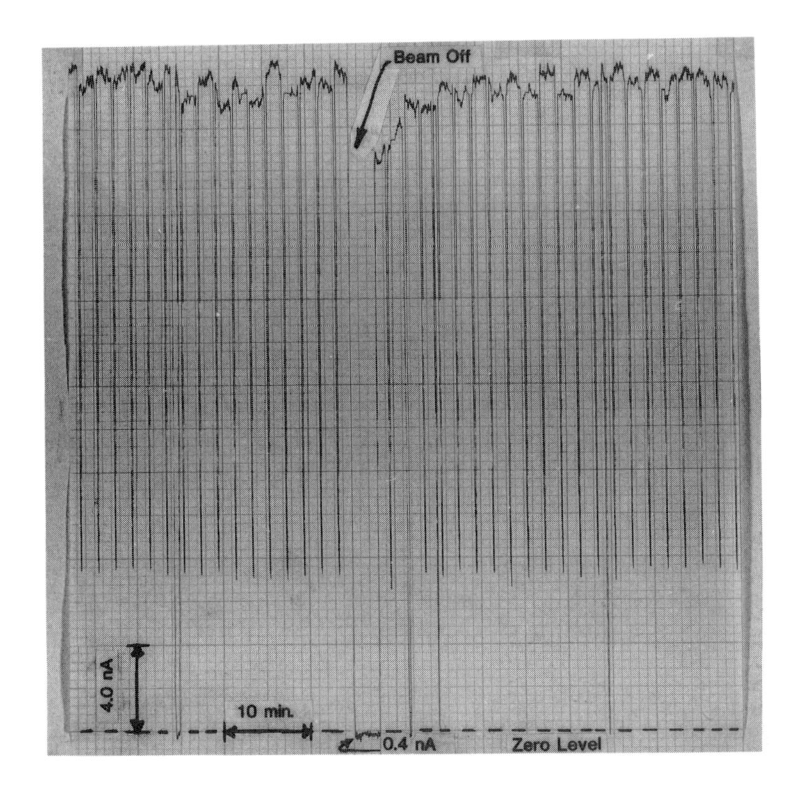

Figure 11 P- Average Current at Receiver Output

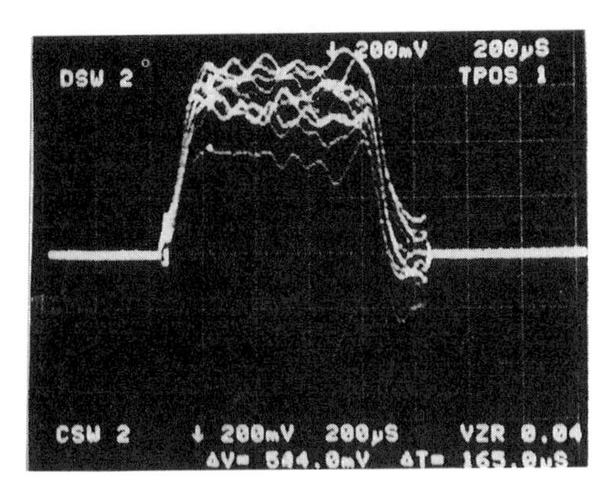

Figure 12 P- Beam 240 nA Peak, No Averaging

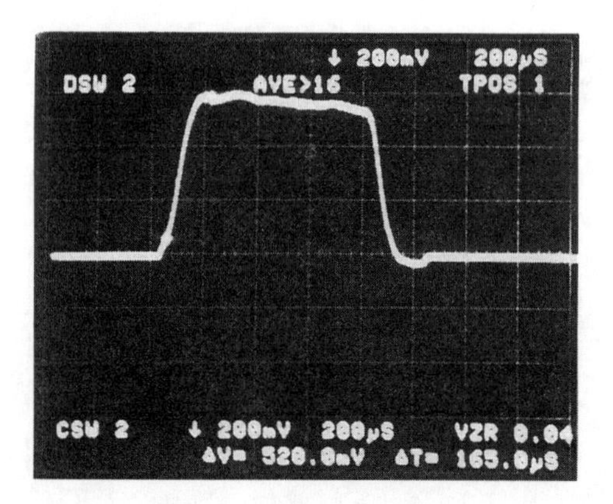

Figure 13 P- Beam 240 nA Peak, Average Equals 16

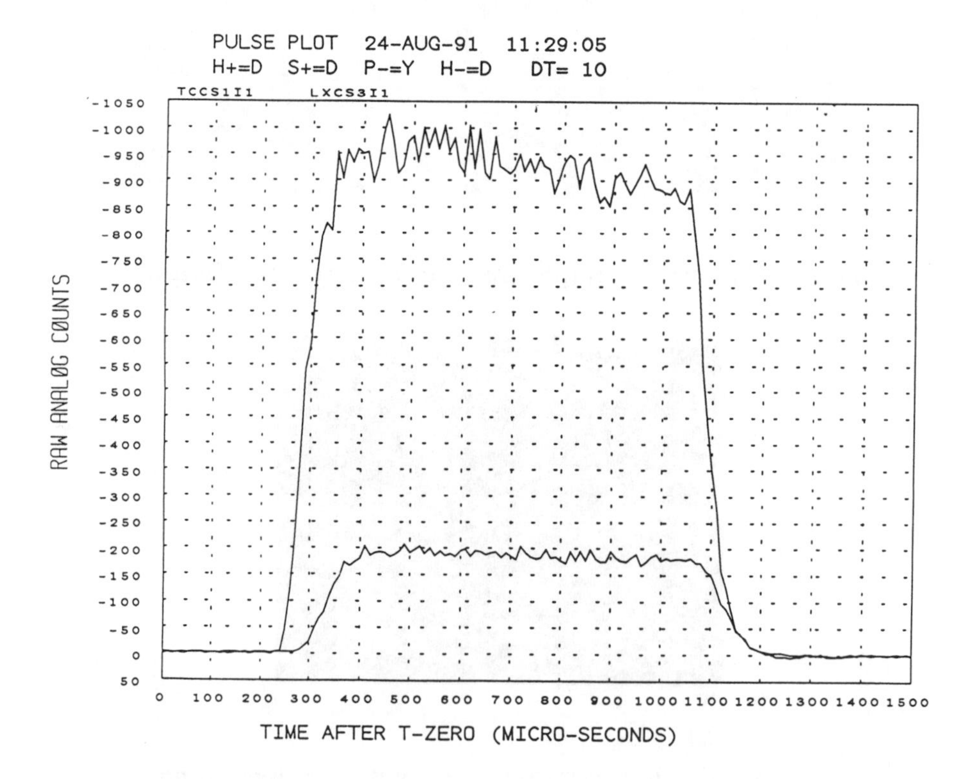

Figure 14 P- Beam 18µA and 3.5µA, Production Units

Conclusions

To assess the usefulness of the HSCM system, it is necessary to examine the benefits and drawbacks.

Benefits:

1) Accuracy and Linearity - The peak current outputs are accurate within 0.3% and linear over full scale of operation (10 volts).

2) High sensitivity - Signal to noise ratio is equal to 1 at peak current indication of 40 nA. Average currents can be measured to the 1 nA level.

3) Wide dynamic range - The device can accurately measure beam currents from 40 mA to 40 nA peak current.

4) Nonintercepting - The measurement of beam current will not interfere with accelerator operation.

5) High gain stability and reliability - Operation over the last 3 years has proved the unit can function reliably with instrument calibration performed once per year.

6) Energy independence - The accurate measurement of beam current is independent of the beam energy.

7) Vacuum - The current monitor is external to the vacuum pipe. This is convenient for maintenance and vacuum contamination considerations (outgassing).

8) Calibration - The current monitor is easily calibrated independent of beam operation.

Drawbacks:

1) AC device - Since the current monitor is an AC device, it will not measure neutral or DC beam currents. If H^- particles are stripped to H^+, the current monitor will measure only the difference between the components.

2) Radiation damage - Since the electronics must be located as close as possible to the current transformer to minimize noise effects, they are exposed to ionizing radiation fields. This may cause the current monitor to fail due to radiation damage to the

electronics. However, this is not normally a problem since the currents we are interested in measuring are low intensity.

3) Bandwidth - The bandwidth of the unit has been limited for noise reduction.

4) Electrical isolation - To prevent measuring currents that flow along the beam pipe, an electrical isolation joint must be provided at one end of the current monitor.

5) Cost - Building a unit today the hardware would cost approximately $20K to $25K.

Over the past three years the HSCM system has become an integral part of the operation and problem diagnosis of the P- ion source. It is also used to calibrate other instrumentation in the P- experimental areas. It has met all specified design criteria except for an SNR equal to 1 at 20 nA peak; the present device has an SNR equal to 1 at 40 nA. This deviation is acceptable. For the applications at LAMPF the benefits far outweigh the drawbacks.

I thank Andrew Browman for the technical guidance he provided on the project design and implementation. Contributors to the mechanical design were Louis Morrison for the prototype and Edgar Bush for the production unit.

References

1. A.A. Browman, "LAMPF Beam Transmission Monitor Systems," Proceedings of the 1981 Accelerator Conference, Washington, DC, March 11-13, 1981.

2. Ott, Henry W., "Noise Reduction Techniques in Electronic System, "John Wiley and Sons, Inc., 1976.

3. "Application Information, OPA101," Burr Brown Integrated Circuits Data Book, 1989, pp. 2-49 thru 2-52.

4. Letzer, S. and Webster,N., "Noise in Amplifiers," IEEE Spectrum, August 1970, pp. 67-75.

5. F.R. Gallegos, L.J.Morrison, A.A. Browman, "The Development of a Current Monitor System for Measuring Pulsed-Beam Current Over a Wide Dynamic Range," Particle Accelerator Conference, Vancouver Canada, May 13-16, 1985.

Advanced Light Source Beam Position Monitor

J. Hinkson
Lawrence Berkeley Laboratory, Berkeley CA 94720

Abstract

The Advanced Light Source (ALS) is a synchrotron radiation facility nearing completion at LBL. As a third-generation machine, the ALS is designed to produce intense light from bend magnets, wigglers, and undulators (insertion devices). The facility will include a 50 MeV electron linear accelerator, a 1.5 GeV booster synchrotron, beam transport lines, a 1-2 GeV storage ring, insertion devices, and photon beam lines. Currently, the beam injection systems are being commissioned, and the storage ring is being installed. Electron beam position monitors (BPM) are installed throughout the accelerator and constitute the major part of accelerator beam diagnostics. The design of the BPM instruments is complete, and 50 units have been constructed for use in the injector systems. We are currently fabricating 100 additional instruments for the storage ring. In this paper I discuss engineering, fabrication, testing and performance of the beam pickup electrodes and the BPM electronics.

Introduction

The beam position monitors for the ALS have been discussed before in a brief paper[1]. In the space allowed here I report on the engineering problems and solutions in greater detail. Each BPM system is composed of an array of beam pickup electrodes, a set of high-quality coaxial cables, a bin of processing electronics, and a controlling computer. These components will be fully treated in the following pages.

The storage ring circumference is 200 meters. It has 12 curved sections each about 10 meters long and 12 straight sections of 6.7 meter length. Two of the straight sections are used for beam injection and RF cavities. The remaining straight sections are reserved for insertion devices. BPMs are installed in the curved sections only (96 total, 8 per section). The booster synchrotron circumference is 75 meters and is composed of four curved and four straight sections. Thirty-two BPMs are installed nearly uniformly around the ring. Table I shows accelerator parameters relevant to beam diagnostics.

Table I. Accelerator Parameters for Beam Diagnostics

Parameter	Storage Ring	Booster	Linac
Energy (GeV)	1-2	0.05-1	0.05
RF Frequency (MHz)	499.654	499.654	2998
Harmonic Number	328	125	-
Minimum Bunch Spacing (ns)	2	8	8
Revolution Period (ns)	656	250	-
Number of Bunches	1-250	1-12	1-20
Repetition Rate (Hz)	-	1	1-10
Maximum Average Current (mA)	400	20	125
Single Bunch Current (mA)	8	3	10
Bunch Length (2σ ps)	28-50	100	30
Tune ν_h, ν_y	14.28, 8.18	6.26, 2.79	-
Synchrotron Frequency f_s (kHz)	13.87	256-44	-

Requirements

The storage ring BPMs are required to make continuous, non-destructive measurements of average beam position during beam injection and stored beam lifetime. Fully corrected beam position data will be delivered to the beam orbit control computer at 10 Hz. In addition, single-turn position data is to be accumulated at the ring revolution frequency and stored in internal BPM memory.

The booster ring BPMs function similarly to those in the storage ring with the performance specifications being less rigorous. Average beam position at a selected time in the booster energy ramp (0-0.4s) is reported at 1Hz. Single turn measurements are made also. The booster ring revolution frequency is higher, 4MHz, making single turn measurement timing a little more difficult.

BPMs in the Linac and beam transport lines measure beam at 1 to 10 Hz. There are 17 BPMs located in these areas. The beam repetition rate is low here, so we have slightly modified the storage ring BPM design to accommodate this lower rate. Table II shows the basic specifications for the BPMs in the different accelerator areas.

Table II. BPM Performance Specifications

Parameter	Storage Ring	Booster	Linac
Resolution (low speed) (mm)	0.01	0.1	0.1
Resolution (high speed) (mm)	0.5	0.5	-
Repeatability (low speed) (mm)	0.03	0.1	0.5
Repeatability (high speed) (mm)	0.5	1.0	-
Data Storage (turns)	1023	1023	-
Low Speed Response (Hz)	20	1	1
High Speed Response (MHz)	5	7	-
Dynamic Range (dB)	40	40	50
Center Frequency (MHz)	500	500	500
Beam Pickup Style	Button	Button	Button and Stripline
Pickup Coupling (ohm @ 500MHz)	0.1	1.0	1.0 and 8.0

Measurement Method

Shafer[2] and Littauer[3] have reported on the various methods used to determine transverse position of a charged particle beam. At the ALS we use the difference-over-sum technique which is most often used in synchrotron light sources. Our beam pickups consist of two pairs of electrodes mounted in the beam vacuum chamber wall. Beam signals induced in these electrodes travel via coaxial vacuum feedthroughs and coaxial cables to the BPM electronics where the signals are processed to reveal the X and Y position of beam center of charge. At nearly all locations in the ALS beam is composed of short electron bunches, between 10 mm and 30 mm, FLHM (Full Length Half Maximum). A 10 mm relativistic bunch corresponds to a pulse duration of about 33 ps FWHM (Full Width Half Maximum). In the booster and storage rings these short bunches are periodic and contain harmonics of equal amplitude extending into the multi-GHz region. Position information is contained in the amplitude of any of the harmonics sensed by the electrodes. By amplifying and detecting the amplitude of a selected harmonic we determine beam position by taking the difference between detected signals from two opposed electrodes, dividing the result by their sum, and multiplying that result by the reciprocal of a sensitivity constant. For example, beam position as measured by two horizontally opposed pickup electrodes would be calculated as follows:

$$X = \frac{V_1 - V_2}{V_1 + V_2} \frac{1}{S_x} \quad (1)$$

where V1 and V2 are amplitudes of detected beam harmonics and Sx is a constant in % / mm.

Storage Ring Beam Pickup Electrodes

We use button style electrodes for BPM pickups in the storage ring. The major reason for using them instead of other pickup devices is because of their low beam impedance. Coupled bunch beam instabilities (a major problem in storage rings with many short bunches and high average current) are driven by energy stored in reactive impedances in the vacuum chamber. With nearly 400 BPM pickups in the storage ring, we were concerned about their impact on a rather limited narrow-band beam impedance budget. Another reason for using buttons is that they may be precisely fabricated and installed with good accuracy.

A button pickup is usually a thin, circular metal disk supported by the center conductor of a coaxial vacuum feedthrough. The button is mounted flush with the wall of the beam vacuum chamber to avoid being struck by the beam. Where synchrotron radiation is present the buttons are configured to avoid being struck by the radiation, or a mask is installed to intercept the radiation. At least four buttons are used in a combined function monitor to determine horizontal and vertical beam position. Since the beam bunch length in electron rings is usually very short, it is common to make the button diameter equal to twice the rms beam bunch length (2σ) in order to collect maximum charge and to have good position sensitivity. Generally, peak current in the beam bunch is high, so signal strength for the BPM electronics is not a problem. A fraction of the beam bunch charge appears as voltage on the button self-capacitance to ground. Usually the capacitance is loaded by 50 Ω, resulting in a high-pass configuration for the beam signal.

Excellent button and coaxial feedthrough devices suitable for ultra-high vacuum service are available from industry today. The feedthroughs typically have well controlled impedance and are free of resonances up to 10 GHz or higher. Feedthrough and button installation in the vacuum pipe is accomplished by either welding the feedthrough into a flange which is later attached to the beam pipe or by welding the feedthrough into a section of the vacuum pipe directly. In both of these cases the exact location of the button face relative to the geometric center of the beam pipe is difficult to determine. Mechanical and electrical offsets in the button array are found by wire antenna measurements when practical or by external measurements. In the ALS storage ring we chose not to use either of these button installation methods. There were two reasons for this. Our unusual vacuum chamber prevented us from fabricating and characterizing a pickup assembly which we could later attach to the chamber. Secondly, we were very concerned with the mechanical precision and stability of the button installation. We felt that a button welded in place would not meet our accuracy requirements.

The ALS storage ring vacuum chamber is not the usual stainless-steel pipe or extruded aluminum chamber most often used in storage rings. The 10-meter curved vacuum chambers are formed by two machined pieces of aluminum which are welded together. Large slabs of aluminum (10.3 x 1.5 x 0.1 meters) were machined by a numerically-controlled, 5-axis mill. Very intricate machining was performed on the top and bottom of the slabs to accommodate magnet pole tips, flanges, diagnostic ports, BPM pickups, photon stops, vacuum pumps, and the electron beam channel. The beam chamber is somewhat diamond shaped and is open to an anti-chamber on the outer radius. The anti-chamber is used for vacuum pumping, photon stops, and photon ports.

Having a machined chamber with thick walls afforded us the opportunity to place holes for the BPM pickup electrodes (buttons) with good accuracy relative to survey points on the chamber itself. Our intent was to position the electrodes precisely in the chamber holes using a machined ceramic spacer and to hold the electrodes in place with a spring-loaded coaxial vacuum feedthrough. In this case neither feedthrough welding or flange placement determined the location of the pickup electrode. RF contact between pickup components was made with "Multilam" RF fingers. The fingers provided enough mechanical compliance to remove transverse stress from the vacuum feedthrough. Figure 1 shows the installation scheme for the pickup and related parts.

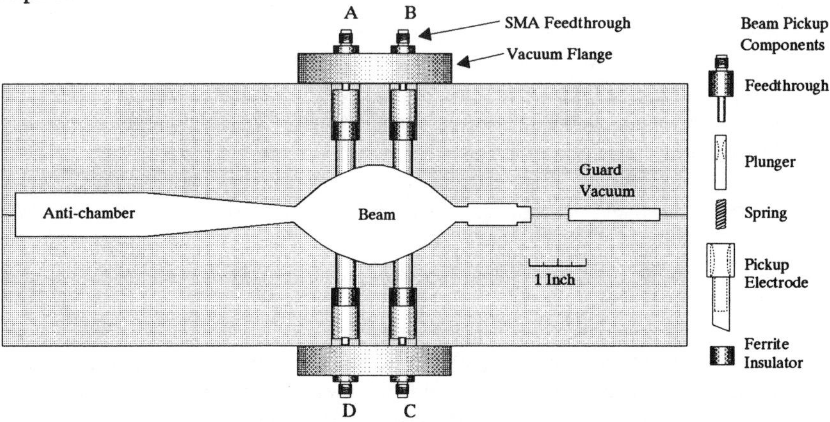

Fig. 1 Cross section of ALS storage ring BPM buttons with exploded view of button parts. Shaded area is a very simplified representation of vacuum chamber.

Figure 1 illustrates the pickups installed vertically with two feedthroughs welded into a single vacuum flange. Notice the pickup face exposed to the beam is beveled to conform to the vacuum chamber wall. This was not our original design. In the prototype full sized vacuum chamber the BPM pickup holes were drilled at 25 degrees off vertical making the flat-faced electrodes fit flush to the beam chamber wall. The prototype chamber construction program revealed that the milling machine could not hold our machining tolerances at a 25 degree angle, so the holes were drilled vertically in the 12 production chambers. This made it necessary for us to bevel the pickup face and install a ceramic key to prevent pickup rotation.

While our BPM pickup design may have had certain mechanical virtues, we faced a number of electrical problems. The pickup is essentially a loss-less capacitor at low frequencies, but in the microwave region it exhibits annoying resonances. These resonances are due to the different radii of the pickup components. The pickup is composed of five short transmission lines of differing characteristic impedance. The end transmission line (the pickup electrode) is open-ended. These transmission lines are resonators at microwave frequencies all of which can be excited by the short beam bunch. We detect these resonances by network analyzer measurements of either S11 at the SMA connector or S21 from the beam side of the pickup. The shunt impedance of these narrow band resonances (5 GHz and up) exceeded our impedance budget. We damped the Q of these resonances by replacing the ceramic spacer and anti-rotation key with NZ31 lossy ferrite. This brought the total narrow-band impedance of all the pickups back within limits and increased the low frequency pickup capacitance to about 25 pF.

Another problem appeared when we placed two feedthroughs on a single flange. The coupling between adjacent buttons rose from a figure of about -100 dB (with feedthroughs on separate flanges) to about -40 dB. There was sufficient space in the tiny gap between the flange and vacuum chamber (due to the metal vacuum seal) for coupling between the feedthrough center-conductors at 500MHz. To determine button electrical center offset due to electrical and mechanical imperfections, we required that coupling between the pickups be between the button faces only. We eliminated the feedthrough coupling by installing RF seals around each center conductor at the flange-to-chamber interface. In spite of the difficulties we have experienced with the storage ring buttons, we feel we have an adequate design. They are accurately placed in the vacuum chamber, and tooling ball fixtures near the buttons allow us to measure their position relative to upstream quadrupole magnet survey points.

Storage Ring Pickup Testing

A small prototype vacuum chamber was fabricated in two halves and fitted with our button pickups. The chamber was placed in a test set that allowed us to install a movable wire to simulate beam and to close the ends of the chamber. Actually, the wire is stationary, and the chamber is moved. Closing the camber ends prevents spurious operation but makes wire installation difficult. Having the chamber in two halves solves that problem. The wire, #30 gauge, is terminated in its characteristic impedance, 290 ohms, to minimize standing waves in the chamber. Part of the 290 ohms is the 50 ohm load on the network analyzer reference (R) port. See figure 2. The wire is driven by the RF output of a HP8753 network analyzer. We move the chamber using a precision X-Y stage and measure the response of the buttons. This is done to determine the sensitivity of the buttons to beam motion, find the beam coupling impedance, and determine the extent of button nonlinear coupling at large beam (wire) displacements. Small sight holes drilled in the X and Y planes near the pickups permit us to position the wire to within 0.01 mm of mechanical center.

It is apparent in Fig. 2 that a large impedance mismatch occurs at the entrance of the test chamber. The source impedance of the network analyzer absorbs reflected voltages. Normalizing the measured button signals to the analyzer R port reduces the effects of the impedance mismatch and the changing Zo of the wire as it is moved far off center. We determined that the storage ring button sensitivity is:

$$S_x = 7.67\%/\text{mm}$$
$$S_y = 5.72\%/\text{mm}$$

The linear equations used to determine beam position are:

$$X = \frac{(V_a + V_d) - (V_b + V_c)}{(V_a + V_b + V_c + V_d)} \frac{1}{S_x} \tag{2}$$

$$Y = \frac{(V_a + V_b) - (V_c + V_d)}{(V_a + V_b + V_c + V_d)} \frac{1}{S_y} \tag{3}$$

where Va, Vb, Vc, and Vd are detected button signals (see figure 1).

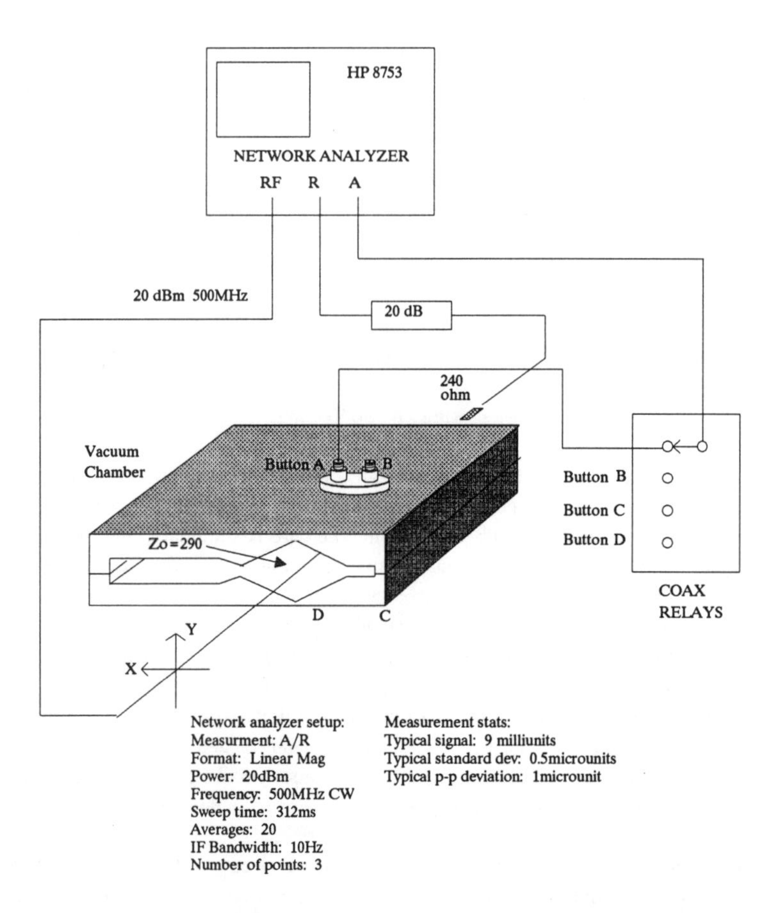

Fig. 2 Diagram of storage ring BPM button test set. The 290 ohm wire is moved in X and Y as the button response is measured by the network analyzer.

Since the buttons are not placed in the measurement axes, we must use all four signals to calculate either X or Y beam position. Equations (2) and (3) are valid (for the 0.03 mm accuracy specification) for deflections less than 2 mm off center. In order to make accurate position measurements at larger displacements and correct for pin-cushion nonlinearities we make use of the method developed by Halbach[4]. In this development the response of the four buttons is determined for any beam position X,Y,

$$V_i = f_i\,(X,Y) \tag{4}$$

where $f_i(X,Y)$ is a Taylor series whose coefficients are uniquely determined for the vacuum chamber cross-section and button locations using POISSON, an electrostatic analysis program. Then the two coupled equations

$$g(X,Y) = \frac{(V_a + V_d) - (V_b + V_c)}{(V_a + V_b + V_c + V_d)} \tag{5}$$

$$h(X,Y) = \frac{(V_a + V_b) - (V_c + V_d)}{(V_a + V_b + V_c + V_d)} \tag{6}$$

are solved numerically for X and Y using a fast converging iterative program resident in the local computer. For our storage ring eight Taylor series coefficients are used. We tested the validity of this approach by taking data from many wire locations as far as 20 mm off center. At 20 mm displacement the calculation had 0.1 mm error. The linear position calculation at 20 mm was off by more than 5 mm.

Additional testing of the storage ring buttons was required. While the buttons may be accurately positioned, significant errors occur due to electrical offsets. The contributors to these errors are unequal button capacitance, connector insertion loss, and coaxial cable attenuation. The button capacitance is nominally 25 pF. We measure 10% capacitance variations between buttons. These variations cause a noticeable offset of electrical zero in a four-button array. There are minor differences in the SMA connector insertion loss. The cables we attach to the buttons are nominally 100 feet long. Their attenuation at 500 MHz is not exactly equal. The effect of all of these offsets is compensated by a measurement of the button-to-button transfer function at our operating frequency, 500 MHz. This measurement is performed twice, once when the buttons are installed in the chamber as a quality assurance step, and again when the cables have been attached. This technique[5] compensates unlike gain in the four channels and some degree of mechanical error. We have determined that mechanical offsets of up to 1mm in a button may be compensated with this technique. Tests on BPM buttons at the Argonne Advanced Photon Source[6] have shown good agreement between offsets measured with a wire and the button-to-button transfer function measurement. Performing these measurements when the buttons and cables are first installed has the additional benefit of providing us a data base of information against which we can compare future measurements. Fig. 3 is shows the setup we use to measure the transfer function between buttons in storage ring BPMs.

Booster BPM Pickups

The booster synchrotron is 75 meters in circumference and is composed of 4 arcs and 4 straight sections. Each arc has 7 BPM pickup arrays and each straight has a single array. Each array is composed of four button style pickups rotated 45 degrees from the horizontal plane. Figure 4 shows the button layout. We fabricated our own button assemblies using MDC type N constant impedance vacuum feedthroughs. Each button has about 25 pF low-frequency capacitance. Up to 3 GHz the buttton assemblies exhibit no resonances. Resonances do occur at higher frequencies but are of no concern because the impedance budget for the booster is more liberal than that for the storage ring. The booster BPM system is not required to perform as accurately as the storage ring system, so minor offsets are not important. However, we have tested the booster buttons and cables using the transfer function method anyway to gather baseline performance data.

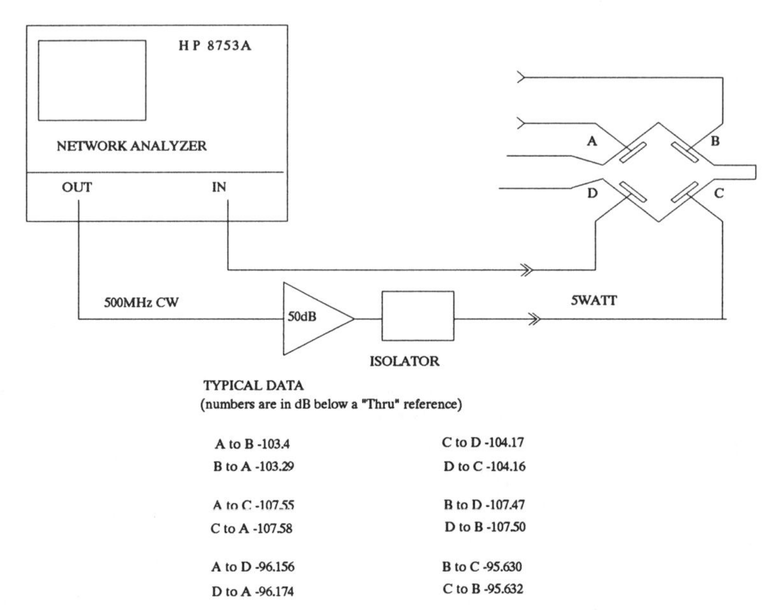

Fig. 3 Diagram of button-to-button transfer function measurement. The power amplifier is required to extend the dynamic range of the analyzer at high values of attenuation.

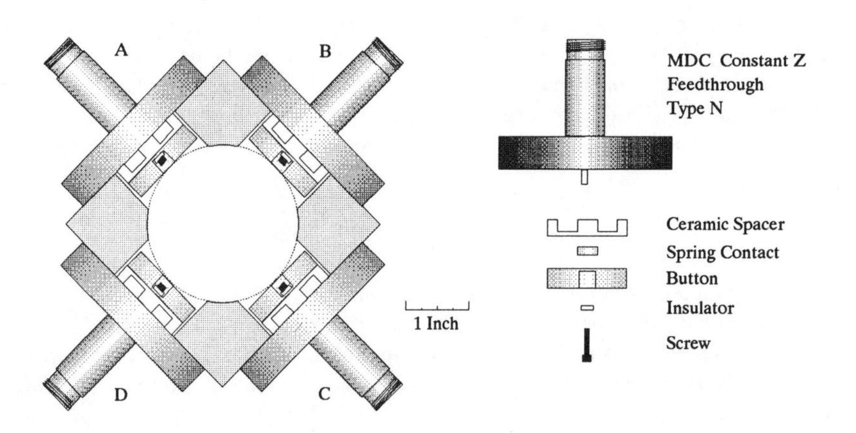

Fig. 4. Booster BPM button arrangement.

Linac and Transfer Line BPM Pickups

Two types of beam pickups are used in these locations. In the Gun-to-Linac (GTL) section where longitudinal space is limited we installed the CERN LEP[7] button. This device is available from Ceramex (MetaCeram) in France. It was carefully engineered for broadband response. The button itself is a thin disk about 1.25 inches in diameter. The disk is supported by a very good coaxial feedthrough that exhibits less that 0.05ρ (reflection coefficient) up to 18 GHz. The button assembly has a 2.75 inch Conflat flange and is ready for installation upon delivery. The four-button arrangement is similar to the booster configuration except that the buttons are not rotated 45 degrees. We could have used the LEP buttons in the booster but we deemed it less expensive to construct our own. In retrospect the costs for either would probably have been about the same. Because of their excellent frequency response, we will use three sets of the LEP buttons as broadband beam monitors in the storage ring.

The remainder of the BPM pickups in the transfer lines are striplines. Fig. 5 shows the layout. These striplines are 150 mm long and have a λ / 4 resonance at 500 MHz. This length was chosen to provide resonant drive for the 500 MHz bandpass filters in the detector electronics RF input section. All symmetrical striplines produce an inverted echo pulse (bunch length << stripline length) at a time 2l / c. With a length (l) of 150 mm, our stripline echo pulse occurs 1 ns after the prompt pulse providing resonant excitation for the filter.

The coaxial feedthroughs we used for the striplines are type N, constant impedance devices from Cermaseal. They have good performance to about 3 GHz. We observe ringing at 5 GHz in the stripline signal when it is excited by the short Linac beam pulse.

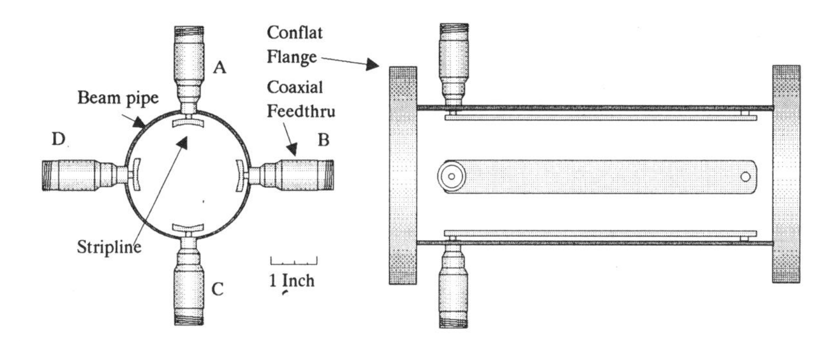

Fig. 5. Beam transfer line BPM striplines.

BPM Electronics

A description of the storage ring BPM electronics follows. The BPMs for the booster and transfer lines are slightly different. As each BPM sub-system is described, the differences will be noted.

The BPM electronics are enclosed in Eurocard bins. The major sub-systems (except for the local oscillator, power supply, and programmable attenuator) are enclosed in 3U, plug-in modules fitted with metal enclosures. The module printed circuit boards are about 8.5 inches long and 4 inches high. The module enclosures are not RF tight. To help shield sensitive electronics we installed RF fingers on the top and bottom bin covers. When the module panel screws are tight and the covers in place, the instrument is relatively immune to RFI. All of the bin components and modules were supplied by Schroff. Figure 6 is a drawing of the bin and modules.

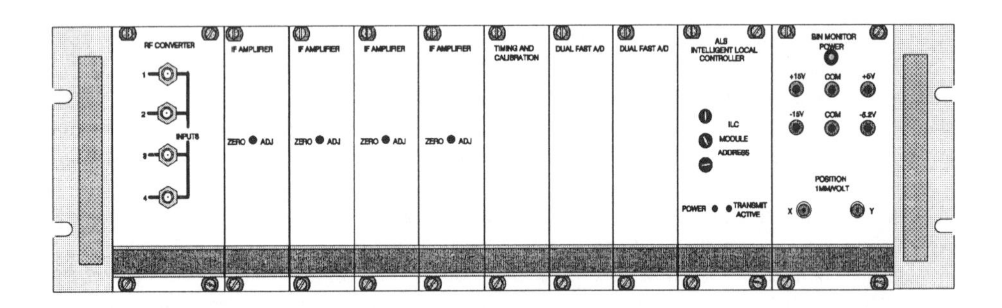

Fig. 6 Drawing of BPM modules and bin. The chassis is 19 x 5.25 inches.

The modules are electrically connected via 96-pin DIN connectors and a six-layer, full-width printed circuit board (back-plane). RF and video signals are routed between modules in coaxial cable and connected via coaxial inserts in special DIN 96-pin connectors. These connectors may have up to 8 coaxial feedthroughs. Nine of the conventional connector pins are lost for each coaxial connection. The coaxial cable connectors (from Palco) mate nicely with Belden 9307 cable, a small semi-rigid 50 ohm coax. Figure 7 shows the basic signal flow between modules.

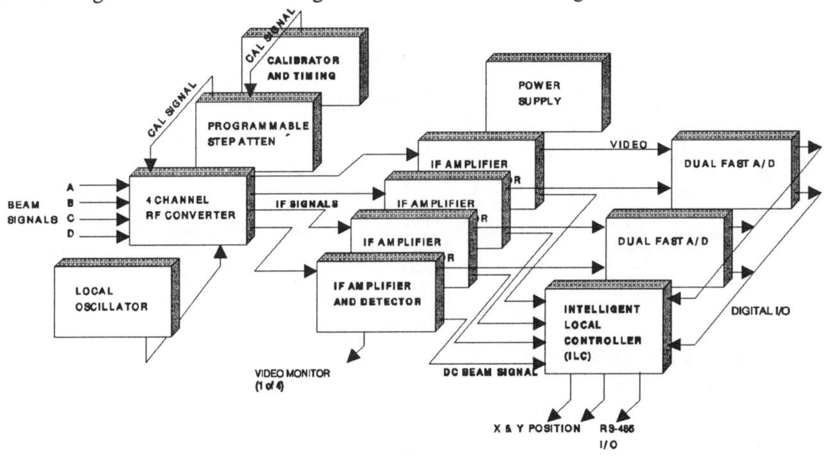

Fig. 7 Basic signal flow between BPM plug-in modules.

Throughout the ALS accelerator complex beam signals are routed to BPM electronics via Andrew FSJ1-50A Heliax cable. This cable was chosen because is has 100% shielding, polyethylene dielectric (more radiation tolerant than Teflon), good flexibility, and long-term stability. The cables terminate near the BPM electronic bins in N-to-SMA bulkhead feedthroughs. We will use .141 semi-rigid coaxial jumper cables to connect beam signals to the four-channel RF converter modules in storage ring BPMs. RG223 is used for the jumper cables in other locations. The connections are made on the front of the RF-modules via SMA connectors. While it would have been more convenient to make beam-signal connections through the BPM bin back panel and DIN connectors, we felt the front-panel SMA connections would be more reliable. All critical signal paths except the beam pickups, Heliax cables, and input connectors are in the BPM self-calibration loop. Therefore, components outside the loop must be stable and reliable.

Each BPM bin contains four super-heterodyne, amplitude-modulation receivers. Beam signals from four pickups are band-limited by the coaxial cables and bandpass filters and are down-converted to the 50 MHz intermediate frequency (IF) in the RF converter module. Four IF amplifier/detector modules process the RF converter output signals. The gain of the IF amplifiers is set by a single DAC output of the Intelligent Local Controller (ILC), the bin computer. The envelope of the IF signals is detected and sent in three paths. For error corrected measurements of average beam position the detector pulse output is peak-detected, filtered and digitized by the ILC. For single turn measurements of beam position the detector output pulse is again peak detected and digitized by a video ADC. The third path for the detector signal is to a monitor jack on the bin rear panel. Since the four IF amplifiers do not have equal gain an ILC controlled calibrator is available for test signal injection into the RF module. The block diagram in figure 8 illustrates the basic design. In the following sections each BPM module is discussed in detail.

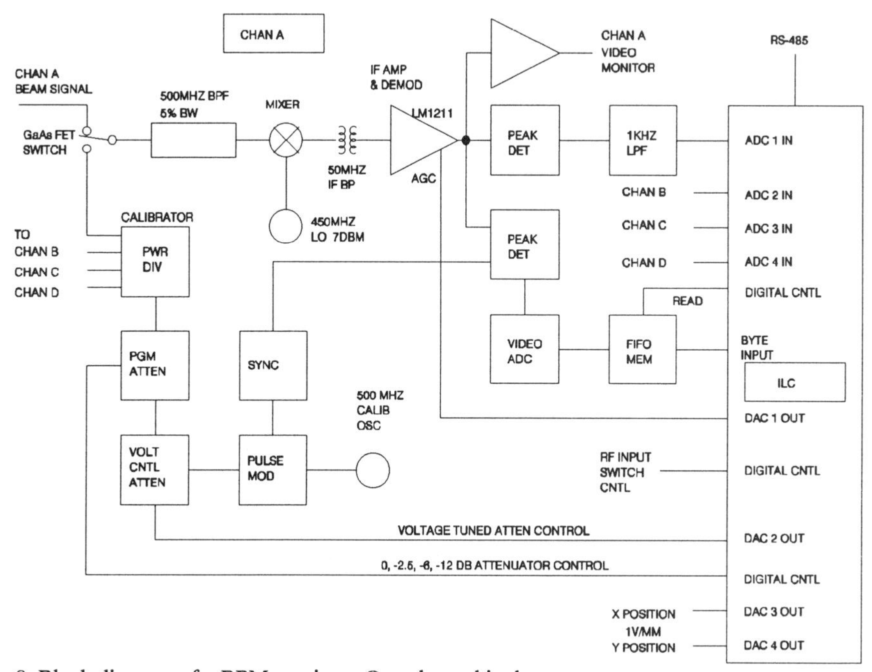

Fig. 8 Block diagram of a BPM receiver. One channel is shown.

RF Converter

The choice of operating frequencies for the RF converter was dictated by the requirement for arbitrary beam bunch-fill patterns in the storage ring. The two extremes of the fill patterns are a single high-current bunch or 328 bunches having equal charge. All patterns in between may exist though we expect the maximum number of bunches will be 250. A single 100 ps bunch rotating at f_0 (1.5 MHz) contains equal-amplitude harmonics of the rotation frequency to 2.5 GHz. With 328 bunches of equal charge and 30 ps duration, the beam harmonics begin at 500 MHz and extend beyond 10 GHz. We set the lower limit for the BPM operating frequency at 500 MHz because lower frequency beam rotation harmonics may be severely attenuated or disappear with certain fill patterns and because the pickups do not have good response below 200 MHz. Considering only harmonics of the RF system and not those of the rotation frequency, we find the upper frequency limit for our BPM system is the beam channel cutoff frequency for waveguide mode propagation. In our case that frequency is about 4 GHz. We were left with eight frequencies from which to choose. We determined that cost essentially scales with frequency, so we chose 500 MHz as our operating frequency. Conventional wisdom dictates that one should not construct a sensitive instrument operating at the same frequency as nearby high-power RF systems. Two factors helped us in this regard. The beam channel will not propagate 500 MHz and the 100% shielding of the BPM signal cables will not admit 500 MHz interference. We have already determined how well these cables perform. The booster injection kicker magnet, a 50 MW, 100 ns pulsed system, does not affect nearby BPMs nor does the 40 KW, 500 MHz CW booster RF system. We believe the 300 kW storage ring RF system will have no impact on BPMs either.

The function of the RF converter module is to heterodyne 500 MHz beam signals down to 50 MHz and to inject calibration signals during the self-calibration sequence. There is no gain in this module. Figure 9 shows the basic circuitry. Standard IF components are available at 50, 60, and 70 MHz. We chose 50 MHz because we found an inexpensive 450 MHz surface-acoustic-wave (SAW) oscillator from RF Monolithics for the local oscillator.

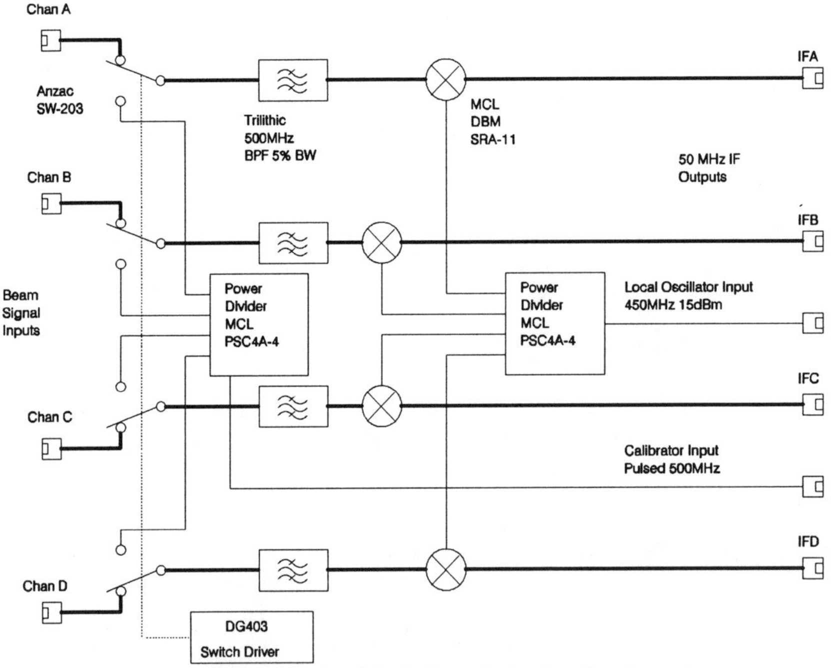

Fig. 9 RF Converter module. Darker lines indicate beam signal path.

We evaluated a number of circuit configurations for the RF converter in an effort to reduce the cost of this module. In our first prototype all components were connectorized. Isolation between channels was excellent and losses were low. Module assembly was difficult, and the unit was relatively expensive. We constructed an RF module having all components mounted on a PC board and interconnected with 50 ohm micro-strip line. Isolation between channels was only 30 dB or so. Our specification for isolation was 60 dB. (We did not try a multi-layer board with shielded stripline.) After a few more iterations we had a hybrid module having only one connectorized component, the bandpass filter. All the other components are pin-connected to the PC board. Connections between components are made via short lengths of Belden 9307 coaxial cable. Isolation between channels is greater than 60 dB at our operating frequency.

The 500 MHz bandpass filter in the RF converter is an important component in the BPM. We tested a variety of filters for impulse response. SAW filters we tested had too much attenuation, long delay, and undesirable pulse feedthrough. Lumped LC filters soldered directly to the PC board exhibited pulse feedthough also (inadequate high-frequency isolation) and relatively high insertion loss. The best filters we found were tubular, 3-section, Chebychevs. They have excellent high frequency rejection and predicable pulse response. Better pulse response may be obtained with linear phase filters, but they are more costly. We found good agreement between SPICE evaluation of a 3-section, Chebychev bandpass filter at 500 MHz and a real filter. Figure 10 shows frequency and impulse response of the SPICE filter. The actual filter performs similarly.

The bandpass filter 3 dB-bandwidth is 25 MHz, much larger than required for single-turn position measurements in either the storage ring or booster. We could have tailored bandpass filter specifications to each application but determined it was less costly to customize the IF amplifiers. All of the RF modules are the same and will function in any of our applications.

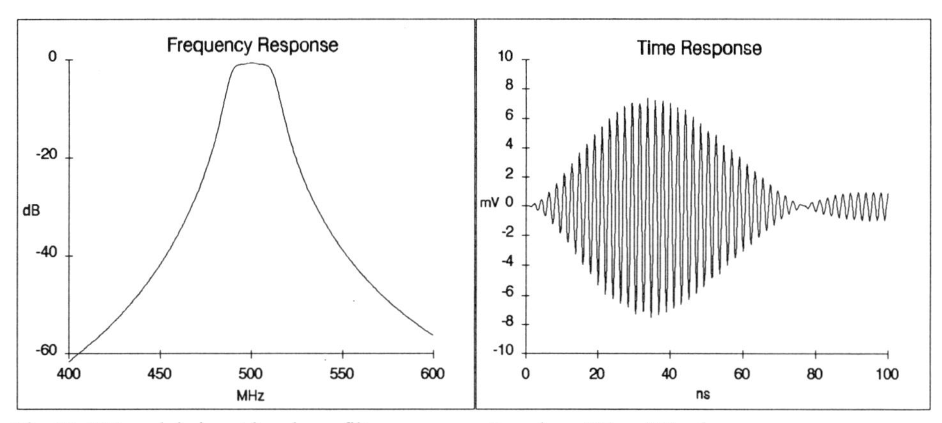

Fig. 10 RF module input bandpass filter response. Impulse - 100ps, 1V pulse.

IF Amplifier

A simplified schematic of the IF amplifier module appears in Figure 11. The RF module mixer output contains not only our 50 MHz beam signal, but many other mixer products. The IF module input filter passes the 50 MHz signals and terminates the out-of-band signals in 50 ohms. This particular IF bandpass filter from MCL makes mixer IF signal processing quite simple. The next component in the signal path is a MCL MAR-6 amplifier. This device has about 20 dB gain and a noise figure of less than 3 dB up to 1 GHz. This amplifier is optional in our design and so far, has not been required. Following the amplifier is a single-ended-to-differential transformer used to match 50Ω to the input impedance of the LM1211. This monolithic integrated circuit from National Semiconductor is a broadband demodulator system originally developed as a receiver for local area networks. It operates between 20 and 80 MHz and has over 40 dB IF gain-control range (AGC). The maximum gain of the IF amplifier section of the LM1211 is established by the input resistors (Rin in figure 11) in the low-impedance common base input stage, the impedance of the IF tuned circuit, and the AGC current. The LM1211 (phase/amplitude) detector is driven by the signal appearing at the IF Out pin.

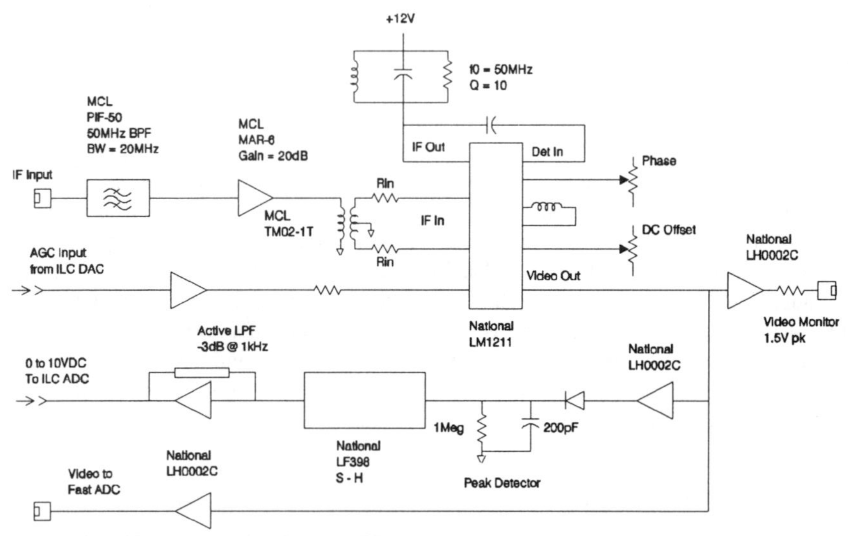

Fig. 11 Simplified schematic of IF amplifier module (1 of 4 in each bin).

The LM1211 detector operates with either AM or FM signals. For beam position monitoring we are interested in the amplitude of the 50 MHz IF signal, so we configured the detector as a quasi-synchronous, amplitude-modulation detector. By quasi-synchronous we mean that the internal phase detector is fed a reference signal developed from the IF signal itself. This of course means that no reference signal is available without an IF signal. A truly synchronous detector would have an unchanging reference signal of the proper frequency and correct phase. A limitation of quasi-synchronous detection is found when the internally-developed reference signal is no longer constant in phase and amplitude. This occurs when the IF signal is too low to saturate the output of the reference signal limiting amplifier. As a result we find a threshold below which the detector output is no longer useful. When this condition occurs, we increase the gain of the IF amplifier via AGC and restore good detector operation. At maximum gain the detector output is quite noisy, and much signal averaging is required in the ILC. At high signal levels we find some gain compression in the IF amplifier output and also some compression in the detector output as it approaches 3.5V.

The signal appearing at the Video Out pin on the LM1211 is a pulse, 0 to +3V, having the shape of the beam bunch pattern (within bandwidth limitations). A single beam bunch generates a waveform similar to Figure 10 (right) at the output of the RF module bandpass filter. After mixing and filtering, this burst of RF is amplified and detected by the LM1211. The detected pulse duration is determined by the bandwidth of the IF amplifier, and for a single bunch is about 150ns. When the beam bunch pattern contains many contiguous bunches, the tuned circuits reach steady-state performance and the pulse appearing at the Video Out pin has the general shape and duration of the beam bunch pattern.

The LM1211 has good carrier suppression at the video output so no additional filtering IF is required. The video output pulses are directed in three paths. A unity-gain buffer amplifier delivers the pulses to a monitor jack on the rear of the BPM bin. The pulse amplitude is halved because of the amplifier's output impedance and cable impedance matching resistor. We use this pulse (and those from the other three IF amplifiers) for oscilloscope monitoring of BPM performance. Another buffer amplifier sends the pulse to a peak detector. When the BPM is monitoring low repetition rate beam (in the Linac for example) the sample-and-hold (S-H) circuit following the peak detector is used to capture the voltage peak. In this case the active low-pass filter following the S-H is simply used as a DC amplifier to scale voltages for the ILC ADC. When storage ring or booster beam is measured, the peak-detector output is essentially DC when the beam orbits continuously. In this case the S-H is used as an impedance buffer and simply tracks the input voltage. The low-pass filter following the S-H removes high frequency noise from the signal before it is digitized. The remaining path for the video output pulses is to a fast 8-bit digitizer which enables the BPM to make single-turn position measurements.

Tuning the IF amplifier and detector is simple. With a pulse modulated 50 MHz signal at the IF amplifier input, the Phase adjustment is set for best detected positive pulse response, and the DC offset is adjusted to remove DC voltage from the base of the pulse. A trim capacitor in the IF tuned circuit is adjusted for optimum pulse response also.

Fast Dual A/D Module (FAD)

There are two FAD modules in the storage ring and booster ring BPM bins. They are not used in the other locations. The function of these modules is to capture a time record of turn-by-turn beam position data for later analysis. A simplified block diagram of the FAD appears in figure 12. The video input of the FAD drives a fast peak detector having a MOS-FET reset switch. The peak detector output buffer amplifier drives the input of an 8-bit flash A/D. The A/D captures the voltage peak (2V maximum) and drives two first-in-first-out (FIFO) memory chips via an 8-bit buss.

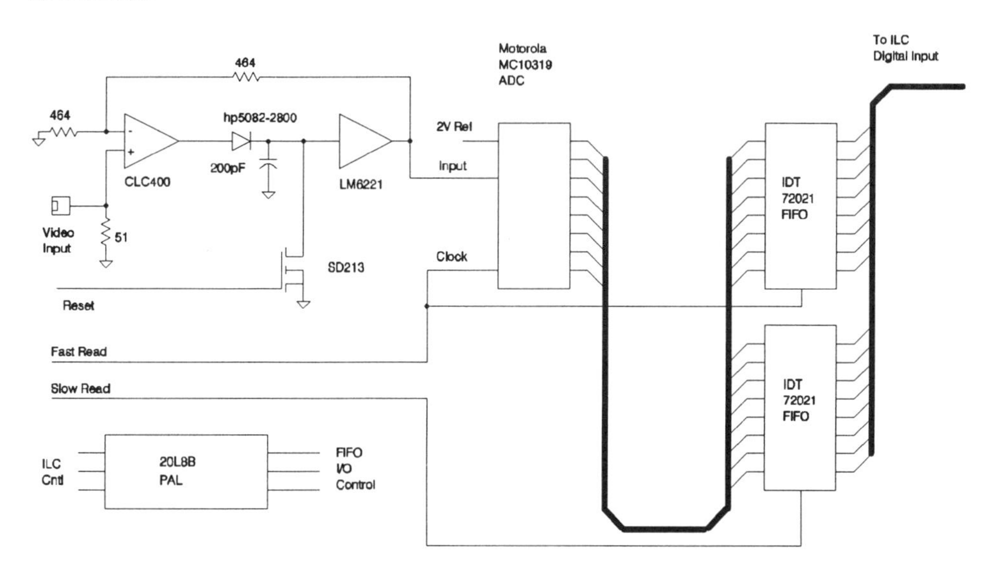

Figure 12. Simplified diagram of BPM Fast A/D module (one channel)

The timing of a FAD sequence is as follows: The ILC sets up FAD operation by clearing any old FIFO data and rearming trigger circuitry. Slightly before beam signals appear at the ring orbit frequency, the reset FET is gated on and the storage capacitor discharged. The FET drive pulse ends and the beam signal appears at the input of the peak detector. Just before the next ring orbit clock pulse arrives the A/D is triggered to acquire the peak of the detected beam signal. After A/D acquisition the FIFOs are strobed to store the data. One FIFO in each channel takes data at the ring orbit frequency. Another FIFO takes data at f_0, $f_0/10$, $f_0/100$, or $f_0/1000$, the rate being set in advance by the ILC via the Timing/Calibration module. This operation continues until an a-synchronous "Halt" pulse (derived in external circuitry) arrives at the bin promptly stopping signal acquisition. All BPM bins receive the "Halt" pulse at the same time. The ILC detects the halted condition and reads the contents of the FIFO memory. Input/output operation of the FIFOs is controlled by a programmable array logic (PAL) chip via commands from the ILC.

In the booster the BPM fast FIFO stores about 250 us of beam orbit data. The slow FIFO may have a time record 250 ms long (albeit with 999 out of 1000 beam signals missing). We are currently using the FAD data from booster BPMs to study beam injection performance. Code has been developed to quickly acquire and analyze the FAD data in the accelerator control system. Fast Fourier analysis of the booster X and Y position data reveals fractional tune. Analysis of data from all BPMs shows beam closed orbit.

Calibration/Timing Module

This module provides timing pulses for the IF and FAD modules. In addition, it generates calibration signals which are used to measure relative gain in the four channels. Input and output operations of the module are controlled by the ILC via a PAL chip. Figure 13 is a simplified electrical diagram of the module.

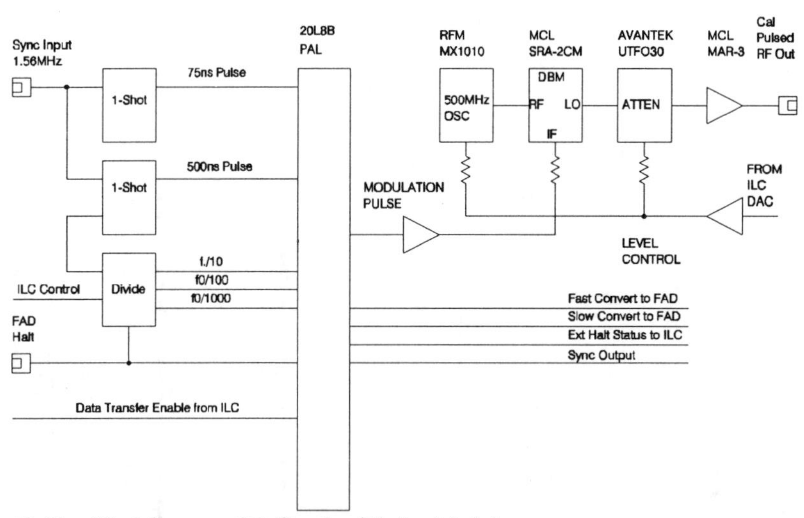

Figure 13. Simplified diagram of Calibration/Timing Module

The calibrator is used to measure differential gain in the RF, IF, and ILC modules. A calibration sequence goes as follows: After the gain of the IF amplifiers has been set by the ILC and beam signals have been acquired, the GaAs switches in the RF module are switched to the calibration input. The calibrator is off at this time as a sample of no-signal offsets is measured.

The beam signals are terminated in 50 ohms in the switches. Switch isolation at 500 MHz is about 50 dB. The calibrator is turned on, and the calibration signal level is automatically adjusted to equal the maximum detected beam signal. The difference between detected calibration signals is measured and gain-offset coefficients calculated. During subsequent beam measurements the measured offsets are subtracted from beam signals and gain coefficients applied to scale the raw beam signal data. With this method we have demonstrated 0.03 mm repeatability over a 40 dB range of RF test signals.

When the BPM measures multi-bunch beam the calibrator pulse length is set to 500 ns. Single bunch beam is simulated by a 75 ns calibration signal. This helps compensate slightly different transient response in the BPM amplifiers.

The calibration signal source is a SAW 500 MHz oscillator. This device has a modulation port which is useful in extending the calibration signal level control range. The RF output of the SAW oscillator drives the RF input of a doubly balanced mixer (DBM) used as an RF pulse modulator. A thin-film voltage-controlled attenuator (Avantek UTF-030) controls the level of the RF pulse at the input of the output buffer amplifier. The block diagram in Figure 8 shows another attenuator in the calibration signal path. This is a programmable step-attenuator used to precisely adjust the calibration signal level in 0.0, -2.5, -6.0, and -12.0 dB steps. When we wish to measure the relative linearity of the IF amplifiers and detectors, we use this attenuator during the calibration sequence to characterize the detectors at four distinct signal levels. This permits us to "linearize" the detectors[8]. Without this compensation, we see beam position errors develop as beam intensity falls. In some applications we may interrupt beam measurements to adjust the BPM gain and re-calibrate, restoring position accuracy. In situations where we may not interrupt the beam measurement for calibration, we rely on the detector linearization technique to give us the accuracy we require. Detector linearization provides about 10 dB dynamic range over which the beam intensity effects are about 0.03 mm.

Local Oscillator (LO)

The function of the LO is to provide the mixing signals required to heterodyne 500 MHz beam signals down to 50 MHz in the RF converter modules. The LO module is installed in the rear of the BPM bin in a small RF-tight enclosure. Figure 14 is a diagram of the this module.

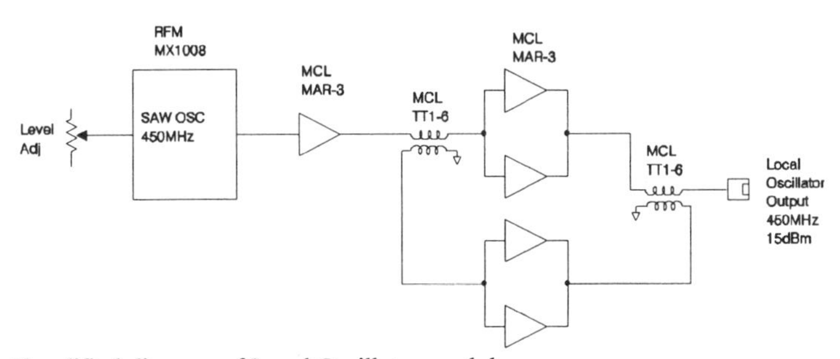

Figure 14. Simplified diagram of Local Oscillator module.

Early in the BPM design phase we considered synthesizing the LO signal from the accelerator master oscillator (500 MHz) and shipping it around the storage ring to all 96 BPM bins. We determined it was less costly to install an oscillator in each bin. Due to the wide bandwidth of the detectors, the LO frequency stability requirements are not severe. The SAW oscillators we are using (from RF Monolithics) are accurate within 100 kHz and their phase noise is acceptably low. (RFM refers to these oscillators as UHF Microtransmitters.) The SAW oscillator power output is adjustable over a 50 dB range up to a maximum of about +10 dBm via an amplitude-modulation port. In the LO we use this port to set the output level for best spectral performance (lowest distortion). We added amplifiers (MCL MAR-3) to boost the LO output to +15 dBm. A hybrid power-splitter in the RF Converter module reduces the LO power to +7 dBm as required by the RF module mixers.

Intelligent Local Controller (ILC)

This module is at the lowest level of the ALS computer control system[9]. Throughout the ALS nearly 600 ILCs will be used to control commercial instruments via RS-232 or the GPIB, run and monitor magnet power supplies, control beam-line components, and run the major systems such as high-power RF amplifiers. BPMs now use 50 ILCs, and 100 more are under construction.

The ILC is the functional equivalent of a Multibus-I or a VME chassis except that it controls fewer devices. It is smaller than these other controllers, fitting in a 3U Eurocard module, and consumes only 5 W. The ILC has three processors; an 80C186, a companion floating-point math coprocessor, and an intelligent serial link controller. These chips share 64 kbytes of dual-ported, battery-backed memory. ILC input/output consists of four channels of 13 bit ADC, four channels of 16 bit DAC, 24 bits of digital I/O, and a 2 Mbit/s serial link to other ILCs and the upper echelon of the accelerator control system. Most of the ILC digital I/O may be configured for either input or output. Additional I/O capability is available with the installation of special daughter-cards connected to an SBX port on the ILC. Through the SBX connector and appropriate cards, we control GPIB devices, RS-232 instruments, or expand the digital I/O by 24 bits. An improved version of the ILC is in the design phase. We expect a 60% increase in speed, four times the memory, 16 bit ADCs, and even less power consumption. Figure 15 is a block diagram of the current ILC architecture.

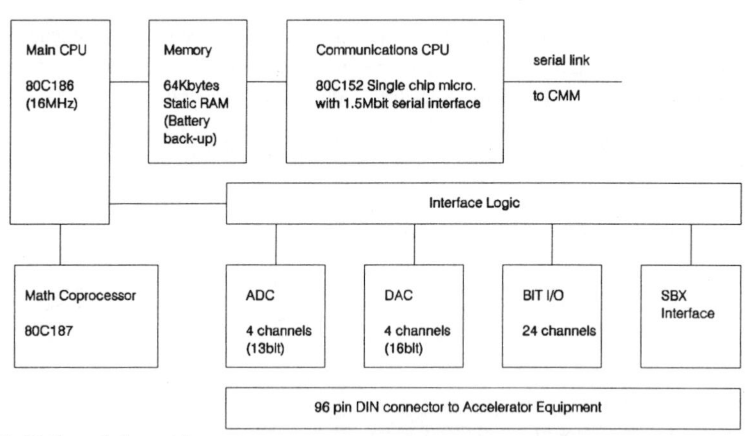

Figure 15. ILC module architecture.

The ALS instrumentation staff developed code for the BPM ILCs in a PC environment using Microsoft Quick Basic 4.5. Programming of the ILC for operation in the accelerator systems is done in C or PLM languages, generally by the Control System staff. In order for us to develop our BPM code we were supplied with Quick Basic libraries, an ILC data base having unique names for each BPM ILC I/O function, and the necessary hardware for PC to ILC communications. We developed a considerable amount of code for testing and developing BPM hardware. Our Basic code ran in the PC, so we did not achieve maximum performance from the ILC. As our code matured, certain functions were programmed into the ILC improving position calculation speed dramatically. In the end we were able to get 30 readings of X and Y position per second with each reading averaged 50 times. Speed is not an issue in the ALS injector BPMs because of the low beam repetition rate. In the storage ring we expect we will achieve about 50 Hz performance with each reading averaged 50 times. Our plan is to synchronize readings in all 96 storage ring BPMs at variable rates set by the machine operator. Total bandwidth for 96 BPM readings and subsequent beam orbit correction will be at least 10Hz and probably higher.

When FAD data are collected and processed by the ILC we see FFTs of X and Y position from a single booster BPM displayed in the control room in a few seconds. Collecting data from all 32 BPMs and displaying data takes a few seconds longer. We are currently working on a real time (1 Hz) display of fractional betatron tune using data from booster BPM FADs. Closed orbit displays will be displayed at 1 Hz also.

Power Supply

Two open-frame linear power supplies are installed in the rear of the BPM bins. We decided not to use smaller switcher power supplies because of the RF noise they generate. The power supplies provide +15, -15, +5, and -5.2 V at about 80 W.

Error Correction

Figures 16 - 18 show how our single-point calibration and multi-point linearization routines improve the BPM performance. The signal source for these tests was a CW generator at 50 MHz connected to the inputs of four IF modules. The the RF input voltage was stepped in 1 mV increments up to 100 mV. Three 180 degree hybrid power dividers evenly split the signal between modules. The gain of the IF modules was set to give full scale voltage at 100 mV into the hybrids. Figure 16 shows the raw DC voltage out of the IF Amplifier/Detector modules and the calculated X and Y positions. The position errors are rather large due to gain differences in the IF modules. Figure 17 shows the effect of a single point calibration at 70 mV input. Notice the detector gains are equal at 70 mV and the X and Y beam position is zero as it should be for equal inputs. Notice also that the calculated position deviates greater that 0.03 mm as the voltage is changed less than a factor of two. Below 10 mV input the calculated position error is large due to the LM1211 detector threshold. Figure 18 shows how the linearization technique overlays the raw data. A straight line is included in the graph for comparison. The position plot for linearized data shows good performance to about 10 dB below full scale.

I stated that we have a method for correcting the pin-cushion distortion we observe in the storage ring BPM response. In order to use this method we must make linear measurements of the button voltage. In our tests of the correction method we used a wire and a network analyzer to measure button response to wire motion. A 20 mm displacement in a single axis resulted in button voltages differing by a factor of 10 or so. When the wire was placed at 20 mm in X and Y, the button nearest the wire developed voltages nearly 100 times higher than the opposing button. To accurately measure voltages over such a dynamic range a very good detector (one essentially as linear as the network analyzer detector) is required. With all of our error correction schemes the LM1211 detector will never be that good.

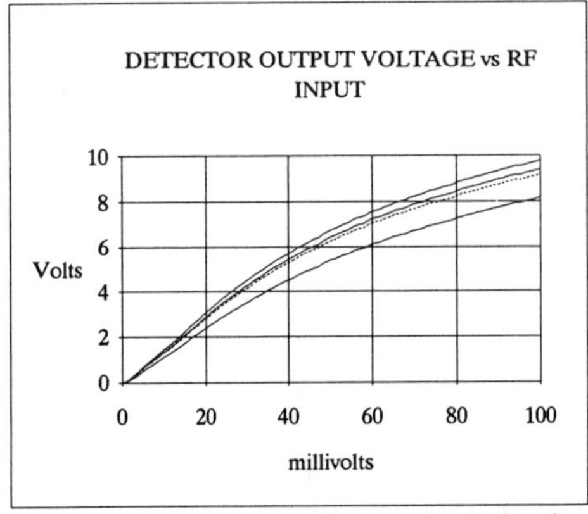

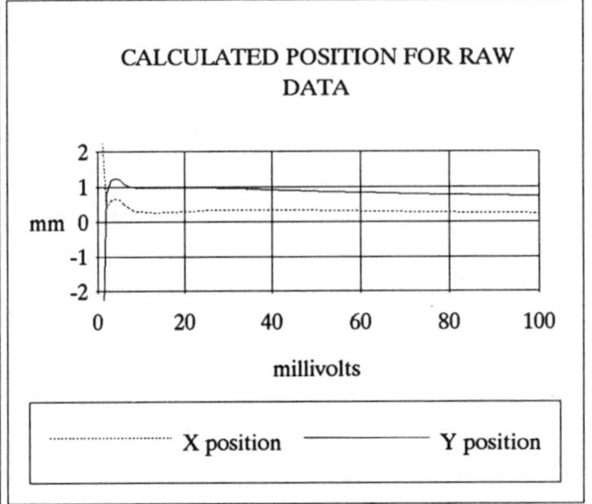

Figure 16. Raw detector voltage and calculated position.

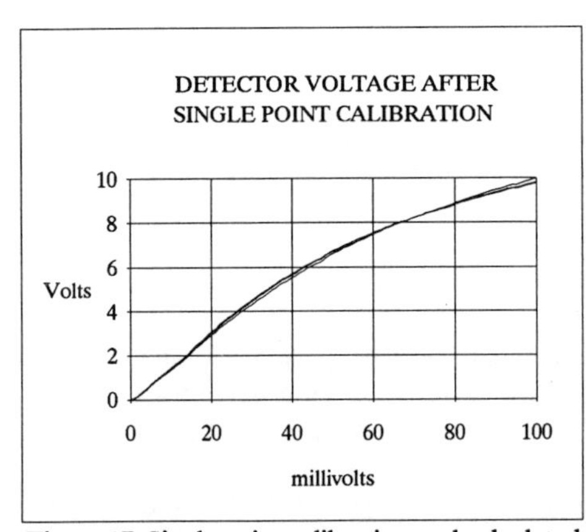

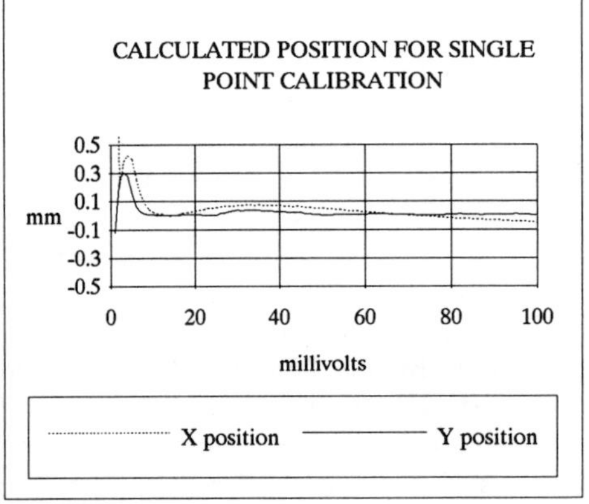

Figure 17. Single point calibration and calculated position.

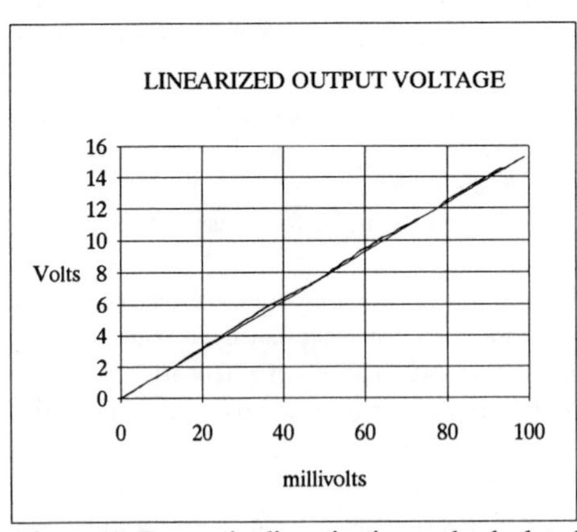

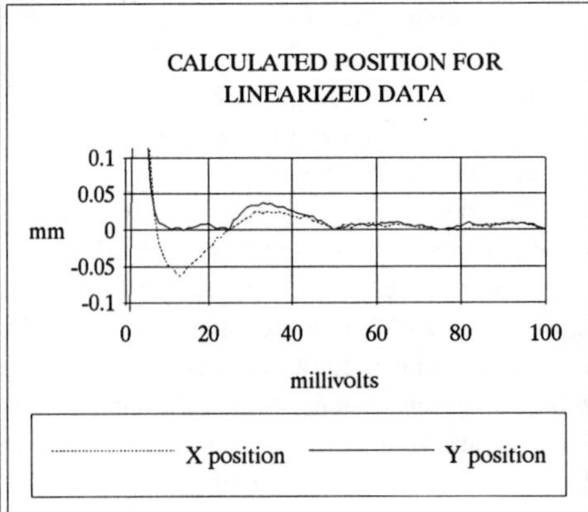

Figure 18. Four point linearization and calculated position.

The ALS storage ring beam is not expected to be off center by more than a few millimeters in normal operation. The vertical aperture will ultimately be limited to 10mm by insertion devices. We expect we will be able to correct for pin-cushion distortion over the anticipated range of beam motion. One may argue that accurate measurements of beam position well off the beam pipe center are not required, that is, we must know only what direction to steer the beam to get it back on center. That may be true, but prudence dictates we should have a method for accurate measurements when the beam orbit is less than perfect.

Conclusion

I have described the ALS beam position monitor and how it is working as we commission the beam injection systems. Having single-turn measurement capability has made the booster BPMs very useful at this early stage. We expect this feature will helpful in commissioning the storage ring also.

The wide bandwidth we require for single-turn measurements made error correction a major task in the BPM development. A single receiver/detector BPM multiplexed to four pickups would certainly have fewer systematic offsets and gain errors requiring compensation. Indeed, we plan to install such systems for beam position monitoring at the entrance and exit of insertion devices. These BPMs (modeled after the NSLS X-ray ring BPMs[10]) will be used for the most accurate and stable monitoring in the storage ring. They will also be part of the active interlock system we will use to prevent vacuum chamber damage due to mis-steered photon beam.

Acknowledgements

My thanks go to a number of people for helping in the development of the BPM. Jim Johnston did much of the circuit design, lab work, and software development. Mike Fahmie coded the PALs. Steve Magyary provided the ILCs and translated BASIC code to PLM. Chris Timossi built and coded the ILC BPM data base. Al Geyer is currently building 100 more BPM bins. I am particularly thankful for an interactive and productive relationship with Tom Henderson, John Meneghetti, and Kurt Kennedy of the ALS Mechanical Engineering staff while we designed the BPM pickups.

References

[1] J. Hinkson, J. Johnston, I. Ko, "Advanced Light Source (ALS) Beam Position Monitor," in Proceedings of the IEEE Particle Accelerator Conference, Vol. 3, p. 1507. (1989)

[2] R. Shafer, "Beam Position Monitoring," AIP Conference Proceedings, 212, E Beadle & V Castillo Ed., p 26, (1989)

[3] R. Littauer, "Beam Instrumentation," AIP Conference Proceedings, 105, M. Month, Ed., p. 869, (1983)

[4] K. Halbach, "Description of Beam Position Monitor Signals with Harmonic Functions and their Taylor Series Expansions," Lawrence Berkeley Laboratory AFRD Report, LBL22840

[5] G. Lambertson, "Calibration of Position Electrodes Using External Measurements," Lawrence Berkeley Laboratory Advanced Light Source Internal Note LSAP05, (1987)

[6] G. Decker, Y. Chung, E. Kahana, "Progress on the Development of APS Beam Position Monitoring System," Presented at IEEE PAC, San Francisco (1991)

[7] J. Borer, et al., "The LEP Beam Orbit Measurement System," in Proceedings of the IEEE Particle Accelerator Conference, Vol, 2, p. 778, (1987)

[8] J. Hinkson, Jim Johnston, H. Lancaster, "Linearizing the Electronics of the ALS Beam Position Monitor," LBL Advanced Light Source Internal Note LSEE061, (1989)

[9] S. Magyary, et al., "Advanced Light Source Control System," in Proceedings of IEEE Particle Accelerator Conference, Vol. 1, p. 74, (1989)

[10] R. Biscardi, J. Bittner, "Switched Detector for Beam Position Monitor," in Proceedings of IEEE Particle Accelerator Conference, Vol 3. p. 1516, (1989)

This work was supported by the Director, Office of Energy Research, Office of Basic Energy Sciences, Materials Sciences Division of the U.S. Department of Energy, under Contract No. DE-AC03-76SF00098.

GROUNDING AND SHIELDING IN THE ACCELERATOR ENVIRONMENT

Quentin A. Kerns
Fermi National Accelerator Laboratory; Batavia, Illinois 60510-0500*

ABSTRACT

Everyday features of the accelerator environment include long cable runs, high power and low level equipment sharing building space, stray electromagnetic fields and ground voltage differences between the sending and receiving ends of an installation. This paper pictures some Fermilab installations chosen to highlight significant features and presents practices, test methods and equipment that have been helpful in achieving successful shielding. Throughout the report are numbered statements aimed at summarizing good practices and avoiding pitfalls

INTRODUCTION

The need to consider shielding was impressed on me 40 years ago when for three weeks a betatron failed to achieve full energy despite checking and rechecking. It turned out the machine had been working all the time; what had happened was that the phototubes being used for beam observation were cut off by the stray field of the magnet about two-thirds of the way up to full energy. I never forgot it. Over the years I found out from experts how to design and use magnetic shields.

Figure 1 is an aerial view of the Fermilab accelerator complex with its 1000 meter radius rings. In the lower center we see the triangular P-bar source and at its left the Booster ring, the Linac and the High-Rise building, facing out over the external beam lines. There are many miles of signal and timing cables encircling the Main and the Tevatron rings and tracing the beam line. A new generation of accelerators replaces many of these cables with fiber optics runs[1]. Here, however, we will discuss the profusion of cables, buses, coaxial lines and waveguides that remain as needed items, not displaced by fiber optics.

Figure 1

*Operated under DOE Contract Number DE-ACO2-76CH0300

Power lines are and will be needed in present and future accelerators. Figure 2 shows the Fermilab master substation. Site power enters on 345 KV high lines on graceful white poles. There are five 40MVA and one 60 MVA transformers, stepping the 345KV down to 13.8KV for underground feeder distribution. Without going into details of all the loads, let me describe two.

Figure 2

The pulsed power load for the main ring is fed separately by XFMR 82B. The P-bar source has its own transformer, 83A.Thus there is some isolation between "noisy" power and "quiet" power. This idea is continued at the 13.8KV to 480V substations; where there is a need to distinguish "quiet" from "noisy" loads, two separate 480V transformers are placed outdoors at a service building. Indoors, the direct 480V and 208-120V stepdown transformers feed the local power loads. Often the loads are SCR or chopper supplies that could cause line noise unless filtered. A good rule is:

(1) Know the power source for your equipment and plan for filters, voltage regulation and non-interruptible power as needed.

EMI GENERATORS

The accelerator environment is not an electronic clean room. We tend to be our own worst enemy. See Table I for some devices that produce conducted or radiated EMI. For brevity, the list includes just a few sources, less than 1% of the total.

Take the first and second columns of Table I as potential EMI generators. Consider that the group of people designing them may not be the same as the group designing the low-level instrumentation. Rule:

(2) Get these groups together, early in the design stage of the high-power equipment. Set goals for permissible radiated and conducted EMI from the high power equipment.

Table I.

A FEW EMI GENERATORS

MACHINE	*GENERA-TOR*	*V, I FREQUENCY*	*CABLE OR TRANS-MISSION LINE*	*GROUNDED OR FLOATING*	*NOTES*	*GROUND FAULT PROTEC-TION*
Tevatron RF	150 KW Tetrode	53 MHz	9 inch coax	Grounded	D.C. flange test	NO
Main Ring RF	150 KW Tetrode	53 MHz	None-Dir. coupled to cavity	Grounded	D.C. flange test	NO
Booster Bias	SCR Supply	0-2500A, 30V, 15 Hz	1/4x4" Bus	Floating	10K Resis-tors to gnd	YES
Booster Guide Field	SCR Supply	67.5A to 970Amps 15Hz	750 MCM water-cooled bus	Floating	Resistors to ground	YES
Booster Kicker MK05	Thyratron	60KV 1.699μsec cable PFN	RG/220U	Supply grounded, kicker fl.	Current waveform monitor	NO
Linac Up-grade Quads	Switched cap banks, 2msec at base	175A, flat top. 870V, 15 pps	3/4" Tri-axial cable	(-)P.S. terminal grounded; mag. coil floating	Triaxial outerbraid grounded, both ends	NO
Linac Up-grade RF	12MW klystron	805 MHz 125μsec 15 pps	WR-975 waveguide	Grounded, both ends	D.C. flange test, RF flange test	NO
Linac Up-grade Modulator	24 MW PFN	125 μsec pulse at 15 pps	Shielded RG/220 pair	(-)Termin. grounded only at klystron xfmr	Triaxial outer braid grounded, both ends	NO
Drift-tube Linac Debuncher	100 KW Tetrode	201 MHz	3 1/8" Rigid coax, 3" Heliax	Grounded, both ends		NO

LOW LEVEL-HIGH POWER JUXTAPOSITION

The central problem is that high power equipment tends to interfere with low level instrumentation. In Figure 3, Typical Accelerator Environment, consider that the box labeled "Power Supply" could be any of the items described in Table 1 columns 1 and 2. The goal is to minimize electrical noise pickup at the instrument racks, where sensitive circuitry may be observing small signals from a detector in the tunnel. The booster (ferrite)

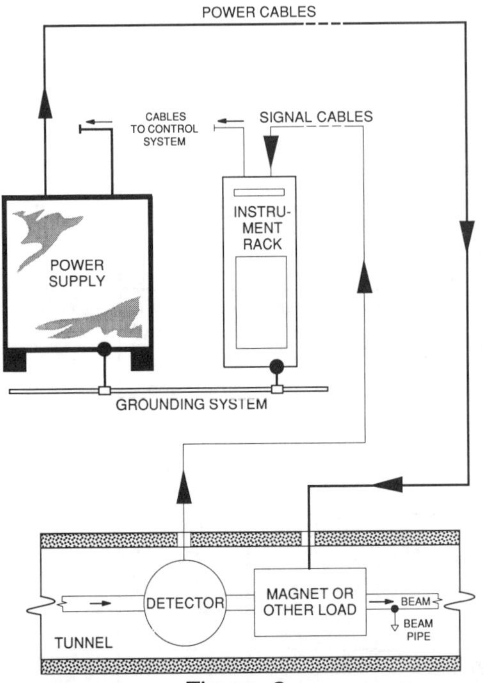

Figure 3

bias supply is an example from Table 1. Figure 4 shows the top of this Ferrite Bias Supply, 30V at 2500 Amps, 15Hz. The black cylinder in the foreground is the 2500 Amp transductor. The 1/4" x 4" busbar pair emerges from the left foreground single-bolt connector and vanishes in the background, heading down the penetration into the Booster tunnel.

The complete current loop of the 2500A supply upstairs in the Booster gallery, down to the ferrite-cored load in the tunnel 25 feet below, is floating. The only connection to ground is a pair of 10KΩ resistors across the power supply terminals to permit ground fault detection. When these soft ground resistors are lifted, the power supply and load can be hi-potted and meggered to ground. The aim of the floating circuitry is to eliminate ground currents. The busbar circuit net enclosed area is kept small to reduce stray magnetic field and the beam pipe is effectively magnetically shielded with a long Permalloy cylinder. The conducted ground current is a displacement current of about 250 microamps, 140 dB below the load current and certainly not troublesome. What can we say about large power supplies?

Figure 4

Large power supplies (*e. g.,* kicker supplies) need these rules:

(3) Reduce magnetic coupling loops in the high current circuits to the minimum.

(4) Provide a separate return circuit for every voltage source, so that you can:

(5) Ground the system at the best point, often with a soft ground like a resistor.

Figure 5 shows the tunnel of the P-bar source with debuncher ring (left) and accumulator ring (right). The bend and quad magnet circuits (see Table I) are floating except for the soft grounding via the 10KΩ resistors used in the ground fault protection circuit Leakage currents of more than a few milliamps would compromise the 10ppm magnet regulation and cannot be allowed. As the picture shows, the laminated iron core of each magnet is grounded by a heavy copper wire to a ground bus encircling the rings just below the lowest cable tray of each ring. The stainless steel beam pipe firmly contacts the magnet laminations and thus is grounded at each magnet. The copper ground bus on the cable tray connects to an array of ground rods buried in the earth outside the tunnel. Each ring has a D.C. Beam Current Monitor (a second-harmonic magnetic modulator type). It is magnetically shielded by multi-layer Permalloy[2] pipes.

Figure 5

Figure 6 shows the surface buldings of the P-bar source. Space does not permit a description of the very interesting RF stochastic cooling ring and kicker systems of the P-bar source, but a tribute is in order. No stack of P-bars has ever been lost because of thunderstorms[3].

Figure 7 is a view down in the Main Ring tunnel showing the row of 18 copper accelerating cavities (53 MHz). The three-stage power amplifiers are clamped directly to the cavities by Marman clamps. All power amplifier sections are tin plated for good RF contact. To check the contact after an amplifier is replaced, D.C. current of 100 amps is run from the top of the amplifier down to the cavity and the voltage drop at each clamp joint checked to verify that it is below 50 μV @ 100 Amps.

Figure 8 shows the top of the 12 MW, L-5859 805 MHz klystron. There is a stack of 4 circular, lead X-ray shields surrounding the isolated collector and additional lead shielding around the top of the solenoid. Both X-ray levels and RF leakage had to be addressed, because the circular shields act

Figure 6

as both X-ray attenuators and as a metallic enclosure to shield collector RF. For improving the RF shielding, it was necessary to remove paint from the contacting surfaces and to apply bolt pressure to force flanges into intimate contact. A later figure shows the D. C. current method of testing joint contact resistance, applied to the stack of lead shields.

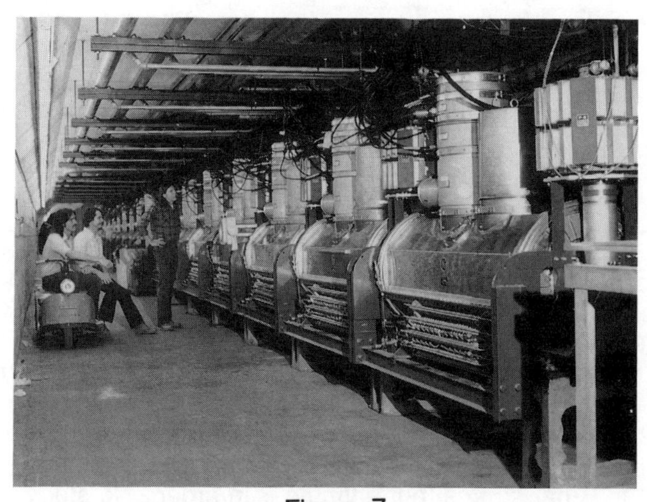
Figure 7

Figure 9 shows the testing of the WR-975 waveguide joints with a 2-point probe. The Tektronix 7104, 1 GHz oscilloscope, directly shows the 805 MHz signal if the flange joint is leaky. At 8MW in the waveguide, every joint exhibits a few millivolts; the test is to see whether a given joint is much leakier than the best joint[4]. If it is, it can be restored by cleaning the flanges, installing a new gasket and retorquing the flange bolts.

One of the most demanding RF installations is the amplifier chain for stochastic cooling, where signals at the shot noise level in the beam are amplified up to watts at the TWT outputs (Figure 10). Even small RF leaks could cause the amplifier chain to become an oscillator. Careful attention is paid to using solid-jacket coaxial cable with good-quality connectors, properly torqued. In some other installations, we have found it best to solder connectors to the copper-jacketed cable (even if the connectors are designed for straight mechanical assembly).

The above experiences suggest another rule:

(6) Have a test plan to determine that the installed shielding actually works as intended.

Figure 8 Figure 9

DETECTOR SHIELDING EXAMPLES

Table II.

DETECTOR	FREQUENCY	APPROX. SIGNAL LEVEL	MAXIMUM ALLOWABLE INTERFER-ENCE	AMBIENT STRAYS	MINIMUM SHIELDING BELOW AMBIENT
D.C. Beam Current Monitor	D.C.	1 mA	0.5μA	10 Gauss... 20 mA on beampipe..	....60 dB ....92 dB
Beam Position Detector	53 MHz	150μV	1μV	1mV/meter	60 dB
Cavity Phase and Amplitude loops	805 MHz	1 Volt	0.5 mV	20mV/ meter	32 dB
Modulator Regulation Loop	125μsec pulse, 15pps	10 Volts	1 mV	2 Volts between widely-separated racks	66 dB

Figure 10

If the EMI generators of Table I are shielded carefully, the ambient noise can be reduced to that of column 5 above. I would consider the levels of column 5 to be a reasonably livable electrical environment. Then, the request to the group designing high-power equipment would be "Shield well enough to meet column 5". The request to the detector people would be, "Shield well enough to live with column 5".

Take a speciall case, the 12 MW klystron of Table I. To reduce 12 MW to 20 mV/meter ambient requires ~131dB of shielding. To reduce 20 mV/meter to 0.5 mV requires only 32 dB.

It may seem unfair to make the high-power people work harder than the low-level group-but it may turn out that way. Keiser[5] discusses "Principles of Electromagnetic Compatibility" in a general context.

SOME TOOLS FOR SHIELDING TESTS

Figure 11 shows some tools for shielding tests. The current transformer[6] on the left has a 3dB passband from 23Hz to 180KHz. It is used to check for current where there should be none (we always find some). Next to it is the Tesla coil, or "cattle prod". The Tesla coil generates ~50KV at its tip. If the tip is not close to another conductor, there is air corona at the tip; the corona spikes form a handy, controllable, radiated, wideband source[7] good from 100KHz to ~4GHz (If the tip is close-sparked to another conductor, the spectrum widens to 22GHz). The Tesla coil is excellent for testing the shielding of certain critical TTL and ECL logic circuits where the error rate must be very small to avoid expensive nuisance trips.
The next item shown is a short dipole antenna formed by stripping back a length of copper-jacketed coax. It is used for hunting corona in high voltage apparatus. A hot-carrier diode detector with 10 millisecond decay time is convenient for scope observation of corona. A second use for the

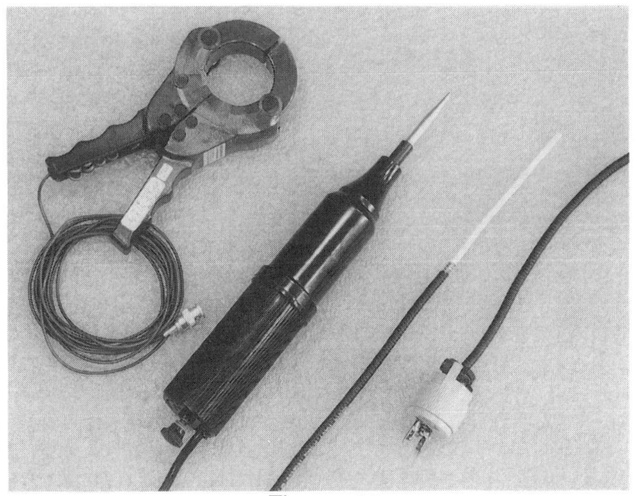

Figure 11

antenna is to radiate RF to test shielding. It is then called an "RF sprinkler". We expect rack instrumentation to work in RF fields up to several V/meter; the RF sprinkler helps verify this. The last thing shown is the 2-point probe that appeared in figure 9, measuring 805 MHz signals across waveguide flanges. The 2 points are very directly connected to a 1/4" diameter solid-jacket 50Ω coaxial cable.

Figure 12 shows the current transformer in use. We found an unwanted external current of ~6 amps peak, shaped like the 125μsec modulator pulse. It was narrowed down to a particular cable. Then it was clear where to add filtering to reduce the stray current to a few milliamps.

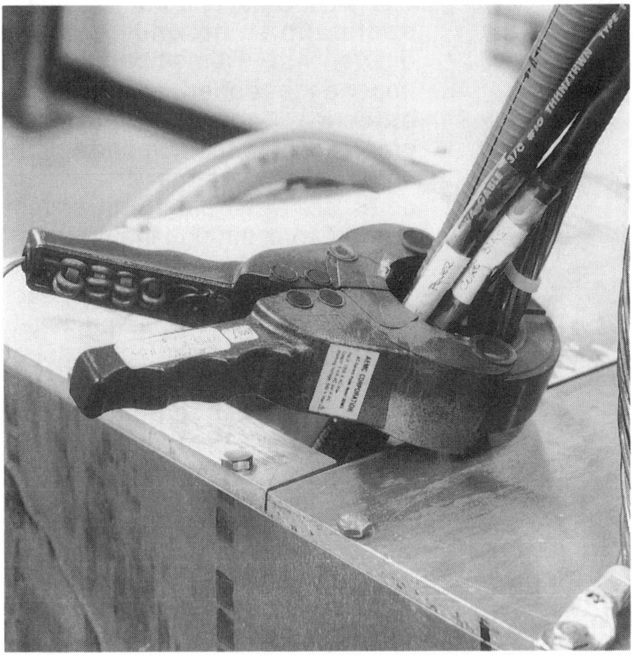

Figure 12

Figure 13 is a typical rack installation. Note the ground wire connected to each rack top. Note also that the shields of all cables are coaxially grounded at the rack top, as is also shown in Figure 17. Sometimes adjacent rack tops are joined by full-width, 10-mil copper jumpers, to create a ground plane.

FIgure 14 shows copper water pipe in use as a conduit (shield) for the primary 117 V.A.C. to the Sola transformer located inside the rack. Power line spikes and RF were prevented from coupling into the rack interior

Figure 13

volume. The rack becomes a first-stage shield. Later, we say more about staging.

Figure 15 shows an earth ground conductor running to a steel building column. It is common to measure milliamps to amps fluctuating currents of 60 and 180Hz A.C. on these conductors.

Figure 16 shows an assortment of ferrite and Permalloy tapewound cores for raising the external impedance of cables, thus reducing coupled noise. Increasing the external impedance lowers the external currents that arise both from B-dot and from source to destination ground voltage differences. Remember that the impedance added to the cable external current path is proportional to the number of turns squared. The larger tapewound cores are often used with several turns of coax through them.

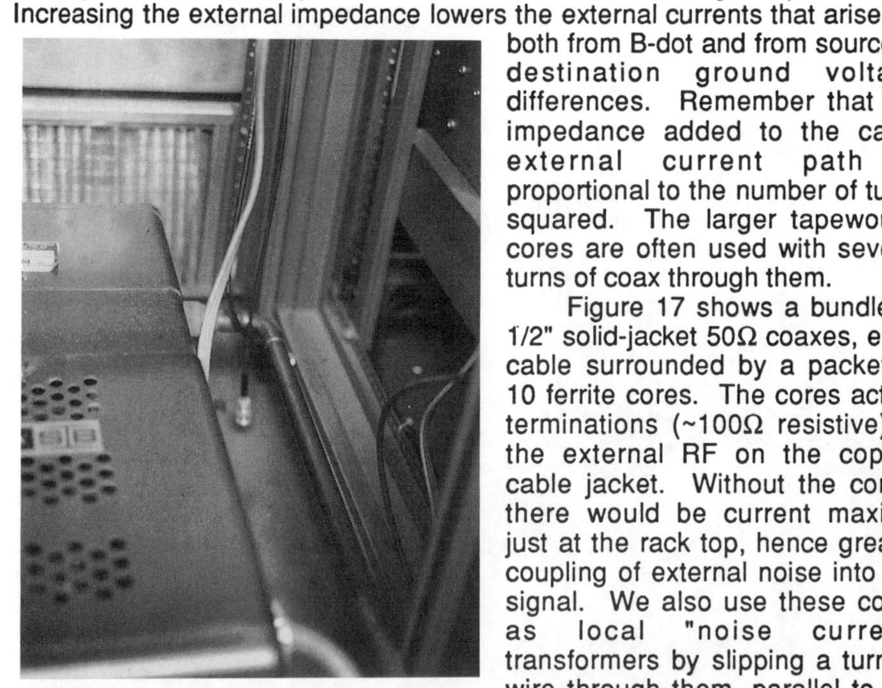

Figure 14

Figure 17 shows a bundle of 1/2" solid-jacket 50Ω coaxes, each cable surrounded by a packet of 10 ferrite cores. The cores act as terminations (~100Ω resistive) to the external RF on the copper cable jacket. Without the cores, there would be current maxima just at the rack top, hence greater coupling of external noise into the signal. We also use these cores as local "noise current" transformers by slipping a turn of wire through them, parallel to the cable, observing via oscilloscope.

Figure 18 shows the vertical omnidirectional antenna for constantly monitoring RF leakage. The klystron drive is inhibited if RF leakage exceeds the set threshold of ~50mV/meter. RF leakage is undesireable for

Figure 15

many reasons. It can cause offsets in operational amplifiers, aliasing in digital sampling scopes and, in larger doses, is potentially hazardous[8].

Figure 19 shows an RF-tight, see-through screen surrounding the X-ray radiation meter. It was observed that very small amounts of RF invalidated the X-ray measurements. The screen was separately tested by beaming RF at it from the horn antenna; it passed the test (as you would have expected).

Figure 20 shows the horn antenna beaming short bursts of RF at the racks. Some of the rack-mounted equipment did not pass the test. There followed a session of removing paint from old panel backs, grounding metal pot shafts and enclosing rack space with blank panels to cover up holes to get the required shielding.

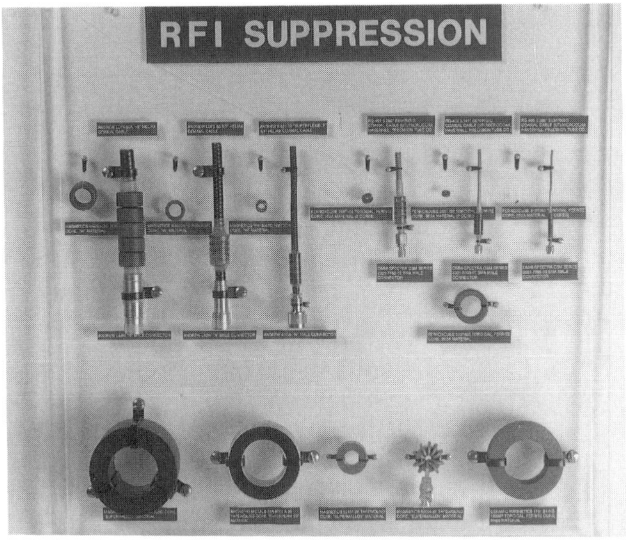

All the racks have power line filters where A.C. comes in at the rack top. All rack top panels are tin-plated copper, bolted to a tin-plated border at the rack top.

Commercially-made, rack-mounted equipment nearly always has painted panels that do not contact the rack adequately as furnished. We have come to expect we will have to improve the contact ourselves.

Figure 16

Figure 21 shows

Figure 17

Figure 18

the copper panels installed in one special rack to make it electrically tighter for its low-level analogue circuitry, controlling the 24MW modulator.

FIgure 22 is a view of racks in the control room Main Ring RF building. Note the blue-painted copper pipes. These pipes carry patch panel cables to the north, east, south and west sides of the control room.

Figure 23 is a close-up of the pipes entering the rack top, which is a tin-plated copper plate attached to a tin-plated border on the rack top.

Figure 24 shows the D.C. current scheme in use to identify and pinpoint poor connections, which, in this case, caused RF leaks from the klystron collector of Figure 8. The bottom ring-to-polepiece connection was the poorest, followed by the top plate. Both connections were later improved.

Figures 25 and 26 show internal shields placed in a NIM plug-in to protect the 9685 comparator from false triggering. In this case, a spurious pulse would shut down the accelerator. To avoid nuisance trips, the shielding was added and bench-tested, up to NEMP levels, 50KV/meter and 5nsec rise time.

Figure 27 shows active noise reduction. The noise voltage difference at power line frequencies and magnet cycling rate between the tunnel ground (*i. e.,* beam pipe) and the rack ground was a problem. In order to complete a run in timely fashion, the circuit shown was assembled. It succeeded in reducing the noise by ~50dB from 3Hz to 1KHz. A

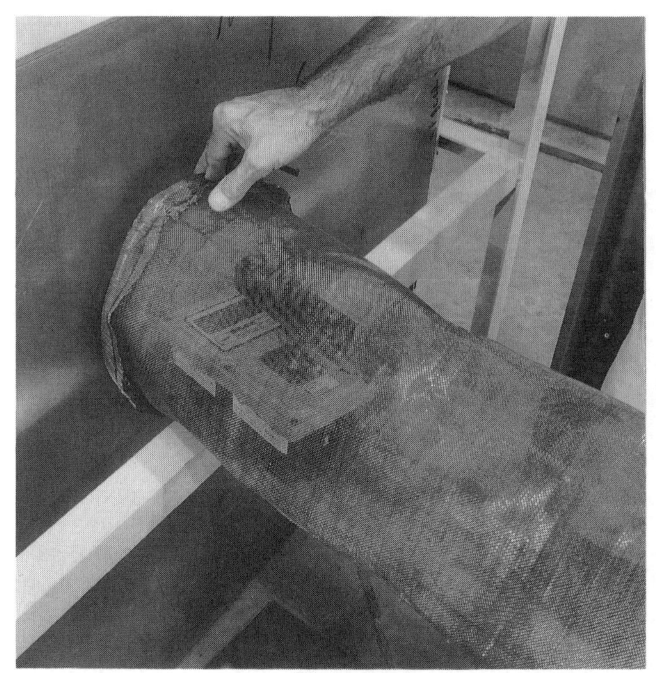

Figure 19

Figure 20

completely passive circuit with a huge core would do a similar thing. Even better would be a detector (or rack receiver) that would eliminate the need for a ground at both ends (we had neglected to follow our own Rule no. 2). It is helpful to design tunnel mounted detectors so they are electrically isolated from the beam pipe. Then, a metal box surrounding the detector is required, of sufficient wall thickness to provide eddy-current

shielding; this box must join the upstream and downstream ends of the beam pipe so that external currents on the beam pipe flow over the box exterior (a Gaussian enclosure). Figure 28 is a tuned, stripline beam position detector. The external beam lines use a number of these; the 53 MHz bunch structure of the beam generates the position signal. Because some beams are low current, the circuitry must be free of interference at even the microvolt level.

An unexpected case of insufficient metal thickness in a shield provides the next example.

SKIN DEPTH EXAMPLES

Figure 29 shows stripline beam position detectors in the beam line, underground at the NØ1 service building.

Figure 21

It was found that there was crosstalk between the detectors in the upper beam and the lower beam. The culprit proved to be a 4-mil Titanium window at the end of the detector, leaking RF because it was too thin-only about 2 skin depths. The solution, shown in Figure 30, was to add a 10-mil sheet of copper to supplement the Titanium. This situation is a corollary to experience with the main ring. The stainless steel beam pipe is not sufficiently thick to exclude all leakage due to wall currents of the beam. Anywhere around the ring, in service buildings, signals of 10-50 μV/meter can be detected and deciphered to show the circulating beam structure. That this is due to beam and not stray RF is shown by shutting off beam and leaving the RF on; the signal goes away.

SUCCESSIVE LAYERS OF SHIELDING AT INSTRUMENTATION RACKS

Standard packaging items can stack up to provide effective shielding. Instrumentation racks, bonded together to form a ground plane at the top as described, are the first layer of shielding. Now the internal volume of the rack has a lower ambient noise level than the outside. Like the onion, more layers follow.

The second layer is the crate, bin or instrument case, with its front panel fully contacting the tin-plated rack frame. A third layer is the plug-in module *per se*. There may be a fourth layer, a separate full-metal enclosure on the

Figure 22

printed circuit board of the module. Walker[9] discusses crosstalk on the board itself.

Power supply wires from one layer to the next are low-pass filtered. Signals are carried from one layer to the next on coax with external ferrite cores. The net effect on noise is that of a ladder attenuator. The rule to make this work is:

(7) Bring all signals in and out of the rack top (the outermost layer of the onion).

A TECHNIQUE TO IMPROVE S/N RATIO

Figure 31 is a view of the massive liquid argon calorimeter at DØ, to be put in the Tevatron colliding beam in 1992. The shielding for the 55,000 twisted-pair signal wires, some of them shown in Figure 32, forms a total system enclosure. Figure 33 is a typical copper duct. Multiply shielded, 150 KVA power transformers feed A.C. power to the platform and three floor-levels of equipment. The grounding of the entire system will be chosen for optimum noise rejection.

Figure 23

Besides careful shielding and grounding, baseline subtraction[10] helps to produce accurate signals. Baseline subtraction here depends on knowing the crossing time of P and P-bar bunches inside the detector (known from RF bucket timing). The signal amplifier's output is sampled and held separately two times, once just before crossing and again just after crossing. A digitized difference is taken to be the true signal, free of drift and low-frequency noise.

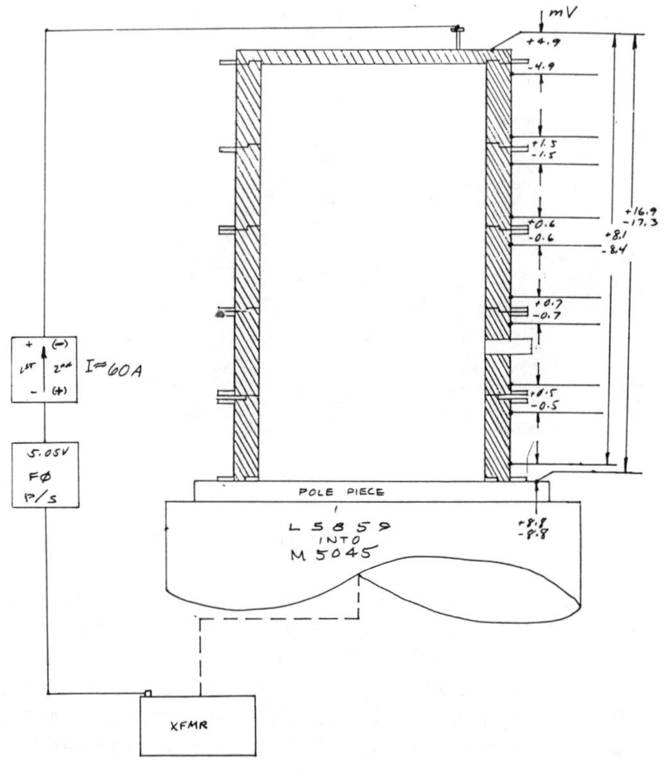

Figure 24

Does shielding have a future? Yes, see Figure 34-the <u>next</u> Fermilab project!

ACKNOWLEDGEMENT

I would like to thank Paul L. Cliff of Fermilab for expert help in assembling this report.

Figure 25

Figure 26

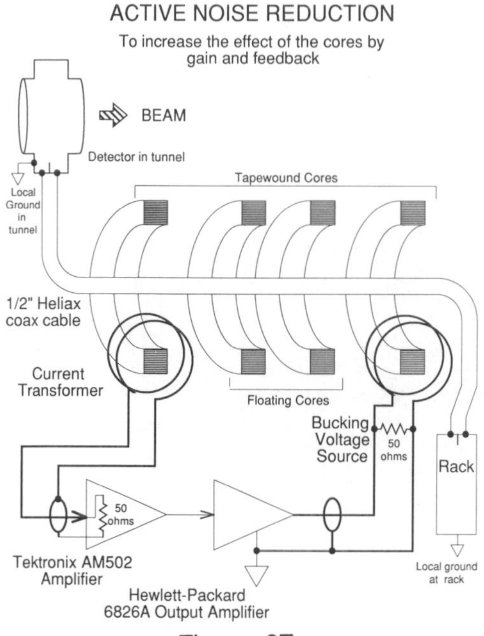

Figure 27

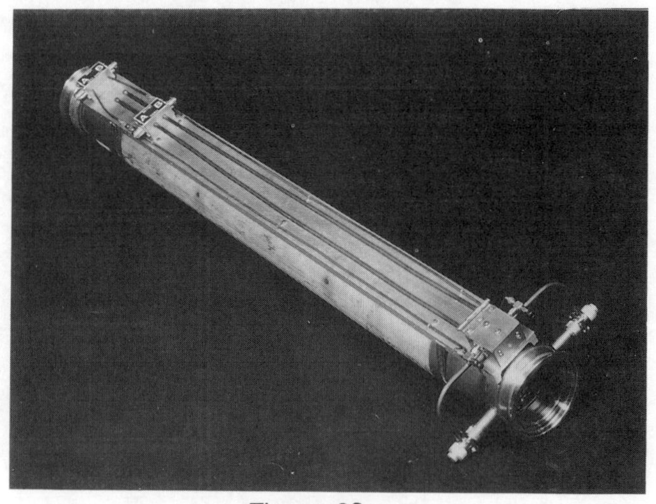

Figure 28

Figure 29

Figure 30

Figure 31

Figure 32

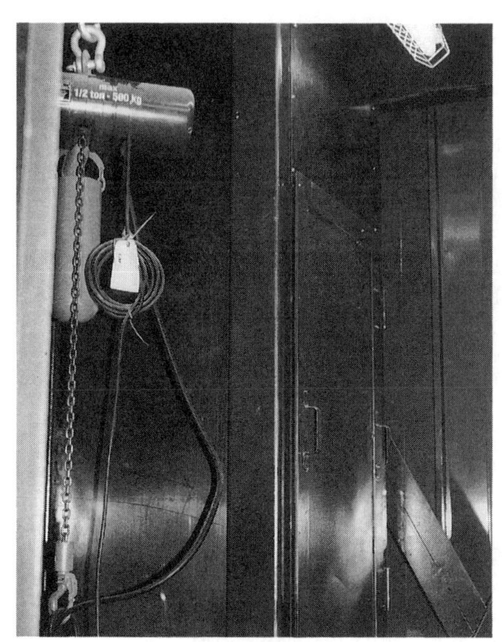

Figure 33

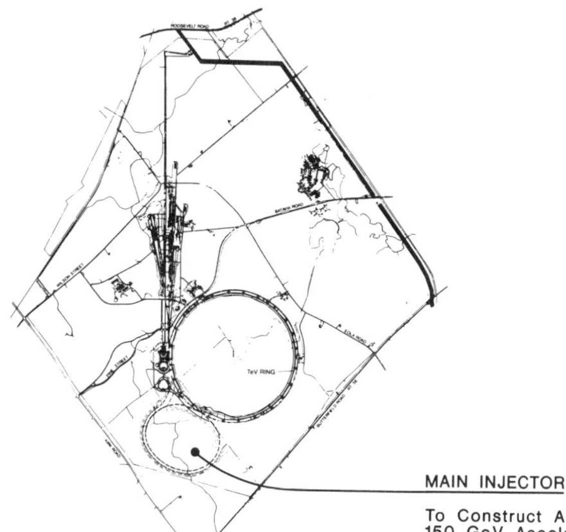

FERMILAB
MAIN INJECTOR

Figure 34

BIBLIOGRAPHY AND NOTES

1. Private communication, from Scott MIller, SSC Timing Section Head, and Sam Crivello, SSC Fiber Group. Fiber optics are cost-effective. Single-mode fibers with pure silica cores are the choice for the 87 Km runs around the Collider, and shorter runs for the HEB, MEB and LEB rings. Radiation effects on fibers and optimum topologies are under study.

2. The name 'Permalloy' was given to the high permeability iron-nickel alloys for low magnetizing forces by Bell Telephone Laboratories. See G. W. Elmen, "Magnetic Alloys of Iron, Nickel and Cobalt", Journal of Franklin Institute, Vol. 207, May, 1929, No. 5. We used 4.75-80 Permalloy; it is marketed under numerous trade names and can be formed, TIG welded and annealed for D.C. μ >50,000.

3. Private communication, Robert Oberholtzer, Fermilab.

4. Aluminum alloy waveguide needs a coating; Spec. MIL-C-5541D Class 3 chromate conversion coating gives suitable electrical contact.

5. Bernhard E. Keiser, "Principles of Electromagnetic Compatibility", 3rd Edition, Artech House, 685 Canton Street, Norwood, MA 02062.

6. AEMC Corporation, Model SD603. We fitted it with a coaxial output cable to a terminated 50Ω scope input.

7. The Tesla coil is a portable impulse generator, handy for comparative tests. Indoors, at distances between 1 foot and 40 feet, the signal can be set by distance over a 60dB range.

8. See IEEE Publications C95.1 (1991) "Standard Safety Levels with respect to Human Exposure to RF, Electromagnetic Fields, 3KHz to 300 GHz" and C95.3, "Recommended Practice for the Measurement of Potentially Hazardous E.M. Fields for RF and Microwaves", avaialable from:

IEEE Service Center
P. O. Box 1331
Piscataway, NJ 08855

9. Charles S. Walker, "Capacitance, Inductance and Crosstalk Analysis", (1990), Artech House *ibid*.

10. Private communication, George Krafczyk, Fermilab.

COUPLED BUNCH MODE INSTABILITIES MEASUREMENT AND CONTROL

D. P. McGinnis
Fermi National Accelerator Laboratory, Batavia, IL 60510*

Introduction

This paper is concerned with longitudinal coupled bunch motion in circular particle accelerators. We will restrict our attention to dipole motion in the absence of Landau damping in order to bring out simple but important concepts of coupled bunch mode motion.The paper will discuss how the motion is excited, methods for detecting the motion, and methods for damping the motion.

In a circular accelerator, beam is accelerated with a RF system that operates at a harmonic of the revolution frequency. As shown in Fig. 1, acceleration occurs when the beam passes over a gap in the beam pipe. The gap in the beam pipe is surrounded with a resonant structure called a cavity. The cavity stores electromagnetic energy. The energy is pumped into the cavity by driving a power amplifier (PA) with an RF oscillator tuned to one of the resonant frequencies of the cavity (usually the fundamental mode). This stored electromagnetic energy will place an electric field across the gap in the beam pipe which will accelerate the beam. The integral of the electric field along the gap is commonly refered to as a "voltage" which can transfer energy to the beam. The RF designer usually tries to maximize this voltage for a given power delivered by the PA. This can be done by raising the impedance of the cavity as seen by the PA. Unless the designer is clever, raising the impedance of one mode in the cavity usually raises the impedance of the other modes in the cavity.

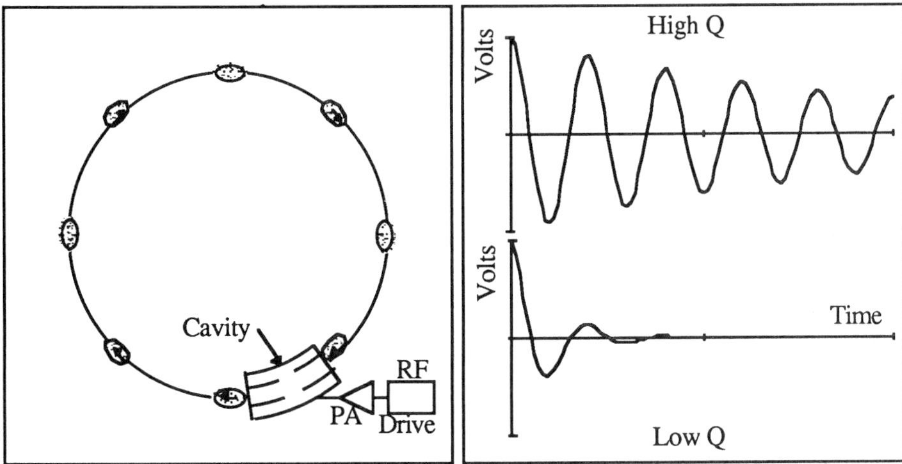

Fig. 1. RF system accelerating bunches in a circular accelerator.

Fig. 2. Impulse response of a cavity.

The maximum number of bunches in the beam is equal to the ratio of the RF frequency to the revolution frequency (which must be an integer and is usually designated as "h".) These bunches reside in RF buckets which are equally spaced

* Operated by the Universities Research Association under contract with the United States Department of Energy.

around the ring and move around the ring at the revolution frequency. When a given bunch of the beam passes through the RF cavity it not only sees the accelerating voltage due to the PA but because the beam can be thought of as a current source, it excites a voltage across the gap due to the impedance of the cavity. If the length of the bunch is short, it acts as an impulse source which not only excites the fundamental mode of the cavity but all the higher order modes as well. The length of time that these modes exist (or "ring") in the cavity is proportional to the quality factor (Q) of the modes as shown in Fig. 2. Usually, the higher the impedance of the mode, the higher the Q of the mode. When the next bunch arrives at the cavity, it sees the accelerating voltage due to the PA plus the voltage due to the higher order modes that were excited by the previous bunch. This extra voltage due to the higher order modes influences the position and shape of the bunch in the RF bucket. Thus, the two bunches are coupled via the wakefields residing in the cavity. This coupling continues all the way around the ring. If the impedance is large enough and the phase of the coupling fields is just right, one can see that this coupling voltage will continue to grow with time resulting in an unstable situation.

Synchrotron Motion and Phase Space

This next section will review the concepts of synchrotron motion and phase space.[1] For simplicity, we will assume that the beam is bunched but not being accelerated. Consider a particle which has a revolution period equal to:

$$T_{ri} = T_0 + \tau_i \tag{1}$$

Since the particle does not arrive at the cavity when the cavity voltage is equal to zero, it will experience a momentum change:

$$d\Delta p_i c = -eV_c \sin(h\omega_0\tau_i) \tag{2}$$

For very small τ_i:

$$d\Delta p_i c = -eV_c h\omega_0\tau_i \tag{3}$$

Because of this momentum change, the particle's revolution period will change according to:

$$\frac{\tau_i}{T_0} = \eta\frac{\Delta p_i c}{p_0 c} \tag{4}$$

where:

$$\eta = \frac{1}{\gamma_t^2} - \frac{1}{\gamma^2} \tag{5}$$

Since the time error τ_i and the energy error $\Delta p_i c$ changed only once every revolution period, the time rate of change of these quantities can be written as:

$$\frac{d\tau_i}{dt} = \eta\frac{\Delta p_i c}{p_0 c} \tag{6}$$

$$\frac{d\Delta p_i c}{dt} = \frac{-eV_c h\omega_0 \tau_i}{T_0} \tag{7}$$

Combining Equations 6 and 7:

$$\frac{d^2\tau_i}{dt^2} + \frac{\eta e V_c h\omega_0}{p_0 c T_0}\tau_i = 0 \tag{8}$$

The solution to Eqn. 8 is:

$$\tau_i = \tau_{0i} e^{\pm j\Omega_s t} \tag{9}$$

where Ω_S is called the synchrotron frequency and is given as:

$$\Omega_s^2 = \frac{\eta e V_c h\omega_0}{p_0 c T_0} \tag{10}$$

According to Eqn.(10), the RF voltage must flip phase by 180° at the transition energy to keep Ω_S real (stable).

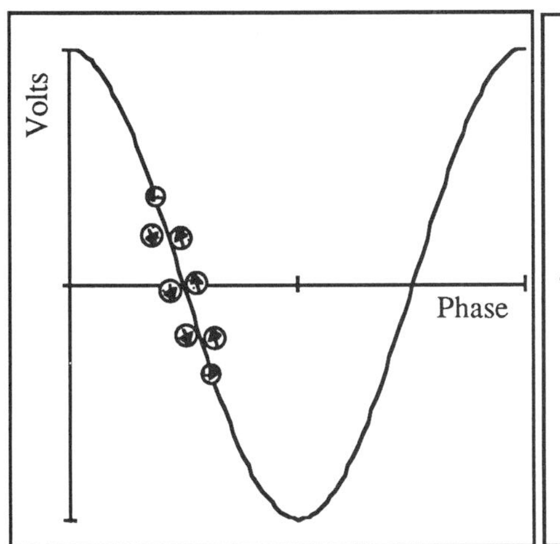

Fig. 3. Time evolution (above γ_t) of the RF phase of a particle as it undergoes a synchrotron oscillation.

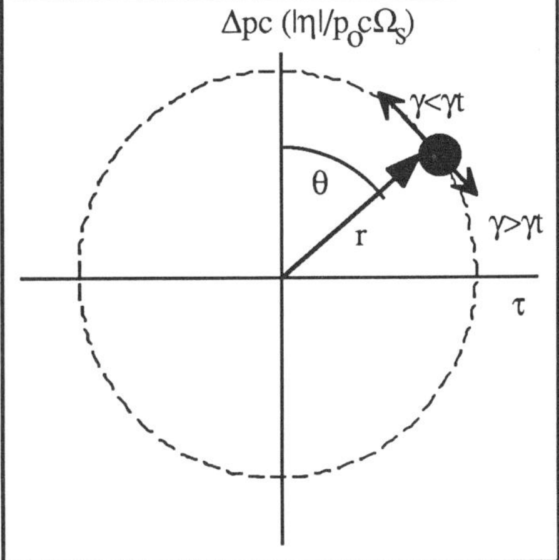

Fig. 4 The momentum error vs. the phase error of a particle undergoing synchrotron oscillations.

The motion described by Eqns. 6 and 7 is shown graphically in Fig. 3. Consider a particle that arrives at the cavity late in time. Assume that the accelerator is above the transition energy ($\gamma > \gamma_t$). The particle will be decelerated and on the next turn arrive earlier in time. The particle will continue to be decelerated until it arrives early in phase with respect to the RF. At this point the bunch will start being accelerated. One can see that the particle's phase will oscillate around the RF zero crossing. Figure 4 shows the momentum error of the particle ($\Delta p_i c$) plotted against the time error of the particle. The trajectory is a circle. Below γ_t, the particle orbits counter-clockwise and above γ_t, it rotates clockwise. This plot is called the phase space trajectory of the particle.

Because the bunch has many particles, it is more convenient to plot the distribution of particles in phase space as shown in Fig. 5. The distribution shown in Fig.5 is described in polar coordinates of phase space (r,θ) where the time error is given as:

$$\tau = \frac{T_0}{2h} r \sin(\theta) \tag{11}$$

and the momentum error is:

$$\Delta pc = \frac{p_0 c \Omega_s T_0}{|\eta| 2h} r \cos(\theta) \tag{12}$$

where θ is some initial angle θ_0 plus a time dependence $\pm\Omega_s t$. Using Fig. 4 as a guide, the top sign is used above transition and the bottom sign is used below transition. Because the distribution must be single valued in θ, the distribution for bunch **i** can be expanded as a Fourier series in θ [2]:

$$g_i(r,\theta) = \frac{N_b \sum_{k=-\infty}^{\infty} c_{i,k} f(r) e^{jk(\theta_0 \pm \Omega_s t)}}{2\pi c_0 \int_0^{\infty} f(r)\, r\, dr} \tag{13}$$

Since $g_i(r,\theta)$ must be a real quantity:

$$c_{i,-k} = c_{i,k}^{*} \tag{14}$$

The function f(r) is the radial distribution of the phase space density. For simplicity, we will assume a "hockey puck" distribution [3] as shown in Fig. 6. The time domain profile of this distribution is found by projecting the distribution on the time axis. The result is a parabolic bunch shape as shown in the lower portion of Fig. 6.

The Fourier coefficients are the strength of the various multipole moments of the beam. For example k= ±1 is dipole motion, k=±2 is quadrapole motion, and k=±3 is sextupole motion. The distribution for monopole, dipole and quadrapole is shown in Fig. 7. Each of the multipole distribution rotates in phase space at the angular rate of Ω_s. The time domain response of the monopole and dipole modes and the response of the monopole and the quadrapole modes is shown in Fig. 8. Dipole motion is characterized by the oscillation of the centroid of the bunch. Quadrapole motion is characterized by the changing width of the bunch.

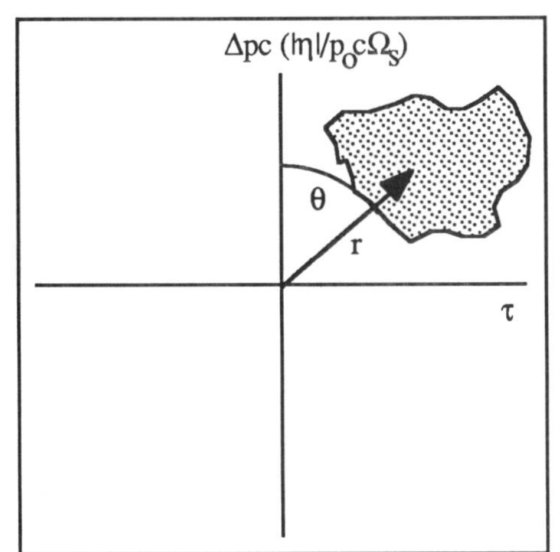

Fig. 5 A distribution of many particles in phase space

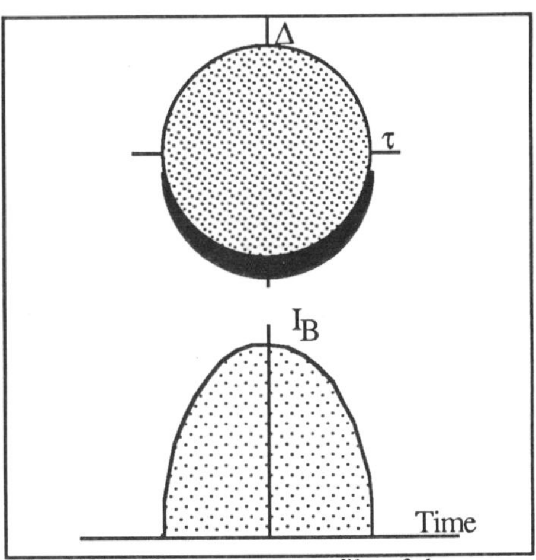

Fig.6 The phase space profile of the "hockey puck" distribution and the time domain projection of this profile.

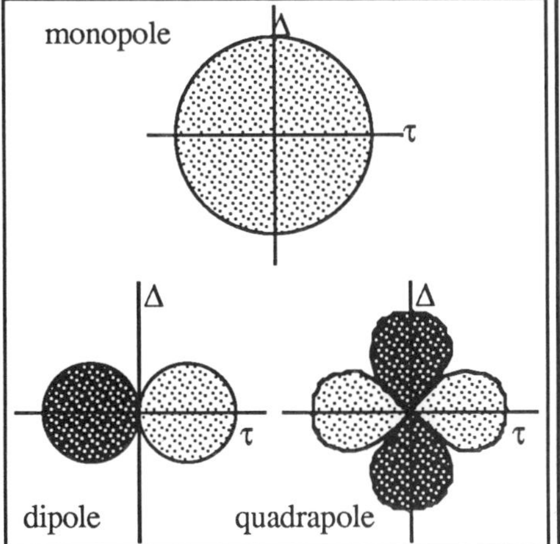

Fig. 7 The phase space distributions of the monopole, dipole, and quadrapoles. The light color is positive charge and the dark color is negative charge.

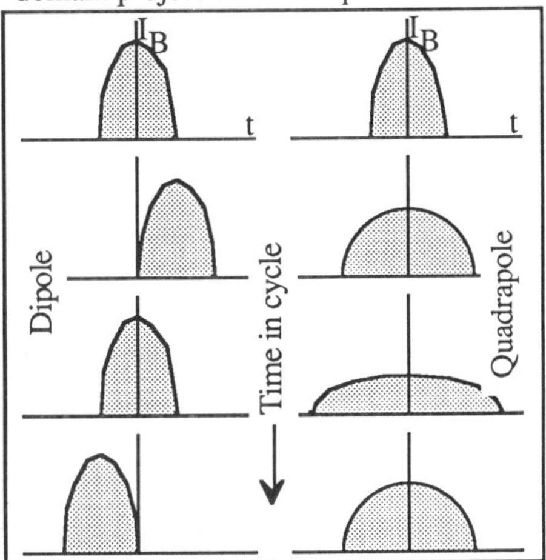

Fig. 8 The projection onto the time axis of the monopole+dipole and the monopole+ quadrapole distributions.

Coupled Bunch Modes

The above section discussed how the bunch distribution in the bucket can be decomposed into multipoles. This section will discuss the relationship between adjacent bunches. For simplicity, we will concern ourselves with dipole motion only. As shown in Fig. 8, dipole motion describes the centroid of the bunch. Consider the bunches shown in Fig. 1 undergoing dipole oscillations. At a given instant in time, the centroid of each bunch with respect to the RF will vary from bunch to bunch. One such pattern is shown in Fig. 9. Because the accelerator is circular, bunch **h+i** is the same bunch as bunch **i**. Thus, the location of the bunch centroids as a function of the bunch number must be periodic with period **h**. Because of this periodicity, the multipole expansion coefficient $c_{i,k}$ in Eqn. 13 can be written as a discrete Fourier series:

$$c_{i,k} = \sum_{m=-h/2}^{h/2} B_{m,k} e^{j2\pi mi/h} \tag{15}$$

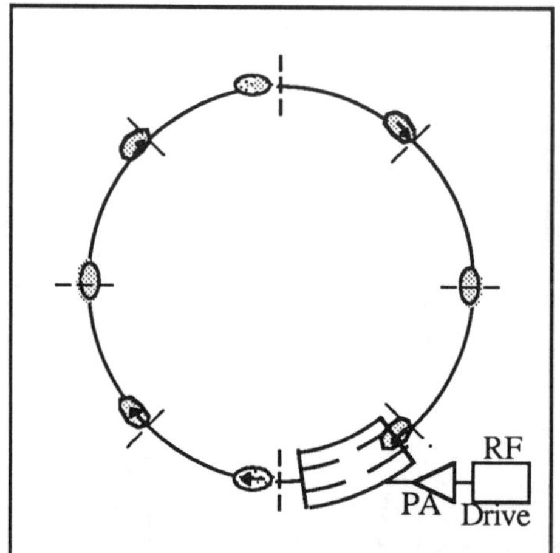

Fig. 9 A snapshot of bunches undergoing dipole synchrotron oscillations for coupled mode number m=±1. The dashed lines indicate the bucket centers.

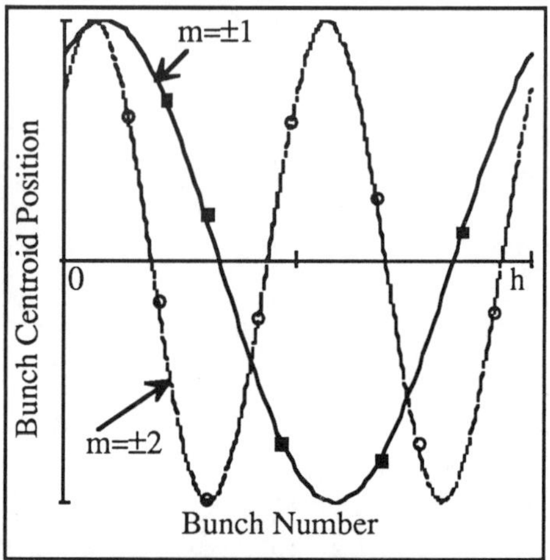

Fig. 10 Bunch centroid position vs Bunch number for coupled bunch modes ±1 and ±2.

Because of the Nyquist sampling criterion, the maximum frequency of the Fourier series for a machine with h bunches is h/2. The centroid in the time domain of bunch **i** is given as:

$$<\tau_i> = \frac{1}{N_b} \int_0^{\infty} \int_0^{2\pi} g_i(r,\theta) \frac{T_0}{2h} r \sin(\theta)\, r\, d\theta\, dr \tag{16}$$

Using the "hockey puck" distribution:

$$<\tau_i> = \frac{-T_b}{3c_{k=0}} \sum_{m=-h/2}^{h/2} b_{m,1} \sin(\xi_{m,1} \pm \Omega_s t + 2\pi mi/h) \tag{17}$$

where $B_{m,1}=b_{m,1}\exp(j\xi_{m,1})$ and T_b is the length of the bunch. Equation 17 shows that the bunch centroid position around the ring is a Fourier sum of all the possible coupled bunch profiles weighted with the coupled bunch mode amplitude $B_{m,1}$. Plotting the bunch centroid position as a function of bunch number for the first two coupled bunch modes is shown in Fig. 10. (Note that the "snap shot" of bunch centroids around the ring shown in Fig. 9 contains coupled bunch mode m=1 only

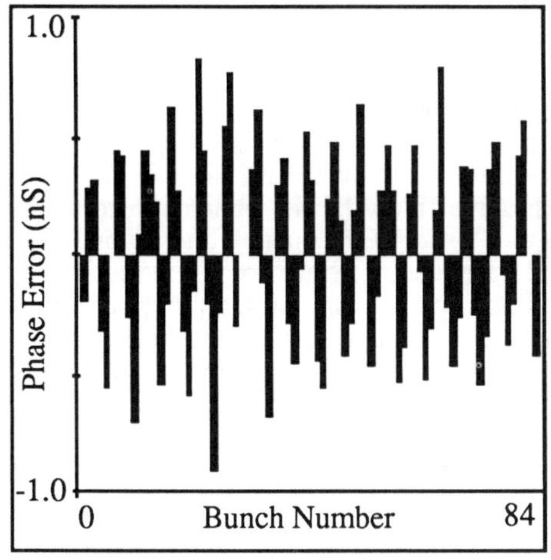

Fig. 11 Measured phase error vs. bunch for the Fermilab Booster.

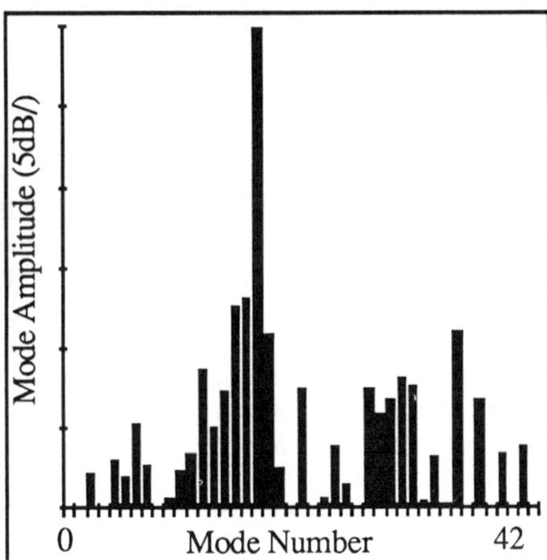

Fig. 12 Fourier (coupled bunch) coefficients of Fig.10

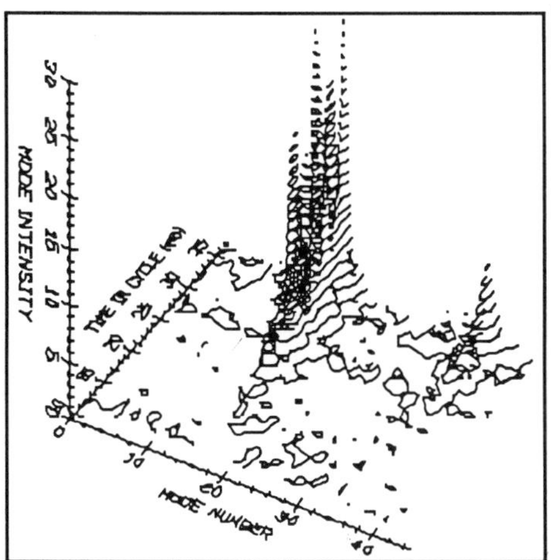

Fig. 13 Coupled bunch mode spectrum through the acceleration cycle for the Fermilab Booster

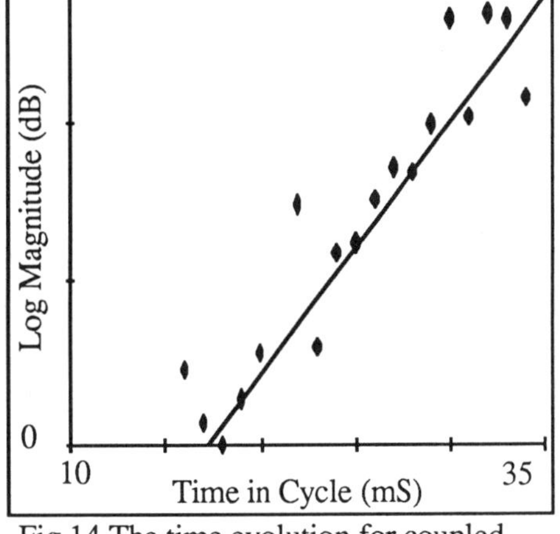

Fig.14 The time evolution for coupled bunch mode ±17 for the Fermilab Booster.

The coupled bunch mode content in the accelerator can be found by taking the Fourier sum of Eqn. 17:

$$B^*_{-n} - B_n = \frac{j6c_{k=0}T_b}{h}\sum_{i=1}^{h} \langle\tau_i\rangle e^{-j2\pi ni/h} \tag{18}$$

The measurement prescribed by Eqn. 18 is performed by recording the instantaneous beam current as a function of time for a single turn in the acceleration cycle with a high speed oscilloscope.[4] The digitizing rate must be able to resolve the bunch length. For each bunch, the centroid is computed by multiplying the bunch profile by time and integrating. This value is normalized by dividing by the total charge in the bunch. In this way variations due to unequal bunch populations are eliminated. The centroid spacing between adjacent bunches will be a combination of the coupled bunch mode oscillations and the phase advance of the RF frequency. The phase advance due to the RF can be eliminated by making a least squares fit to the the bunch centroids and removing the linear increase of phase due to a fixed RF frequency and second order increase in phase due to the slewing of the RF frequency. Higher order corrections are small enough to be neglected. The coupled bunch phase error for a typical turn in the Fermilab Booster is shown in Fig. 11. The Fourier analysis according to Eqn. 18 of the bunch centroid data shown in Fig. 11 is displayed in Fig. 12. Note that since mode B_n and mode B_{-n} have the same number of oscillations, it is impossible to tell them apart according to this method. The above procedure can be repeated to obtain a mountain range display of the coupled bunch mode spectrum as a function of time in the cycle as shown in Fig. 13 . Once the mountain range display is obtained, the evolution of a particular mode as a function of time may be analyzed. Figure 14 shows the evolution of mode 17. The data can be fitted to an e-folding rate of 5.5 mS.

Coupled Bunch Mode Spectrum

As discussed earlier, coupled bunch motion is the result of the beam current exciting higher order mode impedances in the cavity. The strength of the coupling will be proportional to the beam current. To calculate the beam current, consider the current due to a single particle **p** in bunch **i**. The longitudinal position of the particle in the ring is:

$$s_{pi}(t) = R_c \left(\omega_0 t - \frac{2\pi i}{h} + \omega_0 \tau_{pi}(t) \right) \tag{19}$$

For an observer (or cavity) at a fixed location in the ring, the current due to particle **p,i** is:

$$I_{pi} = q\omega_0 \sum_{n=-\infty}^{\infty} \delta\left(\frac{s_{pi}}{R_{pi}} - 2n\pi\right) \tag{20}$$

which can be expanded in a Fourier series:

$$I_{pi} = \frac{e\omega_0}{2\pi} \sum_{n=-\infty}^{\infty} e^{jn(\omega_0 t + \omega_0 \tau_{pi}(t) - 2\pi i/h)} \tag{21}$$

Using Eqn. 11 and the well known identity between exponentials of sines and Bessel functions, Eqn. 21 becomes:

$$I_{pi} = \frac{e\omega_0}{2\pi} \sum_{n=-\infty}^{\infty} \sum_{m=-\infty}^{\infty} J_m(n\pi r_{pi}/h) e^{j(n\omega_0 t + n\omega_0 \tau_{pi}(t) + m\theta_{pi} - 2\pi n i/h)} \tag{22}$$

The current due to all the particles in the bunch is given by integrating Eqn. 22 over all phase space weighted by the phase space density for the bunch given by Eqn. 13.

$$I_i = e\omega_0 N_b \sum_{n=-\infty}^{\infty} \sum_{k=0}^{\infty} \left[c_{i,k} f_{k,n} e^{j(n\omega_0 t - 2\pi ni/h - \pm k\Omega_s t)} + c_{i,k}^* f_{-k,n} e^{j(n\omega_0 t - 2\pi ni/h \pm k\Omega_s t)} \right] \tag{23}$$

where:

$$f_{k,n} = \int_0^{\infty} r\, f(r)\, J_{-k}(n\pi r/h)\, dr \Big/ 2\pi c_{k=0} \int_0^{\infty} r\, f(r)\, dr \tag{24}$$

Using the "hockey puck" distribution, Eqn. 24 simplifies to:

$$f_{k,n} = \frac{1}{\pi c_{k=0}} \int_0^1 u\, J_{-k}(n\omega_0 T_b u/2)\, du \tag{25}$$

Equation 25 is the multipole form factor[2]. A graph of the form factor for the first couple of multipoles as a function of n is shown in Fig. 15.

The total current in the machine is the sum of all the individual bunch currents. Using Eqn. 23 this sum becomes:

$$I = e\omega_0 N_b \sum_{i=1}^{h} \sum_{q=-\infty}^{\infty} \sum_{p=-h}^{h} \sum_{k=0}^{\infty} \sum_{m=-h}^{h} \left[B_{m,k} e^{j2\pi mi/h} f_{k,p+qh} e^{j((p+qh)\omega_0 t - 2\pi(p+qh)i/h - \pm k\Omega_s t)} + B_{m,k}^* e^{-j2\pi mi/h} f_{-k,p+qh} e^{j((p+qh)\omega_0 t - 2\pi(p+qh)i/h \pm k\Omega_s t)} \right] \tag{26}$$

where the sum over the n harmonics has been replaced by a double sum over p+qh where p and q are integers. For simplicity, let us only consider dipole motion (k=1) and since:

$$\sum_{i=1}^{h} e^{j2\pi(m-p)i/h} = \begin{cases} h \text{ for } m=p \\ 0 \text{ for } m \neq p \end{cases} \tag{27}$$

Eqn. 26 collapses to:

$$I = 4\pi I_{dc} \sum_{q=-\infty}^{\infty} \sum_{m=-h}^{h} b_{m,1} f_{1,m+qh} \cos\left((m+qh)\omega_0 t - \pm\Omega_s t + \xi_{m,1}\right) \tag{28}$$

where $I_{dc}= eN_bh/T_o$. Equation 28 shows that the strength of a coupled bunch mode can be determined by examining the amplitude of the synchrotron sidebands of the beam current with a spectrum analyzer. An example spectrum is shown in Fig 16. The spectrum is strongest when the form factor is a maximum. For dipole oscillations, this maximum occurs when the frequency band is near $1/T_b$. Fig. 17 shows experimental data for the Fermilab Booster . The envelope of RF harmonic lines (spaced every 53 MHz) follows the monopole form factor. The coupled bunch mode lines (mode 16 and 36) follow the dipole form factor.

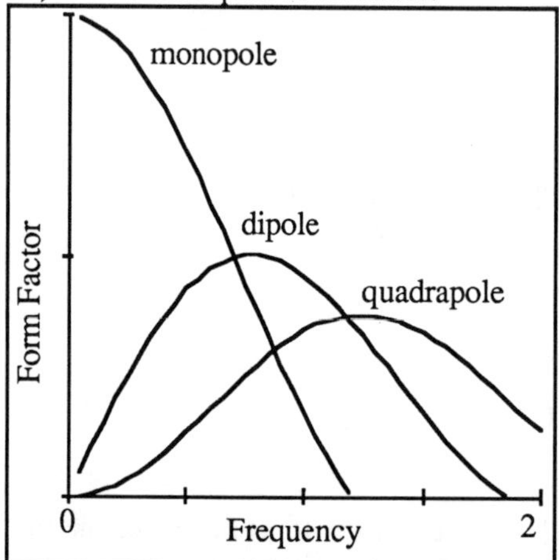

Fig. 15 The form factor for the first three multipole modes. The time axis is in units of 1/bunch length.

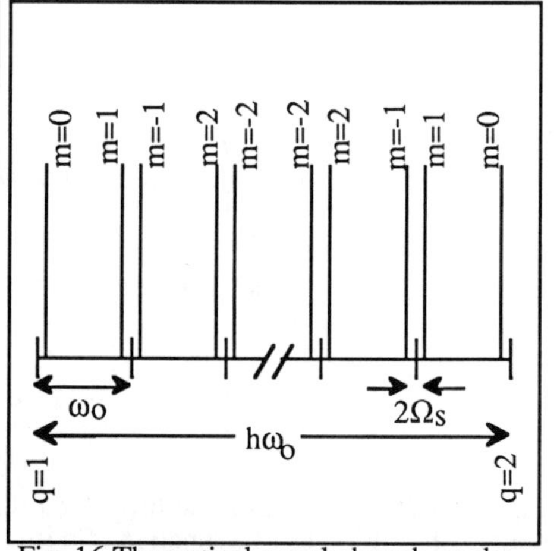

Fig. 16 Theoretical couple bunch mode spectrum between the first and second RF harmonic.

The Coupled Bunch Mode Growth Rate

Figure 13 shows that a coupled bunch mode may become unstable and grow with time. One should not only expect the amplitude of the mode to change, but the synchrotron frequency of that mode should change as well. With these modifications the beam current in Eqn. 28 becomes:

$$I= 4\pi I_{dc} \sum_{q=-\infty}^{\infty} \sum_{m=-h}^{h} b_{m,1} f_{1,m+qh} e^{\alpha_m t} \cos\left((m+qh)\omega_0 t - \pm\omega_{sm} t + \xi_{m,1}\right) \tag{29}$$

where α_m is the amplitude growth rate and ω_{sm} is the new synchrotron frequency of mode m. When the beam current passes through a cavity it will excite the cavity modes and store energy in these modes. The energy transfered from these cavity modes onto bunch **i** is given by Ohms law:

$$\Delta E_i = -eI(t) \otimes Z\ (t)\Big|_{t=\frac{2\pi i}{\omega_o h}} \tag{30}$$

which is the convolution of the beam current and the impulse response of the cavity evaluated at time $t=2\pi i/\omega_o h$. Using Eqn. 29, the energy change can be written as:

$$\Delta E_i = -e4\pi I_{dc} \sum_{q=-\infty}^{\infty} \sum_{m=-h}^{h} b_{m,1} e^{\alpha_{m,1} t} f_{1,m+qh} \left| Z^{\pm}_{m+qh} \right| \cos\left(\Phi^{\pm}_{m+qh} + \xi_{m,1} - \pm\omega_{sm} t + 2\pi mi/h\right) \tag{31}$$

where:

$$\left| Z^{\pm}_{m+qh} \right| e^{j\Phi^{\pm}_{m+qh}} = Z((m+qh)\omega_0 - \pm\, \omega_{sm}) \tag{32}$$

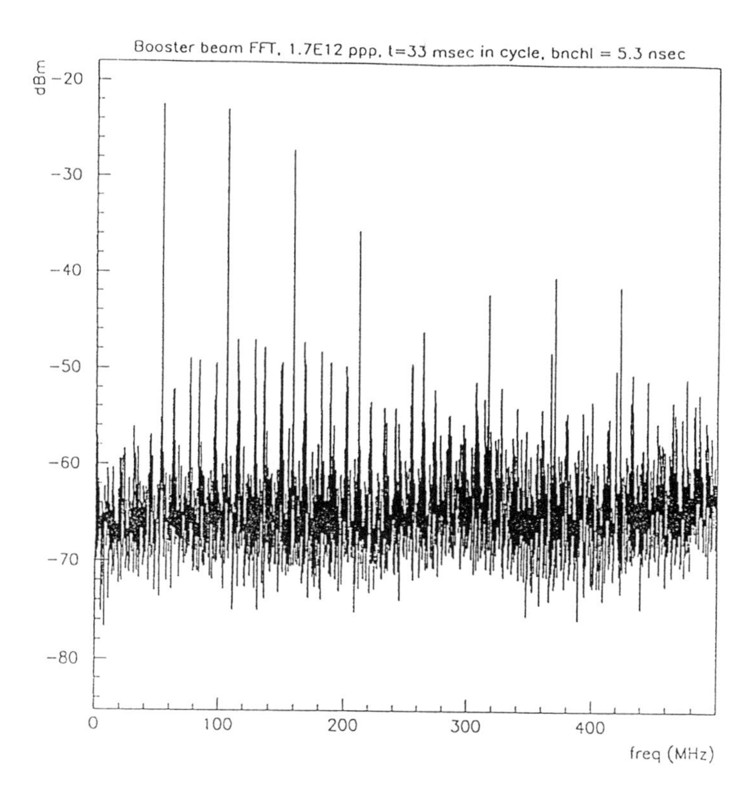

Fig. 17 Experimental data from the Fermilab Booster . The envelope of RF harmonic lines (spaced every 53 MHz) follows the monopole form factor. The coupled bunch mode lines (mode 16 and 36) follow the dipole form factor.

The total change of energy imparted to the bunch is given as:

$$\frac{d\langle\Delta p_i c\rangle}{dt} = \frac{eV_c h\omega_0 \langle\tau_i\rangle}{T_0} + \frac{\Delta E_i}{T_0} \tag{33}$$

Substituting Eqn. 24 for $<\tau_i>$, the growth rate of the coupled bunch mode is:

$$\alpha_m = - \pm \frac{\eta e I_{dc}}{2 p_0 c T_0 \Omega_s} \sum_{q=-\infty}^{\infty} F^{P}_{m+qh} \mathrm{RE}\{Z((m+qh)\omega_0 - \pm\Omega_s)\} \tag{34}$$

and the shift in synchrotron frequency is:

$$\omega_{sm} = \Omega_s + \frac{\eta e I_{dc}}{2 E_0 T_0 \Omega_s} \sum_{q=-\infty}^{\infty} F^{P}_{m+qh} \mathrm{IM}\{Z((m+qh)\omega_0 - \pm\Omega_s)\} \tag{35}$$

where using the "hockey puck" distribution:

$$F^{P}_{m+qh} = \frac{12}{T_b} \int_0^1 J_1((m+qh)\omega_0 T_b u/2)\, u du \tag{36}$$

The form factor F^{P} is negative for m+qh<0 and positive for m+qh>0. The form factor is shown in Fig.18. The maximum effect of the cavity impedance occurs at $f=1/T_b$. Equations 34 and 35 show that the shift in synchrotron frequency is due to the imaginary part of the cavity impedance and the growth of the coupled bunch mode amplitude is due to the real part of the cavity impedance. By examining Eqn. 34 in more detail, one can show that above transition, the upper sideband of the spectrum shown in Fig. 8 is unstable and the lower sideband is stable as shown in Fig.19. Below transition, the reverse statement is true.

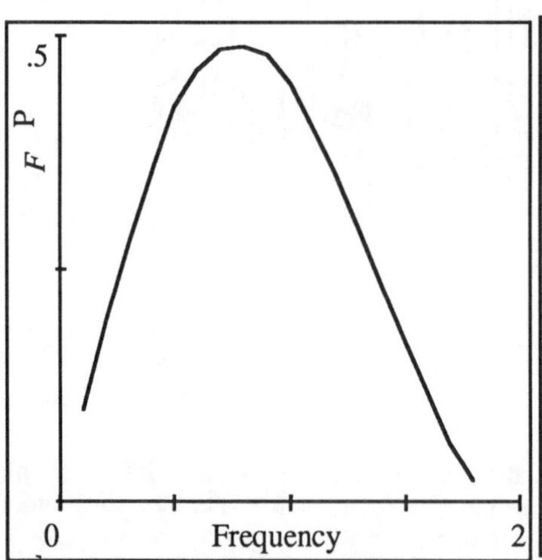

Fig. 18 The growth rate form factor for a dipole coupled bunch mode. The time axis is in units of 1/bunch length.

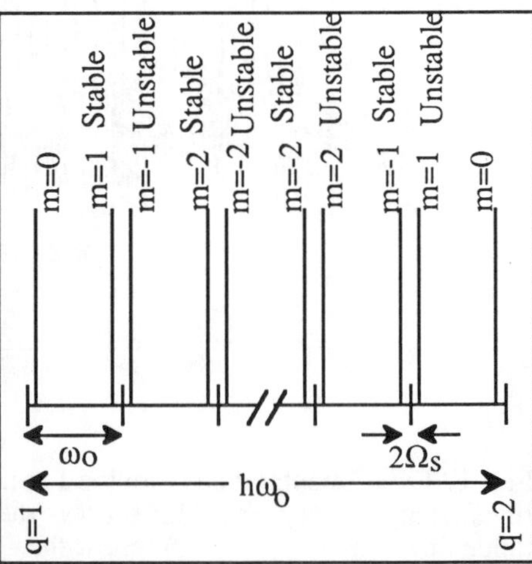

Fig. 19 Stability of the couple bunch mode spectrum between the first and second RF harmonic above transition.

A special case of Eqn. 34 occurs when m=0 and q=+1. The impedance of interest is the fundamental cavity mode. Equation 33 reduces to:

$$\alpha_1 = -\pm\frac{\eta e I_{dc}}{2E_0T_0\Omega_s}F_h^p \mathrm{RE}\{Z(h\omega_0 - \pm\Omega_s) - Z(h\omega_0 \pm\Omega_s)\} \qquad (37)$$

The mode α_1 is unstable above transition (and stable below transition) if the fundamental cavity impedance is resonant above $h\omega_0$. This instability is called the Robinson instability[5].

Equation 34 can be used to estimate the growth rate for a given cavity impedance. For example, in the Fermilab Booster , a total cavity impedance of 20kΩ at 200 MHz would be sufficient for the growth rate of 5.5 ms observed in Fig. 14 given η=.02, p_0c=8 GeV, T_0= 1.6 υS and f_s = 2kHz.

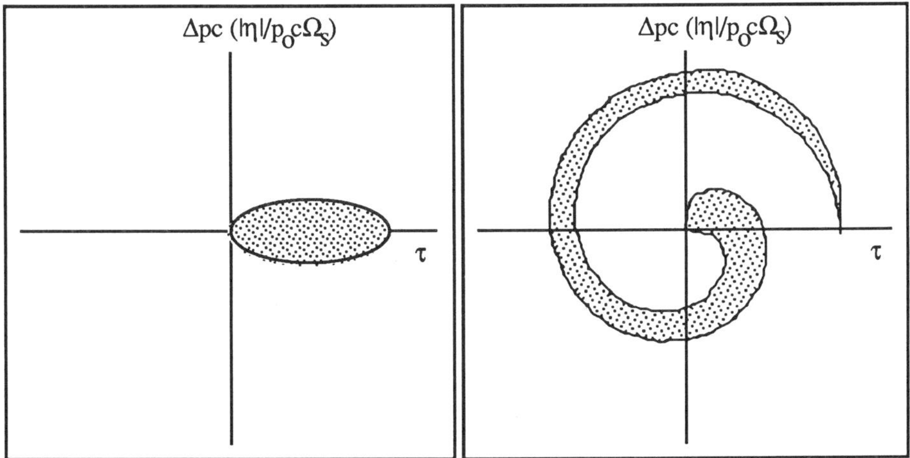

Fig. 20 An initial distribution in phase space.

Fig.21 The distribution of Fig. 20 after a number of synchrotron oscillations.

Landau Damping

In Eqn. 3 we assumed that the RF waveform supplied by the PA could be approximated with a triangle wave instead of a sine wave. This simplification causes the synchrotron frequency of all the particles, regardless of their position in phase space, to be the same.The result of this assumption leads to the conclusion that any cavity impedance, no matter how small, causes an instability. However, for a true sinusoidal RF waveform, the synchrotron frequency is a maximum at the center of the RF bucket and approaches zero near the edges of the bucket. A spread in synchrotron frequencies will cause an initial distribution in phase space as shown in Fig. 20 to smear out and disperse over time as shown in Fig 21.

Another effect due to the spread in sychrotron frequencies is Landau damping[6]. Because each particle has a slightly different synchrotron frequency, the spectrum in Eqn. 28 is not really a collection of delta functions but a collection of frequency bands.The power per Hertz of the frequency bands is diluted with respect to the delta function spectrum. This dilution lowers the driving term of the instability. The threshold for a coupled bunch mode occurs when the growth rate of the instability is greater than the Landau damping rate. As a rough rule of thumb this threshold is met when the growth rate given by Eqn. 34 is greater than the spread in synchrotron frequency[2].

Dampers

A damper system usually consists of a detecting an error signal with a pickup electrode, amplifying and processing the signal with electronics, and delaying the processed signal so that a kicker electrode applies the correction signal at the correct time as shown in Fig. 22. The damper impedance is defined as the ratio of the voltage applied to the beam at the kicker electrode to the beam current sensed by the pickup electrode which can be written as:

$$Z_{Damper} = G_e\sqrt{Z_{pu}Z_k} \tag{38}$$

where Ge is the electronic gain of the damper evaluated at the synchrotron lines, Z_{pu} is the impedance of the pickup and Z_k is the impedance of the kicker. A simple description of a damper is to provide a negative real impedance to the beam to cancel out the positive real impedance of the RF cavities.

If the damper impedance is large enough to cancel the cavity impedance, no coupled bunch mode will grow. If there is no coupled bunch mode, the error signal sensed at the pickup of the damper is zero. Thus, an ideal damper will require no power being dissipated into the kicker. (This is not true for dampers that damp injection oscillations.) However, if there is noise at the pickup (thermal noise, common mode signals, etc.) the noise will be amplified by the damper gain and this amplified noise power will have to be handled by the kicker. Thus, to design a damper system, one must first decide on what gain is needed to cancel out the growth rate of the instabilities. Once the gain is decided on, the amount of noise at the pickup must be measured or determined. The kicker is designed last so as to handle the amplified noise.

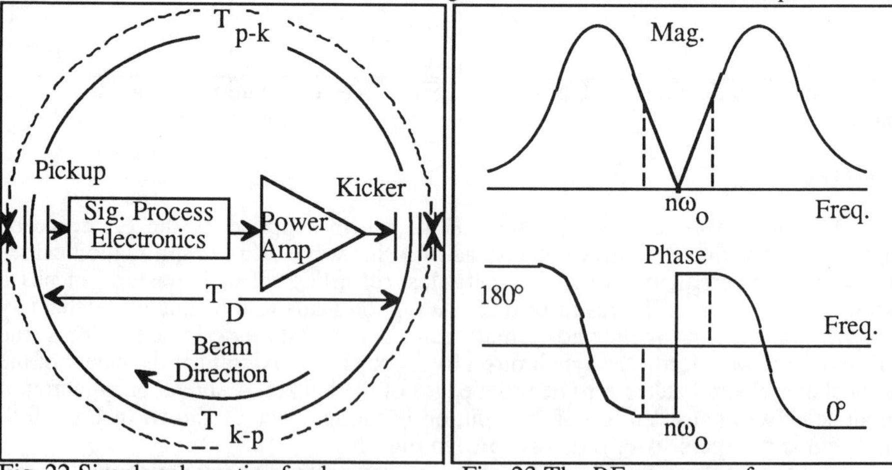

Fig. 22 Simple schematic of a damper system.

Fig. 23 The RF response of a narrow band phase damper above transition.

Phase Dampers

A phase damper detects the phase error of the beam with respect to the driving RF and provides an energy correction to reduce the phase error. A narrowband phase damper works on damping a single coupled bunch mode[7]. To damp all of the coupled bunch modes, there must be h/2 narrow band phase dampers. However, it may not be necessary to build all of the damper systems if only a few coupled bunch modes exist in the machine. Above transition, the narrow band phase damper provides a negative real impedance at the upper sideband and a positive real impedance at the lower sideband. This gain profile is obtained by building a notch filter around the appropriate coupled bunch mode revolution harmonic as shown in Fig. 23.. For a ramping accelerator, this notch frequency would have to track the changing revolution frequency. One such tracking notch filter is shown in Fig 24.

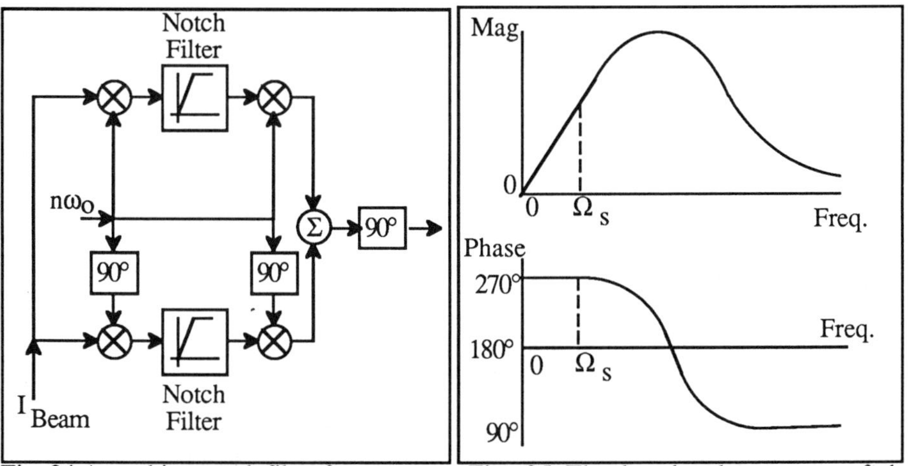

Fig. 24 A tracking notch filter for a narrow band phase coupled bunch mode damper.

Fig. 25 The baseband response of the notch filter shown in Fig. 24.

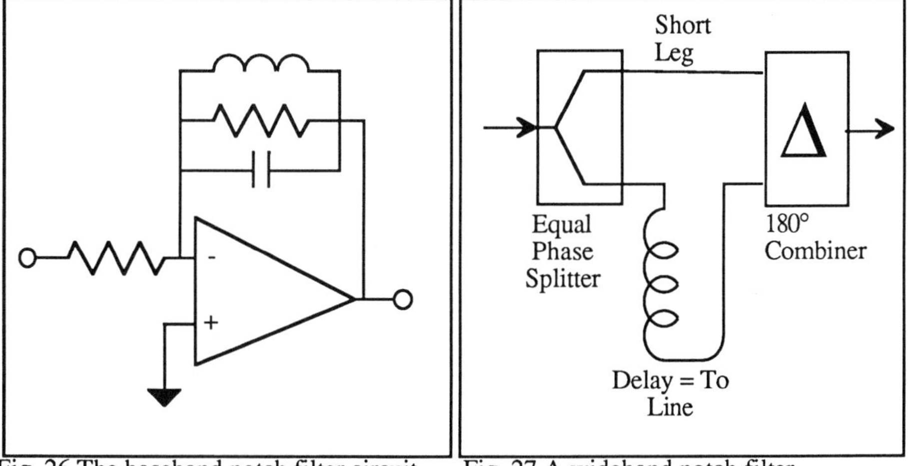

Fig. 26 The baseband notch filter circuit.

Fig. 27 A wideband notch filter.

The baseband filter f(ω) must have a zero at ω=0. Also the filter response should drop off at ω>>0 to minimize noise power. This can be done by placing a pole at a frequency much greater than the synchrotron frequency as shown in Fig. 25. A simple baseband circuit that fulfills the above requirements is a resonant tank circuit shown in Fig 26. Note that a 90 degree phase shift must be added to compensate for the ±90 response at base band. If the resonant frequency or the Q of the circuit is too high, then the gain at the dipole sideband becomes very low. However, if the resonant frequency or the Q is chosen too low, than the phase at the dipole line is becomes non-ideal. The electronic gain of this system is at the dipole lines is:

$$G_e = \frac{\Omega_s}{4} \frac{\partial f(\omega)}{\partial \omega}\bigg|_{\omega=0} \tag{39}$$

For the growth rate of the coupled bunch mode in the Fermilab Booster discussed earlier, the damper gain at the dipole lines of 75 MHz needs to be 70 dB.

A wideband phase damper may be accomplished with a wideband notch filter as shown in Fig. 27. The response of this circuit is shown in Fig. 28. The electronic gain of the filter at the dipole lines is:

$$G_e = \Omega_s/2\omega_0 \tag{40}$$

A wideband system may be though of as a many narrow band channel system, or, in the time domain, a wideband system is a bunch by bunch damper. One can consider the notch filter shown in Fig. 27, as being a comparator of the the phase difference of a bunch with respect to the RF on successive turns. The delay tolerance of the delay line in the notch filter must be less than the ratio of the synchrotron frequency to the bandwidth of the damper. In the Fermilab Booster, the tolerance would have to be less than 0.005%. Also the delay line would have to track the changing revolution period of a ramping accelerator. This might be done with a shift register that is clocked at some multiple of the RF frequency.

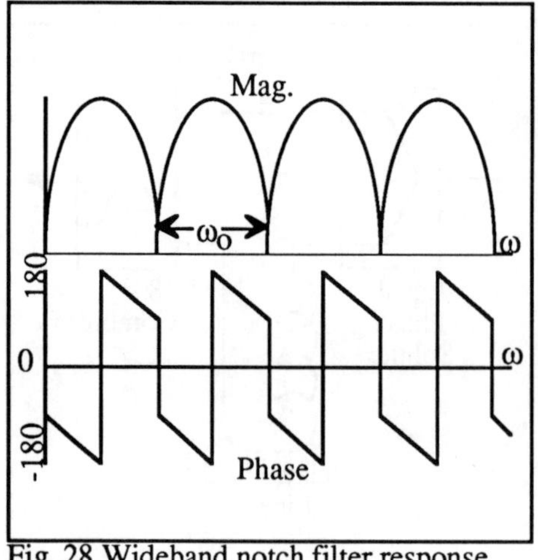

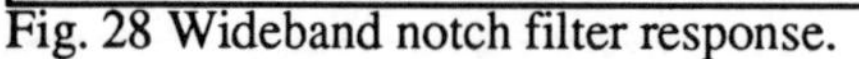
Fig. 28 Wideband notch filter response.

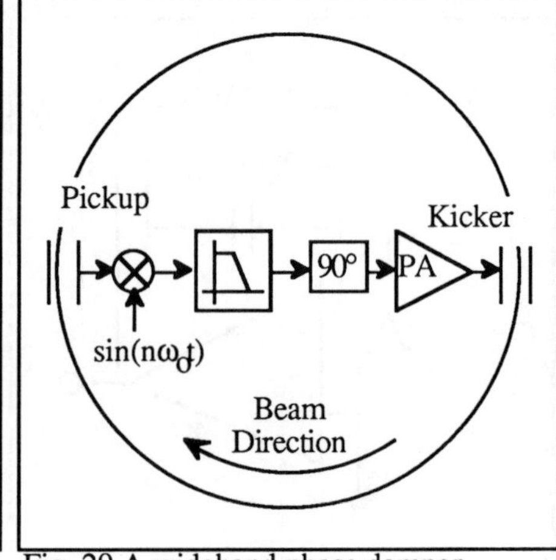

Fig. 29 A wideband phase damper.

A second type of wideband phase damper is shown in Fig 29[8]. The phase of each bunch is compared to a harmonic of the RF frequency. After the low pass filter, the signal is shifted by 90°, amplified by a wideband kicker, and applied to the correct bunch. The bandwidth of the kicker and the low pass filter must be greater than h times the revolution frequency so that the signal from each bunch decays before the next bunch arrives at the kicker. (The decay time of a system is inversely proportional to the bandwidth of the system.) The gain of this system is proportional to the gain of the low pass filter.

Energy Dampers

An energy damper detects the energy error of a bunch and applies an energy correction kick[9]. The energy error is detected by passing the bunch through a horizontal pickup placed in a region of high dispersion in the machine as shown in Fig. 30. The signal received from the pickup is:

$$I_{pu} = \sqrt{2}\, I_{beam} \frac{g_{pu}\alpha_{pu}}{p_0 c\, d} \Delta pc \qquad (41)$$

where α_{pu} is the dispersion function at the pickup, d is the transverse spacing of the pickup plates and g_{pu} is the gain of the pickup. The signal due to particle **p** in bunch **i** is given as:

$$I_{i,p} = \frac{\sqrt{2}\, g_{pu}\alpha_{pu} e\omega_0}{2\pi p_0 c\, d} \sum_{n=-\infty}^{\infty} \Delta p_{i,p} c\, e^{jn(\omega_0 t + \omega_0 \tau_{i,p} - 2\pi i/h)} \qquad (42)$$

Using Eqns. 11-34 as a guide, the decay rate of coupled bunch mode m is:

$$\alpha_m = \frac{\sqrt{2} g_{pu}\alpha_{pu} e I_{dc}\omega_0}{E_0 d} \sum_{q=-\infty}^{\infty} F^{E}_{m+qh} \mathrm{RE}\{Z((m+qh)\omega_0 - \pm\Omega_s)\} \qquad (43)$$

where:

$$F^{E}_{m+qh} = 3\int_0^1 u^2 J_0((m+qh)\omega_0 T_b u/2)\, du \qquad (44)$$

The form factor F^{E} approaches unity when the bunch length and/or the frequency becomes small as shown in Fig. 31. Note that Eqn. 43 is independent of η. Thus the damper does not have to flip phase at transition. Also note that at low frequencies, the damping rate is independent of frequency unlike that of the phase dampers. For a pickup spacing of 5cm and a dispersion function of 2m, the electronic gain needed to damp the mode 16 line in the Fermilab Booster is 90 dB.

The major drawback of the energy damping scheme requires the pickup to have excellent common mode rejection. If the average beam location is not in the center of the pickup, the damper will view the RF harmonics as a large mode m=0 signal. this signal will be amplified essentially as noise power on the kicker. For example, given an electronic gain of 90 dB, a maximum power on the kicker of 2500 W, and a beam current of 120 mA, the common mode rejection required from the pickup is about -50 dB. The pickup itself could provide about -20 dB of this rejection and the rest must be made up with a notch filter.

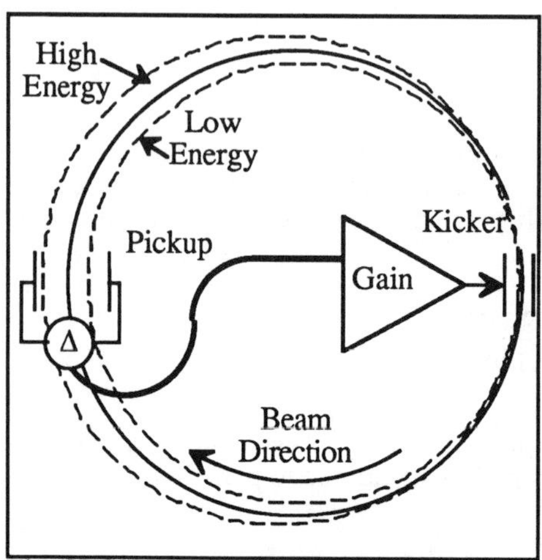

Fig. 30 Schematic of a wideband energy damper.

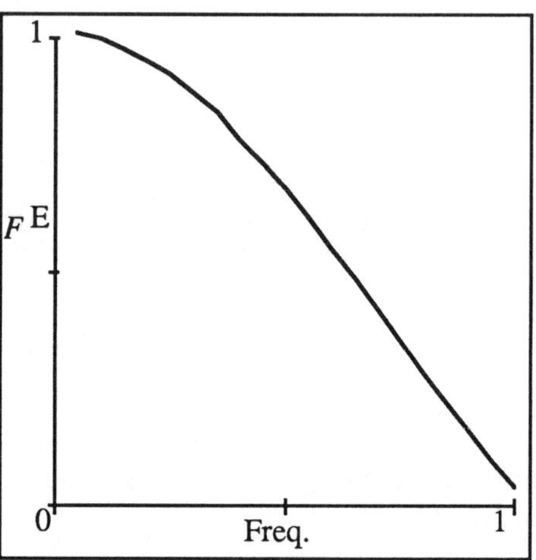

Fig. 31 The damping rate form factor for a dipole energy damper. The time axis is in units of 1/bunch length.

Damper Timing

For all of the above damper systems, it was assumed that the damper delay was phased correctly so as to kick the beam at the right time. For wideband systems, the phase of the damper must match the phase of the beam at each revolution line.Thus, the wideband damper delay must equal the beam delay from pickup to kicker plus an integer number of revolution periods. Since the system is timed for the bunches orbiting at ω_o, long delays add extra phase shift at the synchrotron lines. The phase shift at the synchrotron lines is equal to the synchrotron frequency times the damper delay. Because only the real part of the damper impedance contributes to damping, the damping rate is inversely proportional to the cosine of the phase shift. As the delay approaches 1/4 of a synchrotron period, the damping rate shrinks to zero.

If the revolution frequency of the beam changes during the acceleration cycle, then the damper delay must change accordingly. Changing the damper delay may be done by switching in coaxial cable of different lengths into the trunk line of the damper system during the acceleration cycle. Another technique is to delay the signal digitally by digitizing the signal, delaying it with a shift register and then converting the signal back to analog.

One method of determining the correct delay is to operate the kicker as a both a pickup and a kicker as the same time. Markers are placed on the beam by removing some bunches from the machine. The beam passing underneath the kicker can be detected by the kicker now acting as a pickup as shown in Fig. 32. The gap in the beam signal in Fig. 32 is due to the missing bunches in the beam. With the normal damper pickup connected to the damper system, the signal at the kicker is now composed of the damper pickup signal and the beam signal detected by the kicker as shown in Fig. 33. If the damper is properly timed, then these two signals will overlap. (The damper gain may have to be adjusted so that the damper signal does not obscure the beam signal at the kicker.)

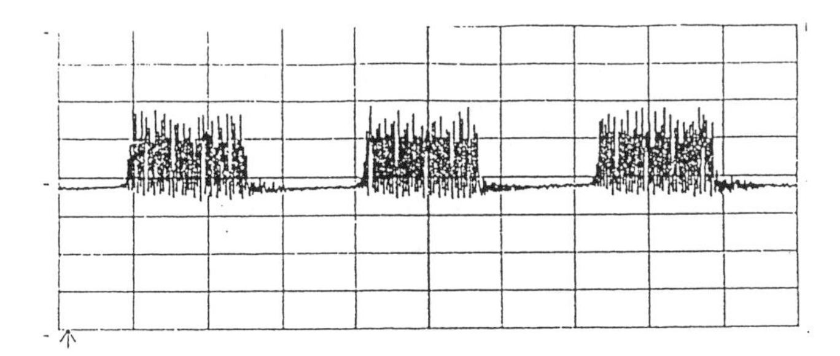

Fig. 32 Beam signal with gap in the beam detected by the damper kicker

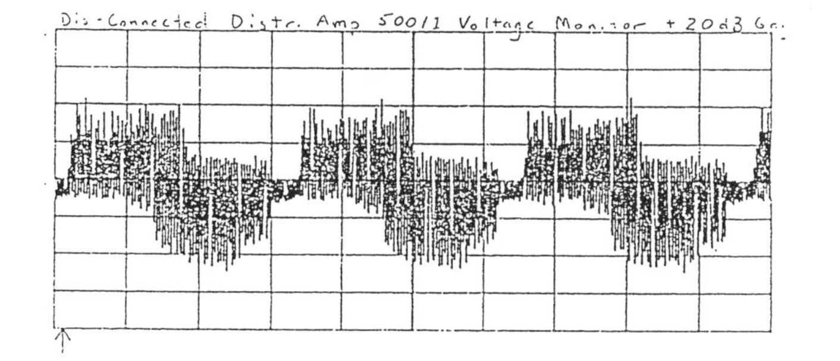

Fig. 33 Beam signal and damper signal at the kicker

For narrow band systems, the above timing technique will not work because the Q of the damper will cause the damper signal to ring for an appreciable period of time after the beam signal has passed. Also for narrow band systems, the group delay of the damper need not match the revolution period of the beam. However, the change in delay of the beam as it accelerates must be matched by a change in phase of the damper. This may be done by placing a variable phase shifter in the trunk line of the damper.

One method for adjusting the phase of the narrowband damper is to measure the open loop gain of the damper at the synchrotron sidebands[10]. The open loop gain is measured by driving the beam at a given frequency and measuring the response of the beam at that frequency with a vector network analyzer as shown in Fig. 34. If the damper is properly phased, then the phase of the network analyzer measurement at both the upper and lower synchrotron sidebands will be 180°.

For fast cycling accelerators, it may be impossible to stop the acceleration ramp to measure the open loop response at a fixed energy. The length of time that a network analyzer must dwell at a single frequency is inversely proportional to the resolution bandwidth of the network analyzer. Thus, if the resolution bandwidth is too low, the network analyzer will not be able to track a fast changing synchrotron line. Expanding the frequency versus time ramp in a Taylor series and keeping terms up to first order, it can be shown that in order for the network analyzer to track a single synchrotron line, the following inequality must hold:

$$n\frac{d\omega_0}{dt} < \frac{\Omega_s f_{IF}}{2} \tag{45}$$

where n is the harmonic number of the measurement and f_{IF} is the resolution bandwidth of the network analyzer.

If this inequality does not hold, the frequency of the network analyzer reference should track a harmonic of the revolution frequency. A circuit that provides such a reference is shown in Fig. 35. In this circuit, a portion of the low level RF of the accelerator is fed into the clock of a direct digital synthesizer. The direct digital synthesizer is programmed by the computer to provide a frequency output that is a fraction of the clock signal. Because of the limited frequency range of the DDS circuit, the DDS output is mixed in a single sideband mixer with the RF VCO output. A single sideband mixer is necessary so that only one RF harmonic is excited. To keep the frequency swing in the reference leg small so that the network analyzer will not lose phase lock, the DDS output is also used for reference leg of the network analyzer. The mixed-up signal is then applied to the beam via the kicker and the response is detected by the pickup. To make the received signal frequency the same as the reference frequency, the pickup signal is mixed down with the low level RF signal where the high frequency component of the mixing product is eliminated by a low pass filter.

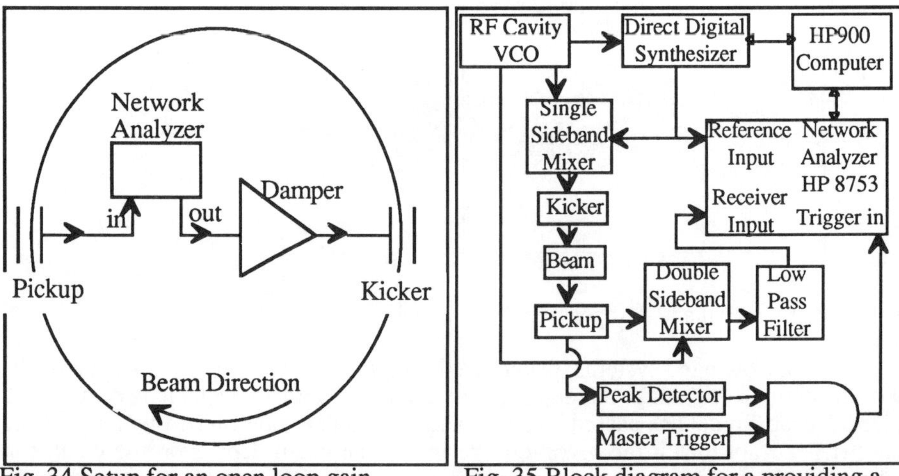

Fig. 34 Setup for an open loop gain measurement of a damper system.

Fig. 35 Block diagram for a providing a tracking network analyzer reference.

Since the reference frequency of the network analyzer is not constant during the measurement, the network analyzer is placed in the Continuous Wave (CW) mode and the amplitude and phase of the ratio of the pickup signal to the reference is displayed as a function of time throughout the accelerator cycle. To trigger the network analyzer at the proper time in the cycle and trigger only if there is beam in the machine, the pickup signal is peak detected and a coincidence gate is formed with the accelerator clock. The measurement described above is performed for only one fractional revolution harmonic. To obtain the response as a function of the fractional revolution harmonic, the computer must increment the fraction, a new batch of beam must be injected into the accelerator, and the network analyzer must be re-triggered. For extra conditioning, the response for a given fractional revolution harmonic can be averaged for a number of beam batches before changing the fraction. This averaging also has the added benefit of reducing signals caused by beam instabilities the are not phase related to the kicker signal.

Since the DDS frequency will not track a multipole line exactly during the acceleration cycle, the width of a given multipole line must be large enough so that the crossing of the DDS frequency through the multipole line will take long enough to fill the IF filter of the network analyzer. By expanding the frequency versus time curve in a Taylor series and keeping terms up to first order, the requirement on the synchrotron line width can be written as:

$$\frac{\Delta\Omega_s}{\Omega_s} > 2\left(\frac{1}{\omega_0}\frac{d\omega_0}{dt} - \frac{1}{\Omega_s}\frac{d\Omega_s}{dt}\right) \Big/ \left(f_{IF} + \frac{1}{\Omega_s}\frac{d\Omega_s}{dt}\right) \tag{46}$$

This inequality usually fails near transition. Figs. 36-37 show the results of a network analyzer measurement performed on accelerating beam in the Fermilab Booster during the last 15 mS of the acceleration cycle. The network analyzer was uncalibrated so the magnitude of the results is un-normalized. The beam intensity was $1\text{x}10^{12}$ protons. The number of points along the time axis is 26. The number of revolution fractions is 220. For this measurement, the number of averages at each revolution fraction was 10. With the Booster operating at a repetition rate of about 4 seconds, this measurement took about 2.5 hours to complete. The measurements were centered around the 74th revolution harmonic. Figure 36 is a three dimensional contour plot of the beam transfer function versus time in the cycle and fractional revolution frequency. The only part of the response that is greater than the noise floor is a set of dipole lines located near the revolution fraction = ±0.004. This revolution fraction coincides with a synchrotron frequency of 2.5 kHz. Figure 37 shows a phase contour of the response for 90°.

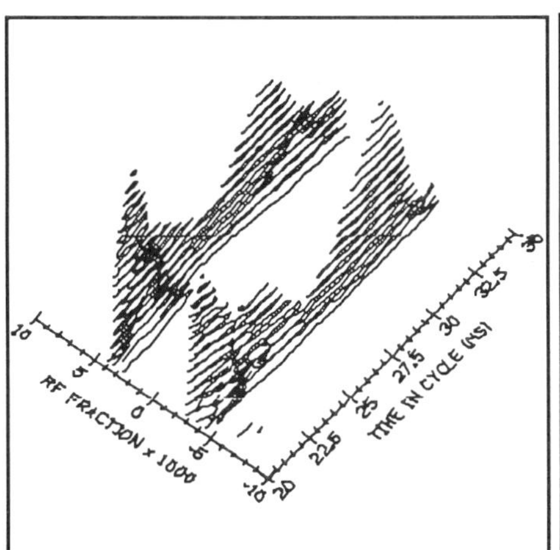

Fig. 36 Mountain range plot of the magnitude of a tracking network analyzer measurement around a revolution harmonic. The RF fraction axis is centered around the 74th revolution harmonic of the Fermilab Booster.

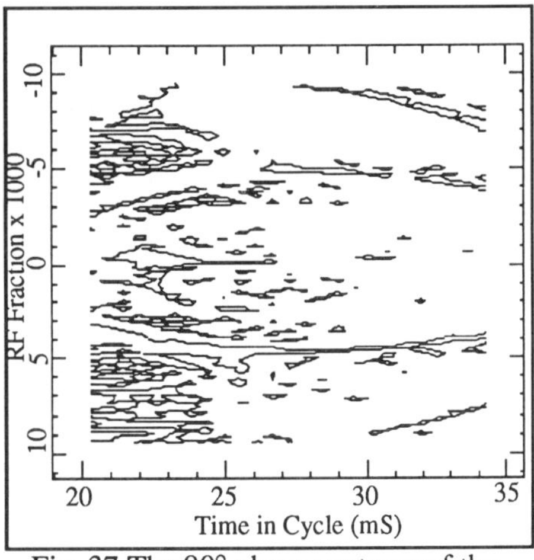

Fig. 37 The 90° phase contours of the network analyzer measurement shown in Fig. 36.

Definition of terms

α_m = growth rate of coupled bunch mode m

α_{pu} = dispersion function at the pickup
$b_{m,1}$ = magnitude of the dipole coupled bunch amplitude for mode m
$B_{m,k}$ = Coupled bunch amplitude of mode m for multipole k
$c_{i,k}$ = Fourier coefficient of mulipole k for bunch i
d = transverse spacing of the pickup electrode

ΔE = energy change due to coupled bunch modes in cavities

Δpc = momentum error
e = Particle electrical charge
F^E = gain form factor for a energy damper
F^P = gain form factor for a phase damper
f(r) = radial distribution in phase space

$f(\omega)$ = filter response of a damper
f_{if} = IF bandwidth of a network analyzer
$f_{k,n}$ = multipole k's form factor

γ = total energy of the particle in units of the rest energy

$g(r,\theta)$ = distribution in phase space
G_e = electronic gain of a damper
g_{pu} = gain of the pickup

γ_t = transition energy
h = RF harmonic number
I_{dc} = DC current in the accelerator
j = $(-1)^{1/2}$
J_m = mth order Bessel function
N_b = number of particles in the bunch
p_oc = ideal momentum

θ = phase coorsinate in phase space
r = radial coordinate in phase space
R_c = Radius of the accelerator
s = position in ring

τ = time error
T_b = bunch length
T_o = ideal revolution period
T_{ri} = revolution period of bunch i
V_c = Peak cavity voltage provided by the PA.

ω_o = ideal radian revolution frequency

Ω_s = synchrotron frequency

ω_{sm} = Sychrotron frequency of mode coupled bunch mode m

$\xi_{m,1}$ = angle of the dipole coupled bunch amplitude for mode m

$Z(\omega)$ = cavity impedance at ω
Z_k = Impedance of kicker
Z_{pu} = impedance of pickup

Acknowledgements

The author would like to thank John Marriner, Jim Budlong, Jim Steimel, Vinod Bharadwaj, and Kathy Harkay for their assistance in preparing this manuscript.

References

1. H. Weidemann, S. Holmes, "Introduction to Accelerator Physics," 1989 US Particle Accelerator School Lecture Notes, pg 57, June 1989

2. F. J. Sacherer, "A Longitudinal Stability Criterion for Bunched Beams," CERN/MPS/Int. BR/73-3, February 1973

3. F. J. Sacherer, "Methods of Computing Bunched Beam Instabilities," CERN/SI-BR/72-5, pg 26,September 1972

4. D. McGinnis, J. Marriner, V. Bharadwaj, "Coupled Bunch Dipole Mode Measurements of Accelerating Beam in the Fermilab Booster," To be published in the Proceedings of the 1991 IEEE Particle Accelerator Conference.

5. A. W. Chao, " Coherent Instabilities of a Relativistic Bunched Beam," 1982 US Particle Accelerator School Lecture Notes, SLAC-PUB-2946, pg 66, August 1982.

6. H. G. Hereward, "The Elementary Theory of Landau Damping," CERN 65-20, May 1965.

7. B. Kriegbaum, F. Pederson " Electronics for the Longitudinal Active Damping System for the CERN PS Booster," CERN/PS/BR/77-9, March 1977.

8. D. Briggs, et. al., "Prompt Bunch by Bunch Synchrotron Oscillation Detection via a Fast Phase Measurement," SLAC-PUB-5525, May 1991.

9. E. Higgins, "Beam Signal Processing for the Fermilab Longitudinal and Transverse Beam Damping System," IEEE Trans. Nucl. Sci. NS-22, 1975, no. 3, pg. 1581

10. D. McGinnis, J. Marriner, V. Bharadwaj, "Beam Transfer Function Measurements of Accelerating Beam in the Fermilab Booster," To be published in the Proceedings of the 1991 IEEE Particle Accelerator Conference.

CEBAF BEAM INSTRUMENTATION*

R. Rossmanith
CEBAF, 12000 Jefferson Ave., Newport News, Virginia 23606, USA

INTRODUCTION

CEBAF, now under construction, is a unique accelerator. The main design goals are: (1) a cw electron beam with energies up to 4 GeV; (2) maximum beam current of 200 μA; (3) relative energy spread no larger than 10^{-4}; and (4) support for up to 3 simultaneous fixed target experiments.

The design is indicated in Figure 1. The beam is accelerated in two superconducting linacs. Because the superconducting structure is expensive, about $200 K per meter, the beam is recirculated several times, in a manner somewhat similar to the many passes in a synchrotron. A detailed cost evaluation showed that the lowest total cost could be achieved with five recirculations.

The cavities, which resonate at 1.5 GHz, are similar to the cavities developed at Cornell University. When they were first produced, the maximum achievable gradient was less than 5 MV/m. With experience, the quality has improved so that several cavities have shown gradients significantly larger than 10 MV/meter.

This unconventional machine requires unconventional instrumentation:

- o With five beams in the accelerating structures at the same time, the position monitors must be able to detect the position of each beam in the presence of the others.
- o Energy spread is closely connected with bunch length, so the bunch length must be measured precisely with sub-picosecond resolution.
- o The high power density in the circulating beam makes necessary a system that can shut the machine down within 20 μsec.

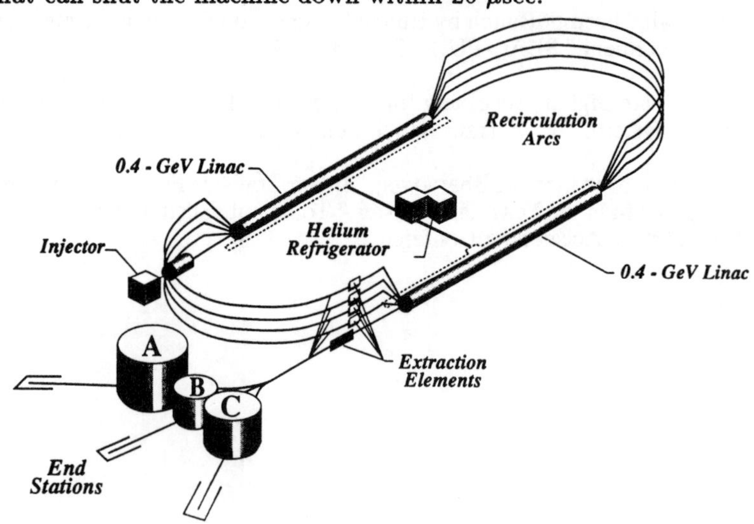

1. The CEBAF accelerator configuration

*This work was supported by the U.S. Department of Energy under contract DE-AC05-84R40150.

BEAM POSITION MEASUREMENT

As can be seen in Figure 1, all five recirculated beams occupy the same vacuum pipe in the linacs, whereas each beam has its own pipe in the arcs. This implies that different types of monitors must be used in the arcs and the linacs. The linac monitors must be able to measure the position of each beam in the presence of four others; the arc monitors have no such requirement.

Linac monitors. In order to measure the position of each beam individually in the presence of other beams, the beam must be marked, and the mark must then be detected. At CEBAF, the marking is amplitude modulation of the beam current at 100 MHz; the modulation is then further altered by changing its sign, or equivalently shifting its phase by 180°, in a pattern that enables the position monitors to suppress all but the desired signal.[1]

To illustrate the coding for a simple case of two beams, assume

$$f(t) = A \sin(\omega t)\, x(t/\tau), \tag{1}$$

where ω is the modulation frequency, τ is the beam circulation period in the machine (4.2 μs for CEBAF), and $x(t/\tau)$ is a function with values ± 1, changing at integer values of its argument. Consider the sequence

$$x = 1, 1, -1, -1, 1, 1 - 1, -1, \ldots \tag{2}$$

If the signal is mixed with itself, the resulting autocorrelation function is

$$f(t)f(t) = A^2 \sin^2(\omega t), \tag{3}$$

representing the detected signal for the first pass. The time average is nonvanishing and is equal to $A^2/2$. If $f(t)$ is delayed by τ and then mixed with the detected signal, however, the correlation is

$$f(t)f(t-\tau) = A^2 \sin^2(\omega t)\, x(t/\tau)\, x(t/\tau - 1). \tag{4}$$

The time average is then

$$\frac{1}{T}\int_0^T f(t)f(t-\tau)\,dt = \frac{A^2}{T}\int_0^T \sin^2(\omega t)\,dt\,(1 - 1 + 1 - 1 \ldots) \approx 0. \tag{5}$$

Thus, the correlation function in this case picks out the signal from the first pass and suppresses that from the second. To suppress the first and observe the second, the signal is mixed with the delayed sequence $f(t-\tau)$.

The sequence $x(t/\tau)$ is not sufficient for three or more passes, but it is easy to extend the argument. We need only to generate a sequence with values ± 1 such that the autocorrelation function with delay will vanish. Such sequences are already well known in coding theories, where they are known as pseudorandom sequences (PRSs).

To illustrate the technique more carefully, define the correlation function as

$$R(x, y; n) = \frac{1}{T}\int_0^T x(t/\tau)\, y(t/\tau + n)\,dt, \tag{6}$$

where $x(t/\tau)$, $y(t/\tau)$ are pulse sequences with values ± 1, T is the total time of the sequence, and $n\tau$ is the time delay. We are particularly interested in the autocorrelation, $x(t/\tau) = y(t/\tau)$, which should satisfy the orthogonality condition in order for the sequence to act as a filter:

$$R(x, x; n) = \frac{1}{T} \int_0^T x(t/\tau)\, x(t/\tau + n)\, dt = 1 \text{ if } n = 0, \qquad (7)$$
$$= 0 \text{ if } n \neq 0.$$

Perfect sequences, which satisfy Eq. 7 exactly, are hard to find and implement. Shift register sequences, however, which will satisfy Eq. 7 to within an arbitrarily small residual error determined by the length of the sequence, can be easily implemented by shift registers with appropriate feedback connections. Programmable array logic (PAL) devices can produce shift register sequences in a single device that needs only a clock pulse. A length-63 PRS generator is indicated in Figure 2; we have used a similar register in our evaluations.

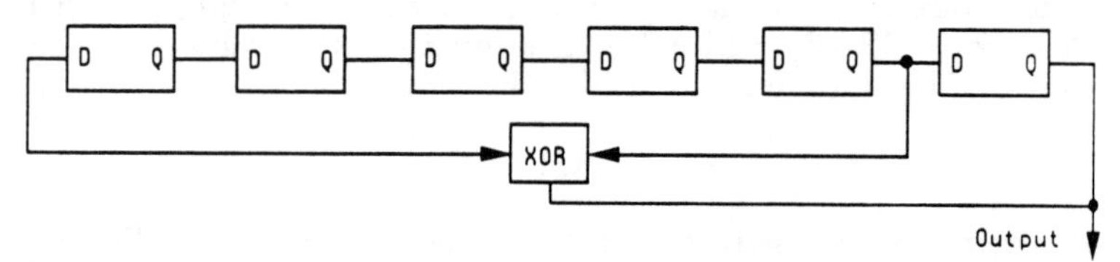

2. Shift register pseudorandom sequence generator of length 63

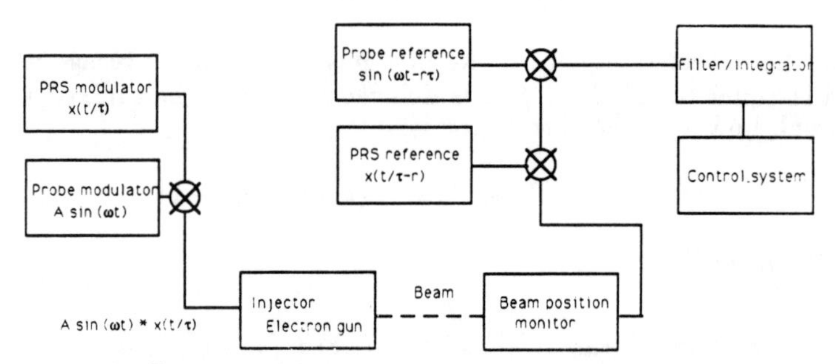

3. Beam position monitor processing

The block diagram for the modulation is shown in Figure 3. The current is modulated with the frequency ω and by a pseudorandom sequence $Ax(t)$,

$$I = I_0 + A\, x(t/\tau)\, \sin(\omega t), \qquad (8)$$

where $x(t/\tau)$ is either $+1$ or -1. For five beam passes through the detector, the output signal at the probe is

$$S_d = \sum_{r=0}^{4} \left(k_r A\, x(t/\tau - r)\, \sin(\omega t - r\tau)\right) + \text{noise}, \qquad (9)$$

where k_r is a constant for each beam which depends upon its position. Because noise is generally quite system-specific and cannot be easily handled analytically, we will henceforth neglect its effect; as usual, however, noise will determine the processing time needed for a given signal/noise ratio. When the detector signal is mixed with the input delayed by j circulation periods, the output is

$$S = \sum_{r=0}^{4} k_r A^2 x(t/\tau - r)\, x(t/\tau - j) \sin^2(\omega t). \tag{10}$$

Integration over m complete sequences will remove the high-frequency time-varying terms, so the signal average will be

$$\overline{S} = \left[\frac{1}{2}k_j A^2 - \frac{1}{2N}\sum_{r\neq j} k_r A^2\right]\left[1 - \frac{\sin 2\omega T}{2\omega T}\right], \tag{11}$$

where ωT is equal to mN; it is seen that the last term in the second brackets is $O(1/T)$.

A position monitor of the type used to observe the beam in the linac regions is the four-wire loop monitor shown in Figure 4.a.

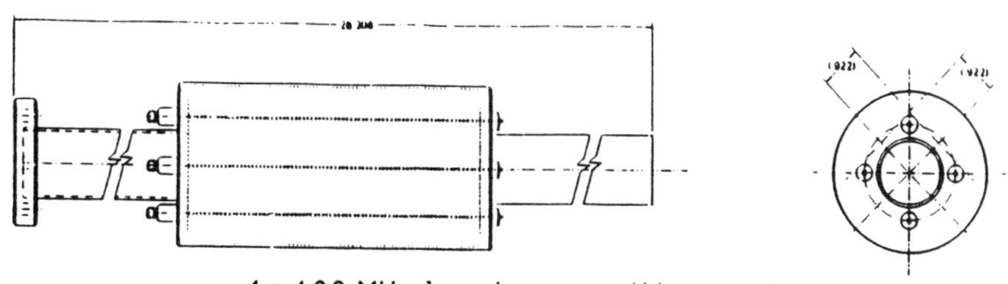

4.a 100 MHz loop beam position monitor

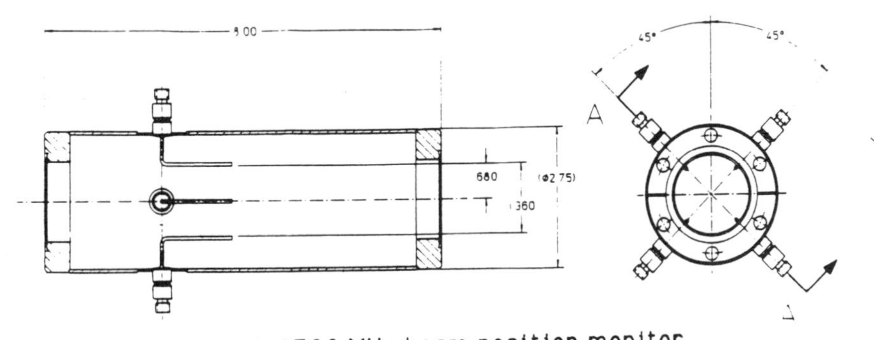

4.b 1500 MHz beam position monitor

Arc monitors. In the arcs, where each beam has its own beam pipe, the position monitors need not respond to the modulation, so they can be optimized to detect the 1.5 GHz rf signal of the beam. An arc monitor is illustrated in Figure 4.b. The two types of monitors are variations of one basic type, the thin-wire pickup. The general design is shown schematically in Figure 5, where the symbols are defined. The significant differences are found in the termination Z_2, which is ∞ for an open line (1.5 GHz monitor) and 0 for a shorted line (100 MHz monitor).[2,3,4]

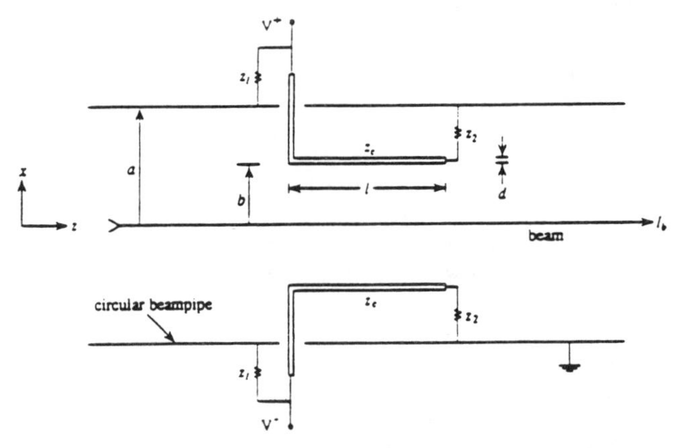

5. General thin-wire beam position monitor

For the monitors, the basic quantity of interest is the transfer impedance Z_T, defined for a single pickup as the ratio of the pickup output voltage for a centered beam to the current I_b. Standard transmission line analysis gives for the transfer impedance[2]

$$Z_T(\omega) = \frac{g\, Z_1 \parallel Z_c(1 - e^{-j\,2\omega l/c})}{1 - \Gamma_1 \Gamma_2 e^{-j\,2\omega l/c}}, \tag{12}$$

where g is a geometric coupling factor, and

$$\begin{aligned} Z_1 \parallel Z_c &= \frac{Z_1\, Z_c}{Z_1 + Z_c} \\ \Gamma_1 &= \frac{Z_1 - Z_c}{Z_1 + Z_c} \\ \Gamma_2 &= \frac{Z_2 - Z_c}{Z_2 + Z_c}. \end{aligned}$$

The amplitudes of the transfer impedance for three pickup types are shown in Figure 6. The plots are normalized to gZ_1 and carried out to $4\omega_0$ to emphasize the relative sensitivities and multiple bandpass characteristics of the pickup.

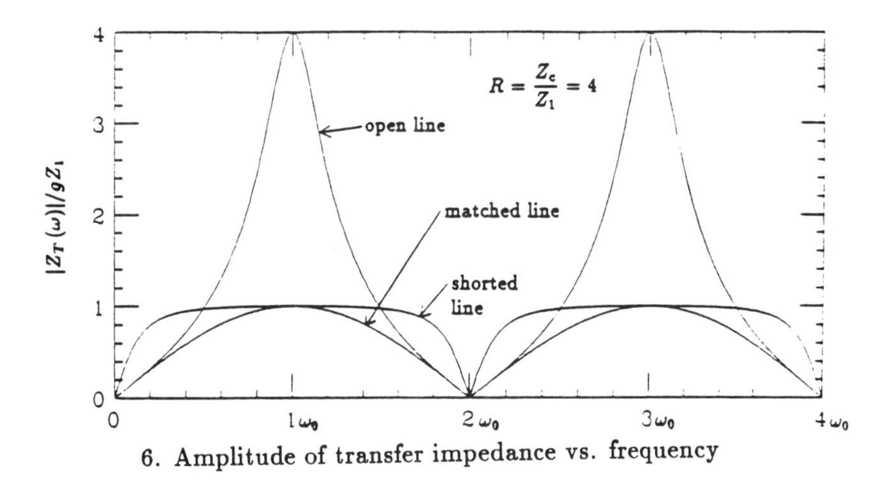

6. Amplitude of transfer impedance vs. frequency

The arc detectors are striplines tuned to the fundamental machine frequency of 1.497 GHz. The magnitude of the beam current, $2I_{av}$, can range from 2 μA to 400 μA peak. The arc monitor receivers synchronously demodulate the signals from the pickups to DC in two stages. As the arcs present almost no bandwidth restrictions, very high accuracy with low beam currents is possible.

A system diagram of the 1.497 GHz arc monitor electronics is shown in Figure 7.a. The system uses a downconverter and phase-locked loop to detect the amplitude of each of the four monitor signals. The basic system components are the tunnel board which downconverts the signal to 1 MHz, and the detector board which contain the phas-locked loops and analog-to-digital converters. The control computer uses the ADC data to calculate the beam position, using the standard difference-to-sum voltage ratio technique.

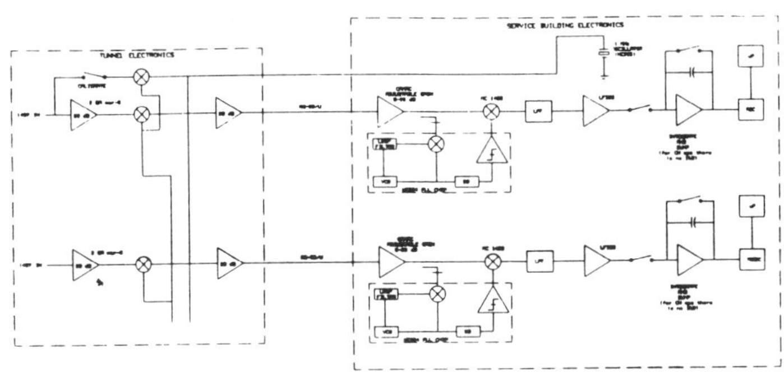

7.a 1497 MHz receiver system diagram

A schematic of the tunnel board is shown in Figure 7.b. Two amplifiers on the front end of each channel amplify the signal approximately 17 dB prior to downconversion. This board is also used to synthesize and inject the calibration signal during autocalibration.

The 100 MHz detectors in the linacs use synchronous detection to convert the 4.2 microsecond current modulation bursts detected by the pickups into pulses that are read with an analog-to-digital converter. The basic system components are: matching network and tunnel line driver board, synchronous detector board, and integrate-and-dump and microprocessor board.

The tunnel board is shown in Figure 7.c. The board amplifies the signal from the pickups by 20 dB. It also has two diode switches that are used to switch the autocalibration signals into the position monitor.

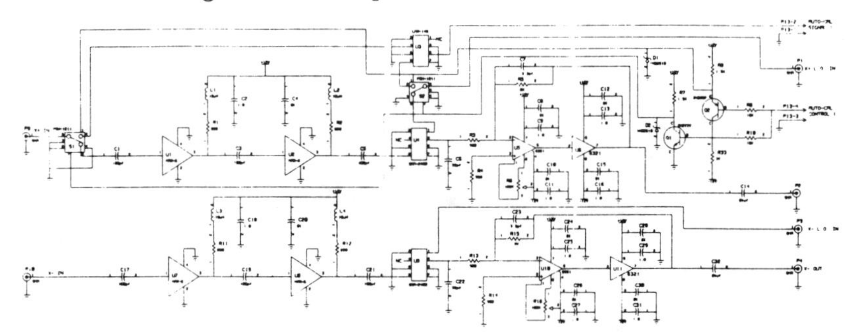

7.b 1497 MHz tunnel electronics

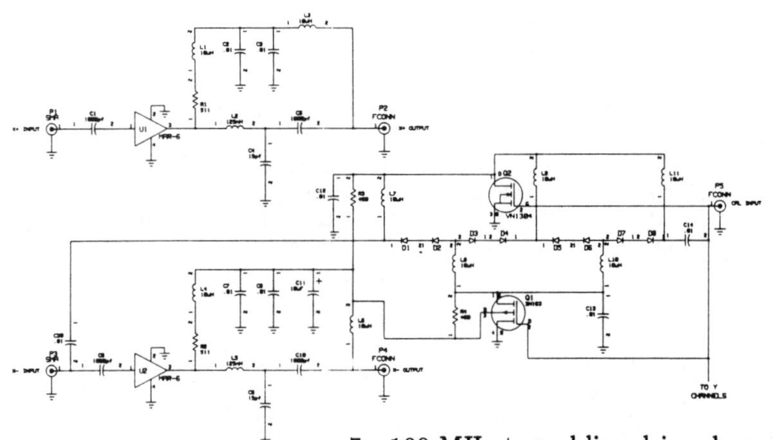

7.c 100 MHz tunnel line driver board

A schematic diagram of the coherent (synchronous) detector board is shown in Figure 7.d. The board contains five basic parts: RF amplifiers, voltage-controlled phase shifters for the local oscillator signals, quadrature phase detectors, downconversion mixers, and baseband amplifiers. It converts the 100 MHz signal to baseband to be read by an analog-to-digital converter. The gain of the baseband amplifiers is adjusted so that the total gain of the board from input to output is 75 dB.

The output signal from the phase detector is passed to the Integrate-and-Dump board, shown in Figure 7.e. In this circuit, the baseband signal from the 4.2 microsecond modulation burst is integrated. Its other functions are: analog-to-digital conversion of the phase detector signals for each channel, digital-to-analog conversion of the phase-adjust signals for each channel, CAMAC interface to the control system computer, and timing circuits for the integrators and converters on the microprocessor board.

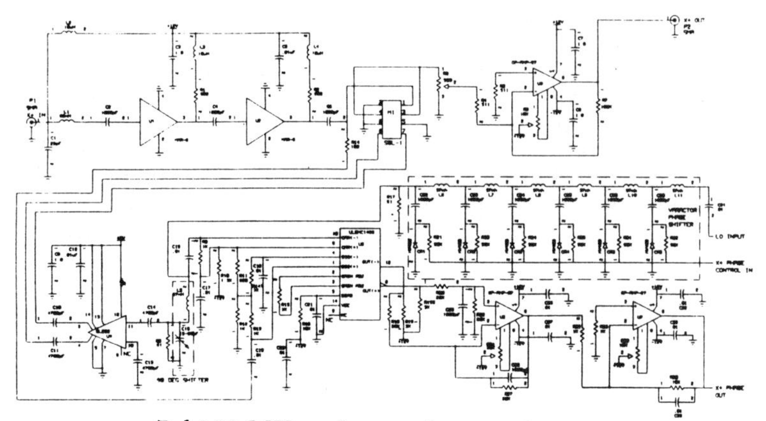

7.d 100 MHz coherent detector board

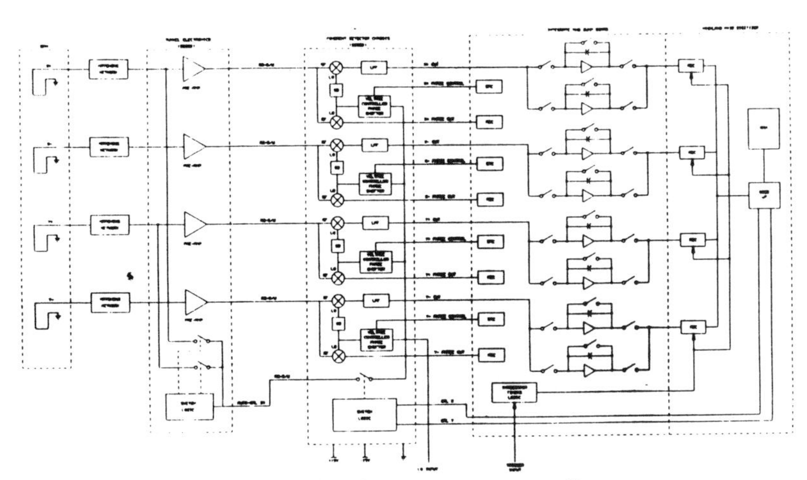

7.e 100 MHz pulsed receiver system diagram

ORBIT CORRECTION TECHNIQUES

Using simulations,[5] we have performed orbit corrections for the full five-pass accelerator. The orbit correction of the lowest energy beam is found by making the beam offset for the first pass zero in each position monitor. However, the corrector settings are not necessarily optimum for the higher passes. A perfect orbit correction cannot be obtained simultaneously for all beams.

Since the linac design uses an alternating position monitor and corrector pattern, we can compute the upstream corrector strength required to make a downstream monitor reading zero for the first pass. The necessary kick $\Delta x'_k$ is calculated from the beta functions and phase advance:

$$\Delta x'_k = -\frac{\Delta x_m}{\sqrt{\beta_k \beta_m}\,\sin(\phi_m - \phi_k)\sqrt{p_k/p_m}}. \tag{13}$$

Figure 8 shows the corrected low-energy beam, and the higher order passes.

Further correction of the higher passes is limited because any change applied to a corrector will disturb the low-energy beam. The following methods can be used for the correction of the higher energies: (1) Optimization of the injection angle and displacement, (2) the use of beam bumps in the low-energy line, which act as kicks for the higher energies, (3) a combination of these, (4) a least-squares fit to 26 variables (26 corrector dipoles with 26 monitor readings), and (5) 36 variables (26 correctors and 10 initial conditions). All of these have been tried in our simulations. As an example of what can be done, Figure 9 shows the results of optimizing the entrance angles and positions of the recirculated beams.

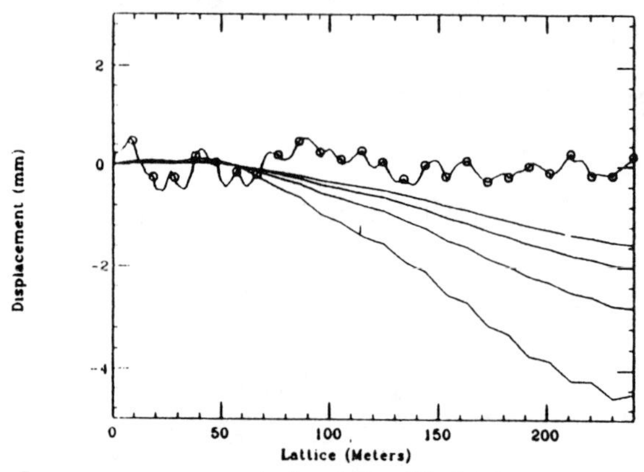

8. Low-energy corrected orbit with higher-energy passes

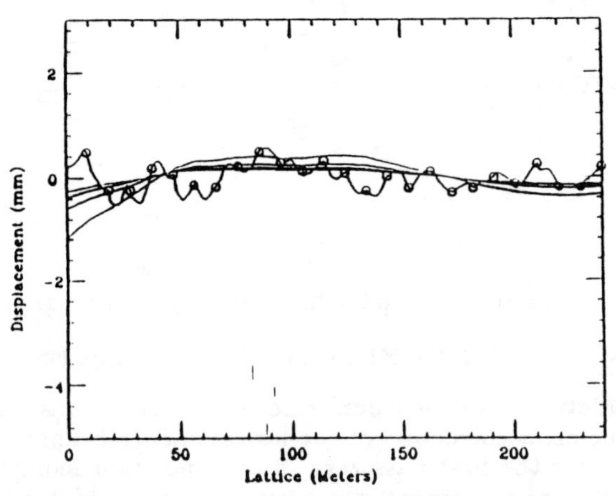

9. Optimization of injection parameters

To aid in establishing a useful beam throughout the accelerator, the beam predictor program OLE has been augmented and improved. In this program, a data base containing relevant numbers from each magnet and accelerating cavity in the machine is kept in shared memory with the operating system itself, providing OLE with instant updating of all information needed to calculate the orbit (betatron function). If an operator changes a magnet setting, for example, a completely new betatron function can be found in less than one second. The program then displays the new function, enabling the operator to simulate steering the beam through the accelerator with no danger to the equipment. An example of the display of the envelope is shown in Figure 10. Other functions also can be called for display by the operator.

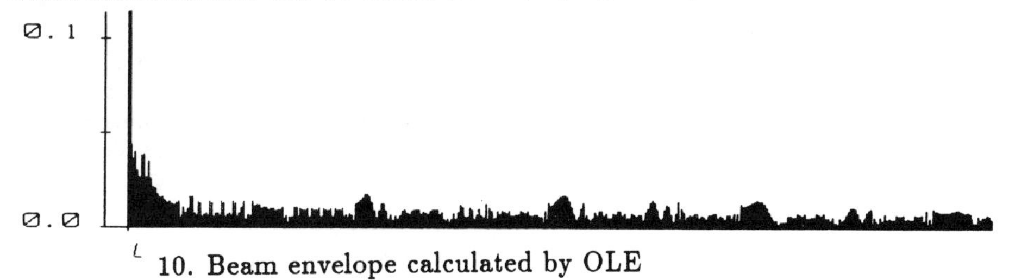

10. Beam envelope calculated by OLE

BUNCH LENGTH MEASUREMENT

The bunch length in the CEBAF electron beam, less than 1° in phase or 2 ps in time when fully relativistic, is too short to be measured by direct timing techniques. We have proposed two methods for obtaining subpicosecond resolution:

1. To measure the self-correlation function of the transition radiation emitted when the beam strikes a thin sheet of metal foil.
2. To observe the change in the energy spread that is introduced when the phase between the beam and one or more rf cavities is altered.[6]

The cavity phase technique is straightforward. The bunch usually rides at the crest of the rf wave in order to maximize the acceleration and to minimize the energy spread. By altering the phase of one or several cavities, the energy spread can be increased by an amount that is proportional to the bunch length. By measuring the energy spread for several phases, the bunch length can be determined with subpicosecond resolution. The method has been successfully tested at the injector, but can later be used at any part of the accelerator.

The cavities in the injector, shown in Figure 11, are driven by two klystrons. The first cavity raises the beam energy to about 2.7 MeV, which is nearly fully relativistic ($\beta = 0.98$). The phase of the second cavity can then be altered to increase the energy spread. Calculations have indicated that the phase length of the bunch at the entrance of the second cavity can be measured with 0.2° precision; this figure represents the limit of validity of the computation.

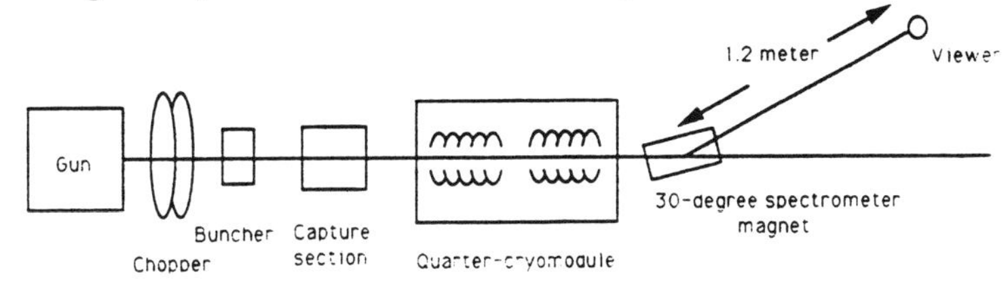

11. Schematic layout of CEBAF injector

To a first approximation, the energy spread produced in the beam by the second rf cavity because of phase spread is given by

$$\delta E = E_0 \sin \phi \, \delta\phi, \tag{14}$$

where E_0 is the energy an electron would gain in traversing the cavity at optimum phase, ϕ is the phase difference of the bunch centroid from optimum, and

$\delta\phi$ is the phase spread in the bunch. The bunch length l is related to the phase spread by

$$l = \lambda\beta\,\delta\phi/(2\pi), \tag{15}$$

where λ is the rf wavelength, and $\beta\,(= v/c)$ is very nearly equal to 1. A closer approximation for δE can be gained by using the beam transport code PARMELA.

We can characterize the beam by its energy spread δE and its phase spread $\delta\phi$. Their input and output values can be connected; for small dispersions,

$$\begin{aligned}\delta E_o &= \left(\frac{\partial E_o}{\partial E_i}\right)\delta E_i + \left(\frac{\partial E_o}{\partial \phi_i}\right)\delta\phi_i \\ \delta\phi_o &= \left(\frac{\partial \phi_o}{\partial E_i}\right)\delta E_i + \left(\frac{\partial \phi_o}{\partial \phi_i}\right)\delta\phi_i.\end{aligned} \tag{16}$$

The transfer coefficients $\partial E_o/\partial E_i$, $\partial E_o/\partial\phi_i$ are computed by PARMELA.

The input energy spread δE_i and the input phase spread $\delta\phi_i$ can be found by least squares, yielding as the best fit the following values:

$$\begin{aligned}\delta E_i &= 0.0386 \text{ MeV}, \\ \delta\phi_i &= 0.247^\circ.\end{aligned} \tag{17}$$

The measured bunch length is nearly 450 fs.

Walter Barry has proposed another way of measuring the bunch length, valid for very short bunches, using transition radiation. When an electron bunch hits a conducting foil, radiation is produced in consequence of the boundary conditions on the electromagnetic fields at the surface of the foil. This radiation has both coherent and incoherent parts. The light produced by individual electrons is incoherent, because no phase relation exists among the electrons. The bunch as a whole also radiates, however, and this is coherent for wavelengths that are roughly equal to the bunch length. The coherent portion can be brought into interference in a Michelson-type interferometer. Measurement of the coherence length in the interferometer reveals directly the bunch length. A schematic diagram of the interferometer is shown in Figure 12.

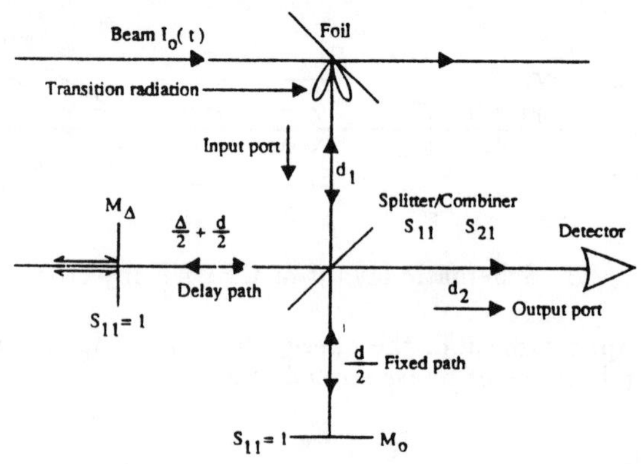

12. Michelson interferometer for autocorrelating coherent transition radiation

ACCELERATOR PROTECTION AND BEAM SHUTDOWN.

Because of the high power density of the beam, if it is steered into any solid object, it will burn a hole in a very short time, typically of order 0.1 millisecond or less. As a design constraint, it is assumed that the burnthrough time is 50 μs. The beam already launched at the injector can require 25 μs to clear the machine, and another 5 μs may be necessary for transit time to get the shutdown signal to the injector. We thus have approximately 20 μs available for response of the protection system.

To detect beam loss, we have installed a system of counters that detect increases in background radiation.[7] If the background counting rate at any counter increases to more than a preset limit, a signal is sent to the fast shutdown[8] (FSD) system, telling it to turn off the beam at the gun. The turnoff is accomplished in two steps: a large negative voltage is rapidly applied to the grid, and then, somewhat more slowly, the high voltage power is removed from the gun. The shutdown is hardwired separately from the computer controls, although it sends messages to the computer. This means that a computer crash will not disable the FSD system.

The system continually accepts and passes a series of 5 MHz square or trapezoidal waves that must be received to keep the gun active. It is fail-safe in the sense that a power failure emulates an FSD trigger. If the permissive signal does not arrive at any level of the safety logic, that level will initiate shutdown within less than 5 cycles. By comparison, a trigger produced by an actual fault condition will initiate shutdown within 3 cycles, a difference that is equivalent to only one more level of logic.

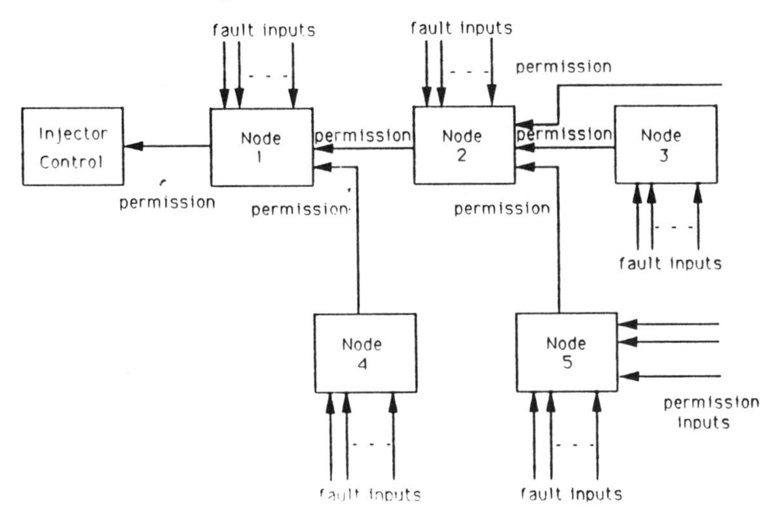

13. Fast shutdown tree structure

While the system was being designed, it became apparent that fault conditions other than beam loss should also shut the beam down. In time, the number of triggers became rather large, so large in fact as to affect the choice of architecture used in implementing the shutdown. To accommodate the large and increasing number of triggers, we have adopted a tree logic (see Figure 13). Up to seven input signals can be received and combined by each node. A positive fault signal on any of the inputs will result in an output fault signal within 3 cycles, or 600 ns. The output signal is then passed on to the next logic level for its action. Since each level introduces its own 600 ns delay, the number of levels cannot be allowed to increase without limit; but the practical limit, corresponding to a total delay of 10 μs, is so large that it does not impose any real constraint.

VIEWERS AND PROFILE MONITORS

Several conventional diagnostic tools are also installed in the CEBAF beamline. They include about 120 actuators that are distributed around the accelerator, and that can insert or withdraw targets from the beam. The targets can be:

Standard fluorescent screens

Aluminum foils to produce transition radiation

Cherenkov cells.

In addition, we have several profile monitors, wire scanners driven by stepper motors.

The beam current is measured by a Faraday cup at low energy, or it can be monitored on-line by parametric current transformers.

In the past several months, we have integrated some of the devices into the control system. The viewscreen images, for example, are digitized and processed. Although they were originally intended to be used solely as qualitative instruments for observing beam position, the computer processing has allowed them to be used for quantitative measurements that are consistent with other methods. An example of a processed image is shown in Figure 14.

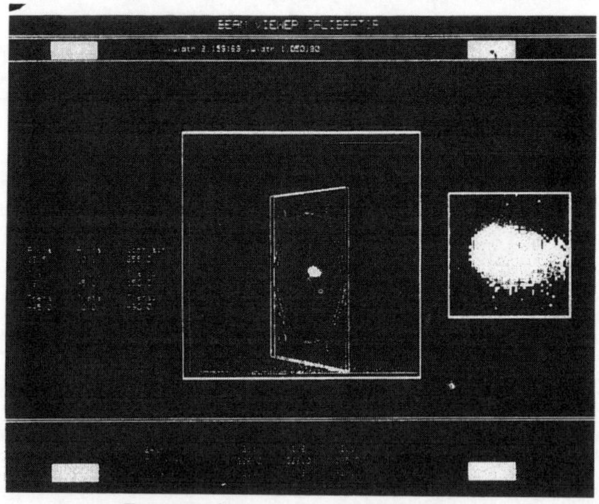

14. Beam viewer processed image

Another example is the harp, a two-wire device for measuring beam profiles. In order to use it in both pulsed and cw operation, the program must control its stepper motor drive in both modes. It must be able to measure the profile of cw beams as low as 1 microampere. For pulsed beams, the peak current of 50 μA is maintained for 30 microseconds, at a repetition period of 60 msec; The average current is then 25 nA. A block diagram of the harp controls is shown in Figure 15. The essential success of the program, as well as the operation of the harp itself, is demonstrated in the profile that is shown in Figure 16.

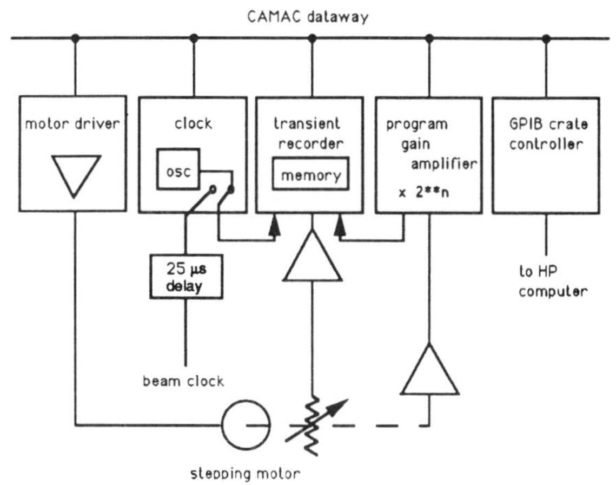

15. Beam profile monitor block diagram

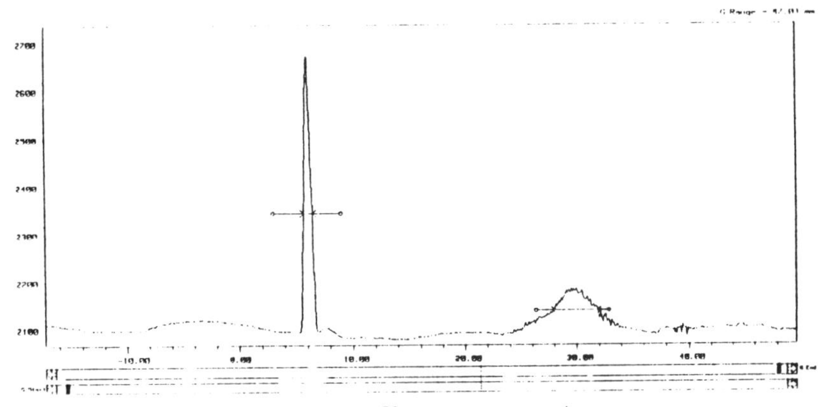

16. Beam profile measurement

BENCH TESTS

Each position monitor and its associated electronics must be individually qualified for service in the accelerator before it is installed. Because of the large number of monitors, we have decided to automate at least part of the process. To this end, Philip Adderley, Karel Capek, and Martice Wise have built a test stand that will measure the response of a monitor to an input signal as a wire that simulates the beam is moved transversely to the axis. The system is sketched in Figure 17. The monitor to be tested is clamped onto the table, and the wire is passed down its axis. The computer then takes over, moving the monitor relative to the wire in a programmed manner, while the network analyzer measures the relevant transfer impedances. The measurements are transmitted to the computer, which calculates the calibrations and generates plots of the data.

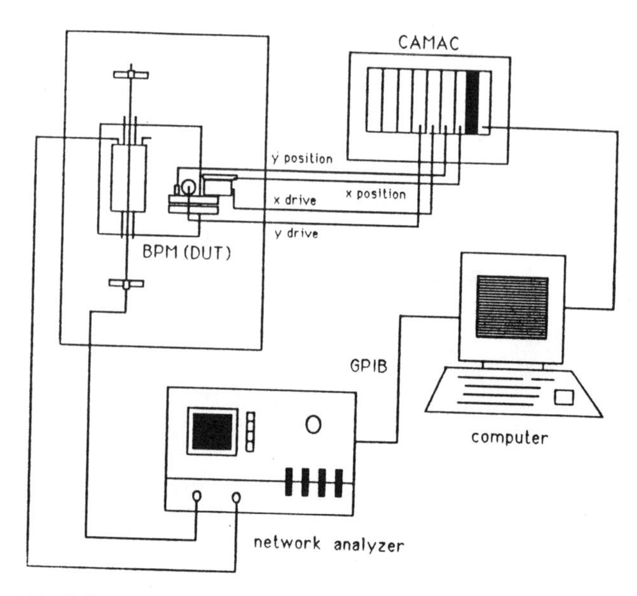

17. Schematic layout of RF position monitor test stand

REFERENCES

1. W. C. Barry, J. W. Heefner, G. S. Jones, J. E. Perry, and R. Rossmanith, "Beam position measurement in the CEBAF recirculating linacs by use of pseudorandom pulse sequences," *Proc 2nd European Particle Accelerator Conf*, 12-16 June 1990, Nice, pp. 723-5. CEBAF-PR-90-009.
2. Walter Barry, "A general analysis of thin wire pickups for high frequency beam position monitors," to be published, *Nucl Inst Methods in Phys Res.* CEBAF-PR-90-024.

   P. Adderley, W. Barry, J. Heefner, P. Kloeppel, and R. Rossmanith, "A beam position monitor for low current dc beams," *Proc 1989 Particle Accelerator Conf*, 20-23 March 1989, Chicago, pp. 1602-4. CEBAF-PR-89-004.
3. W. Barry, R. Rossmanith, and M. Wise, "A simple beam position monitor system for CEBAF," *Proc 1988 Linear Accelerator Conf*, 3-7 October 1988, Newport News, VA, pp. 649-651. CEBAF-PR-89-015.
4. W. Barry, J. Heefner, and J. Perry, "Electronic systems for beam position monitors at CEBAF," *Accelerator Instrumentation: Second Annual Workshop; Batavia, IL, 1990*, pp. 48-74. Edited by Elliott S. McCrory (New York: American Institute of Physics, 1991). CEBAF-PR-90-023.
5. A. Barry, B. Bowling, J. Kewisch, and J. Tang, "Orbit correction techniques for a multipass linac," *Proc 6th Linear Accelerator Conf*, 10-14 Sept 1990, Albuquerque, NM, pp. 444-6. CEBAF-PR-90-016.
6. W. Barry, P. Kloeppel, and R. Rossmanith, "Technique of measuring bunch length by phasing an rf cavity," *Proc 6th Linear Accelerator Conf*, 10-14 Sept 1990, Albuquerque, NM, pp. 719-720. CEBAF-PR-90-014.
7. John Perry, "The beam loss monitors at CEBAF," *Accelerator Instrumentation: Second Annual Workshop; Batavia, IL, 1990*, pp. 294-299. Edited

by Elliott S. McCrory (New York: American Institute of Physics, 1991).
8. J. Perry and E. Woodworth, "The CEBAF fast shutdown system," *Proc 6th Linear Accelerator Conf*, 10-14 Sept 1990, Albuquerque, NM, pp. 484-6. CEBAF-PR-90-015.

Polarization Measurements

R.Schmidt
CERN-SL, GENEVA, Switzerland

1 Abstract

Polarization monitors are an essential tool for the commissioning and operation of those accelerators where polarized beams are required. Experiments with polarized beams were suggested for many accelerators : For electrons/positrons as well as protons machines, for beams with an energy of some hundred MeV up to TeV colliders such as the SSC and future linear colliders. In this paper some aspects of the accelerator physics with polarized beams are discussed. Then an overview of various types of polarimeters is given. A more detailed description of the LEP polarimeter follows and the paper will conclude with a presentation of some results from the LEP 1991 run.

2 Introduction

The degree of polarization is defined as : P = (N1-N2)/(N1+N2) with N1 the number of particles with the spin parallel to a given direction, and N2 the number of particles with the spin in the opposite direction. For an ensemble of particles the classical equations of motion for a magnetic moment $\vec{\mu}$ can be used to describe the spin motion [1]. The rotation of a magnetic moment of an electron in a magnetic field is given by the BMT equation [2] (the electric field is not considered) :

$$\frac{d\vec{\mu}}{dt} = -(e/\gamma mc^2)((1+a)\vec{B}_{\parallel} + a\gamma\vec{B}_{\perp}) \times \vec{\mu} \tag{1}$$

γ is the relativistic factor, a is the anomalous magnetic moment for electrons or positrons, $\vec{B}_{\parallel}$ and $\vec{B}_{\perp}$ the component of the field parallel and orthogonal to the particle momentum. For protons a is replaced by G, with G the anomalous magnetic moment for protons.

We assume a particle travelling in a magnetic field and deflected by the angle α. If the polarization vector is orthogonal to the magnetic field it will rotate around the field axis by an angle $\gamma a \alpha$ for electrons and positrons, and by $\gamma G \alpha$ for protons.

For high energy electron positron storage rings like LEP the spin rotates much faster than the particle. At an energy of 46.5 GeV the number of spin turns for one revolution, the spin tune, is γa=105.5.

3 How to get polarized beams

Protons : sources of polarized protons with a polarization level up to 100 % are available [3]. During the acceleration process the polarization has to be conserved.

Because of the presence of many depolarizing resonances the optimization of the machine to operate with polarized beams is a lengthy and tedious process.
Antiprotons : polarized antiproton beams are not yet available. However, a proposal to polarize antiprotons in a storage ring was made [4].

Electrons : for Linacs and Synchrotons polarized beams have been accelerated at various machines. A polarized source is required, the polarization level of the sources in use is limited to about 50 % [7]. New types of sources with higher level of polarization are still limited in beam current. The acceleration in linacs is easier than in synchrotrons because no or only small deflecting magnetic fields act on the spin. For electron synchrotrons similar techniques to accelerate polarized particles are necessary as for proton synchrotrons.

For e+e- storage rings the phenomenon of self polarization by the so-called Sokolov-Ternov effect is well known [5]. The polarization level increases exponentially to a maximum possible level of 92.4 % if no depolarizing effects are present. The polarization direction is parallel to the magnetic field. The time constant is $\tau = 1.63(R/\rho)\rho^3/E^5$ with R the average radius and ρ the dipole bending radius [1]. For DORIS at 5 GeV/c: $\tau = 5$ min, for LEP at 46 GeV/c: $\tau = 300$ min.

4 Depolarizing mechanisms

In a circular accelerator depolarizing resonances are always present. It is well known that to operate a machine betatron resonances have to be avoided in order not to loose the beam. For the operation with polarized beams the situation is similar : the spin tune must not be an integer value. Not only integer resonances have to be considered, in general the resonance condition for resonances which might lead to depolarization is :

$$\gamma a = k + lQ_x + mQ_z + nQ_s \tag{2}$$

k,l,m,n are integer values and Q_x, Q_z, Q_s the betatron tunes. The integer resonances $\gamma a = k$ appears all 440.65 MeV/c for electrons and positrons (for protons a must be replaced by G).

In the following some examples of polarization measurements are given.

- At the Brookhaven AGS polarized protons were accelerated to 22 GeV/c. Fig.1 shows the optimization of the machine during a period of 3 weeks. Polarized particles are injected and accelerated to a momentum of 6 GeV/c. The depolarizing resonances were compensated until a polarization level of 80 % was achieved. Then the momentum ramp was increased to accelerate the particles to 8.6 GeV/c and the polarization again optimized. After three weeks of continuous optimization a polarization level of 45 % was achieved at a momentum of 18.5 GeV/c.

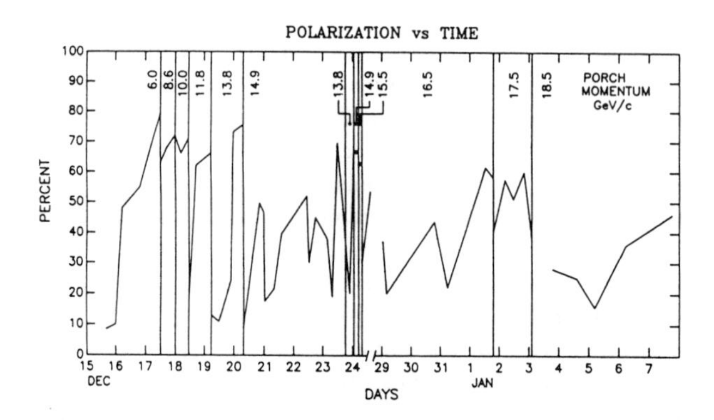

Figure 1: Optimization of the Polarization in the Brookhaven AGS [6]

- A very nice illustration of the existence of depolarizing resonances comes from SPEAR [9]. The polarization level was measured as a function of the momentum between 3.52 GeV/c and 3.76 GeV/c (fig.2). Around the integer spin resonance $\gamma a = 8$ no polarization was observed. Between the resonances the polarization level was close to the maximum possible level of 92.4 %. In between many depolarizing resonances were seen. Most of them can be clearly identified using eq.2. The strongest resonances are : $\gamma a = n - Q_y$ and $\gamma a = n - Q_x$.

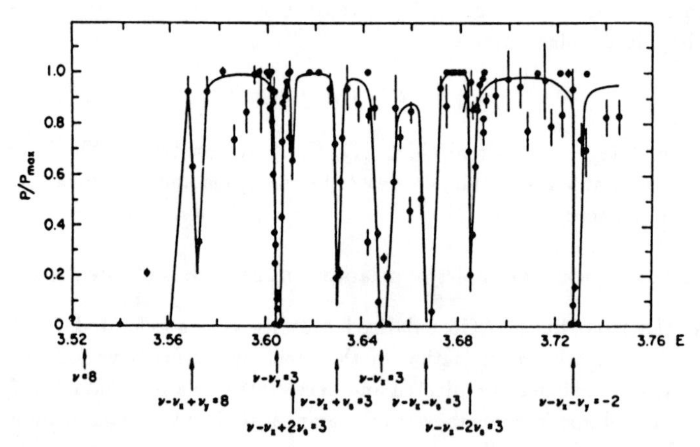

Figure 2: Polarization level as a function of machine energy at SPEAR [9]

- Table I shows a list of electron positron storage rings where beam polarization was measured. In all machines with lower energy a high level of polarization has always been observed. At PETRA a polarization level of 80 % was achieved at a momentum of 16.5 GeV/c after applying corrections which reduced the strength of the most important depolarizing resonances. At 21.5 GeV/c the level of polarization did not exceed 40 %. At LEP the maximum level is about 16 %. TRISTAN reported a polarization level of 40 % and it was observed that the polarization axis was tilted, therefore a longitudinal component was measured. This is a surprising result, because one expects the polarization to be parallel to the magnetic field.

TABLE I : Observed polarization level and polarization time constant for various electron positron colliders [13, 12, 24]

Machine	Energy [GeV]	τ_p[min]	Polarization in %
VEPP	0.65	70	80
ACO	0.53	160	90
VEPP-2M	0.65	60	90
SPEAR	3.7	15	70
VEPP-4	5	40	80
DORIS	5	4	80
CESR	5	300	80
PETRA	16.5	18	80
PETRA	21.5	5	less than 40
TRISTAN	30.0	2	40
LEP	46.5	300	15

The polarization level can be calculated to first order with the computer code SLIM [10]. Although various effects such as energy spread and nonlinear resonances are not considered, the program gives a useful indication of the polarization behavior. For LEP, the results of a simulation are shown in fig.3. An improved polarization optics was assumed and the known imperfections such as orbit deviations, coupling and spurious vertical dispersion were taken into account. The expected polarization level for a spin tune of 105.5 was about 20 %. Repetitive measurements of the polarization at this momentum have been done, a polarization level between 5 and 16 % has been observed during different runs.

It has been always observed that the polarization level is high for small machines, whereas for machines of the size of LEP and TRISTAN the level is low. To obtain a high level of polarization in these colliders optimization procedures are mandatory.

To operate future proton colliders such as LHC and SSC with polarized beams polarized particles have to be injected. Since more than thousand depolarizing resonances are crossed during the acceleration process insertions like Sibirian snakes [11] will be required to maintain the polarization of the injected beam.

To operate an accelerator with polarized beams sophisticated instrumentation is required : The need for a polarimeter which measures the polarization as fast

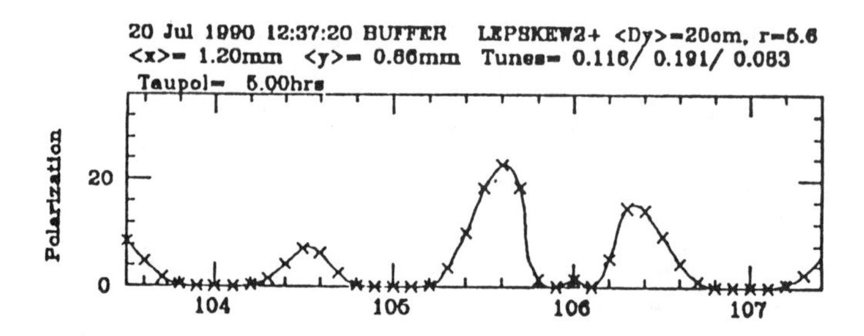

Figure 3: Polarization level from a simulation expected for LEP as a function of the spin tune [8]

as possible is obvious. Many depolarizing mechanisms strongly increase with increasing vertical rms closed orbit. A powerful closed orbit measurement and correction scheme is therefore required. For LEP the rms value of the vertical closed orbit is about 0.7 mm. It is doubtful if a much better precision, say 0.2 mm, can be achieved in an accelerator of 27 km length.

5 Polarization as a tool

To operate an accelerator with polarized beams is difficult and requires both money and manpower. This can be justified by :

- The interest of the high energy physics community to perform experiments with polarized beams.
- The precision calibration of the energy of the beam particles in electron positron storage rings.

Whereas high energy physics experiments require in general longitudinally polarized beams with a polarization level of more than 50 %, the energy calibration can be done with transverse polarization of some percent.

6 Energy Calibration

An energy calibration of the beams is of interest in order to precisely measure the mass of particles as Υ', Z_0 and many others. Fig.4 shows a result from DORIS. The cross section for the Υ' is shown as a function of the center of mass energy (which is twice the particle energy). A precise knowledge of the beam energy is needed to measure the particle mass. In particular for LEP this is of great interest to measure the mass of the Z_0 resonance as well as its width. Different methods have been employed to measure the absolute energy :

- The measurement of the magnetic fields in a reference magnet. This method is limited to a precision of about $\delta E/E = 10^{-3}$.

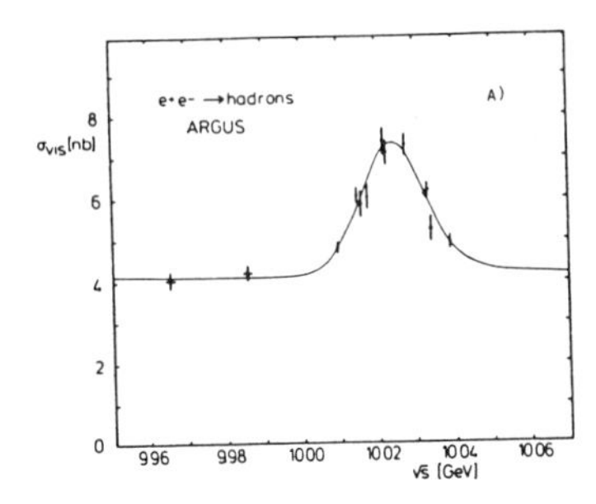

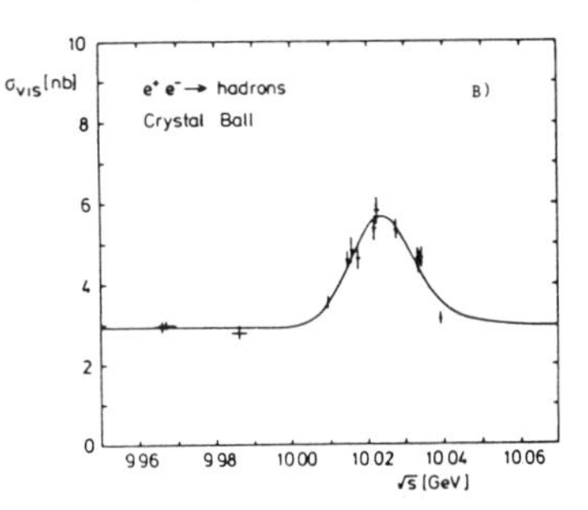

Figure 4: Cross section of the Υ' resonance measured at DORIS [16]

- At LEP the beam momentum has been calibrated using protons [14]. Protons are circulating along the same trajectory as positrons at injection with about 20 GeV/c. Whereas the positron speed is very close to the speed of light with $\gamma = 40000$, the protons are not highly relativistic ($\gamma = 20$). A measurement of the speed difference between the positrons and protons gives the beam energy with a precision of about 10 MeV.

- Depolarization of the beam by applying an external magnetic field [15]. The spin tune is given by $\gamma a = \overline{\gamma a} + n$, with $\overline{\gamma a}$ the noninteger part of the spin tune. The integer n is known by the approximate knowledge of the energy from magnetic field measurements. To depolarize the beam a time dependent magnetic field in the horizontal plane is applied. If the frequency of the field is in phase with the spin rotation frequency around the main dipole field, the spin is rotated around the depolarizer field axis in a resonant way. The rotation leads either to depolarization or to a spin flip. Both cases have been observed. The γ factor and therefore the energy can be simply calculated by:

$$\gamma = \frac{f_{dep}/f_{rev} + n}{a} \tag{3}$$

  with f_{rev} the revolution frequency and f_{dep} the frequency of the depolarizing magnetic field. In fig.5 the polarization level as a function of the depolarizer frequency for DORIS is shown [16]. Below and above the resonance no depolarization is observed. The resonance width depends on the strength of the depolarizing field and can be as small as $\delta E/E = 10^{-5}$ which determines about the limit of this method for energy calibration.

7 Polarimeter : Generalities

Within this report it is not possible to describe all types of polarimeter in detail. Therefore I discuss some general principles common to most polarimeter designs. Then a more detailed description of polarimeters for electrons and positrons follows.

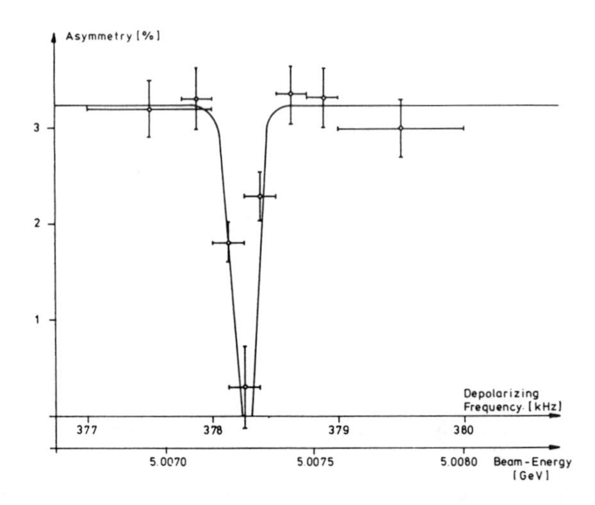

Figure 5: Example of a depolarization curve. Horizontal error bars indicate the sweep range of the frequency generator [16].

All polarimeters are based on the electromagnetic or strong interaction of the polarized beam particles with a "target". The target might be another beam or a piece of material. In some designs the target needs to be polarized. Either the scattered beam particles, the target particles or both are detected (see fig.6).

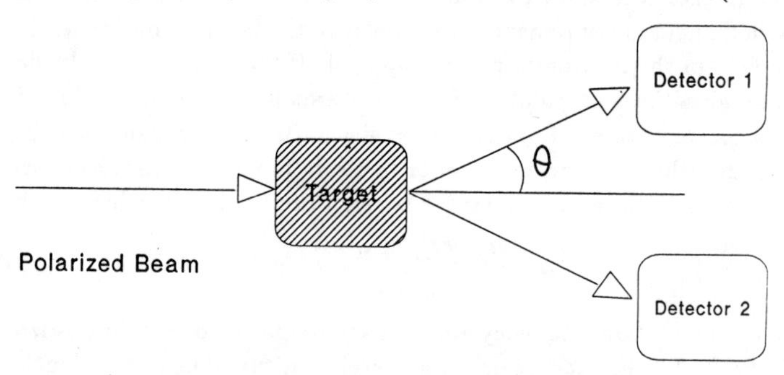

Figure 6: Typical layout of a polarimeter

In the LAB system the particles scattered under an angle θ are detected. The detector measures either the number of particles or the deposited energy. If the beam is polarized the signal is different for detector 1 and detector 2. The polarization level is given by :

$$P = 1/A \times \frac{N1 - N2}{N1 + N2} \tag{4}$$

with A the analyzing power of the polarimeter, N1 and N2 the number of particles or the deposited energy in the detectors. Another possibility is to use only one detector and to measure the differences for positive and negative beam polarization.

Another important quantity is the cross-section for a given scattering process. The larger the cross section, the more particles are scattered. The precision of the measurement is usually limited by statistics, therefore a high rate reduces the time for a measurement. A common problem are possible systematic errors. If detector 1 and detector 2 have different sensitivities even with a nonpolarized beam a signal as expected from a polarized beam is measured. To reduce systematic errors it is common to change the direction of either beam or target polarization. Another estimate of the systematic errors is to measure with unpolarized beams.

In Table II various types of scattering processes in polarimetry are shown. For electron positron polarimeter Møller, Compton and Mott scattering have been used. All cross sections can be calculated analytically with QED. They are valid in a wide energy range. E.g. Møller polarimeter were used for beams with a momentum of some 100 MeV/c up to 50 GeV/c. In high energy physics experiments polarization is observed as well due to the dependence of the cross section for many processes for particle production on the beam polarization [17]. In general these effects are not adequate for polarimetry because the cross sections are too small.

TABLE II: Polarimeter for Electrons [7] [20] [19] [26] and Protons [18]

Polarimeter for electrons Target	Effect	Comments
circular polarized light	Compton scattering	A some 10 % at LEP
polarized e- in ferro-magnetic target	Møller scattering	A about 6 % at SLC
High Z nucleus	Mott scattering	few eV to some 100 keV
polarized e- in the same bunch	Møller scattering (Touschek effect)	less than 1 GeV
positrons in e+e- colliders	different effects	
Polarimeter for protons		
protons in CH_4 or Carbon	p-p elastic scattering	classical method to some 10 GeV
high Z nucleus (Pb)	Primakoff effect : $p+Z = p+\pi+Z$	for high energies
scintillating target	Coulomb-nuclear interference	for high energies

For most proton polarimeters effects based on QCD are used. No analytic calculation of the cross section can be done. Proton-Proton scattering is cited to be the classical method for measuring the polarization but is limited to a beam momentum of some 10 GeV/c [18]. Other techniques are under development, but for the time being they are medium size high energy physics experiments and not adequate for polarimeters built within beam instrumentation groups.

8 Electron-Positron Polarimeter

Møller polarimeter : They have been used to measure the polarization in Linacs and Synchrotrons [21, 19, 20]. Møller polarimeters are destructive for the beam: The beam hits a ferromagnetic foil in a magnetic field of some 100 G. The e- polarization in the target is about 0.08. In the cms system the asymmetry is maximum for scattering angles of $\theta = 90^o$ (fig.7). In the LAB system two electrons with half the beam momentum are observed at an angle of $90^o/\sqrt{\gamma}$. For the SLC the angle is about 5 mrad. Fig.8 shows a measurement done at SLAC : The Møller peak sits on the background from Electron-Fe Mott scattering. The asymmetry was measured with only one detector by reversing the beam polarization.

At the BONN synchrotron two detectors were used in coincidence which largely reduced the background from Mott scattering [19].

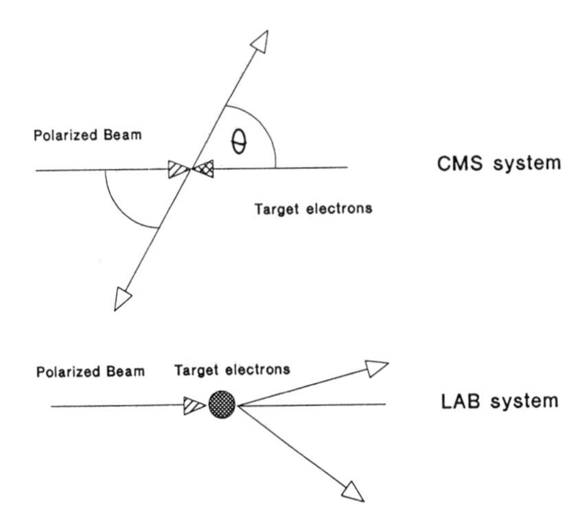

Figure 7: Principle of Møller polarimeter

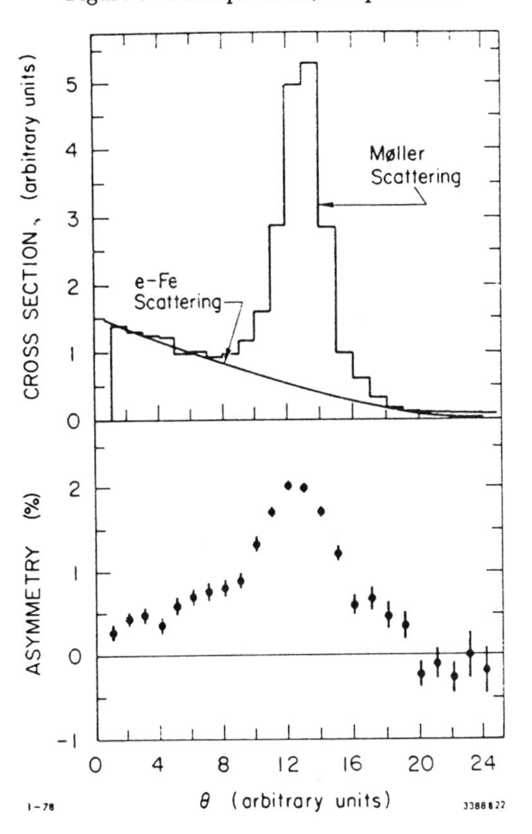

Figure 8: Møller polarimeter : Results from SLAC [21]

Compton polarimeter : Circular polarized light is scattered off the beam. This technique is nondestructive and therefore used to measure the beam polarization in storage rings. In the rest system of the electron the asymmetry is maximum for scattering angles of about 90°.

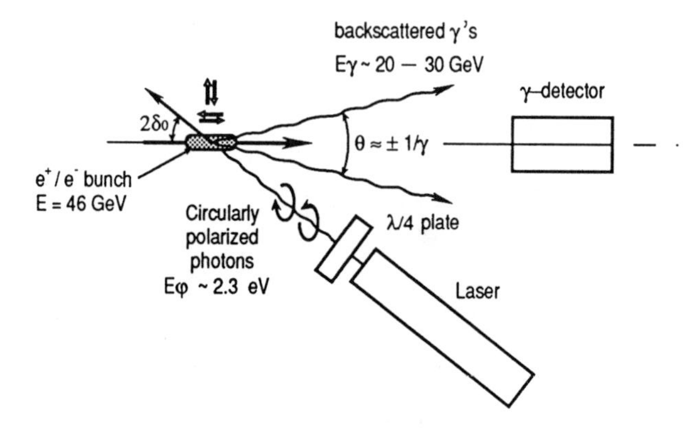

Figure 9: Principle of Compton scattering

In the LAB system the photons are scattered into a narrow cone around the electrons with an opening angle of $\theta = 1/\gamma$, for LEP at 46.5 GeV/c this angle is 10 μrad (see fig.9). Because of the energy transfer of the electron to the photon the momentum of the photon is in the order of the momentum of the beam particles. The photons are detected downstreams after some 10 ... 100 m. The rms width of the vertical distribution is about $(1/\gamma) \times L$, with L the distance to the detector.

At LEP the "mean shift" δy between the distributions of scattered light with positive and negative helicity is observed (fig.10). δy is proportional to the level of polarization, for a fully polarized beams $\delta y = 500$ μrad (for LEP with a beam momentum 46 GeV/c [22]).

The polarized light comes from a laser or from the synchrotron radiation as done in Novosibirsk [23]. Most polarimeter use lasers. A laser polarimeter has the following parts :

- Laser and optical system to generate circular polarized light.
- A system to transport the laser light to the electron beam, including mirrors etc.
- The scattered photons travel to the detector. In the first deflecting magnet after the interaction region the electrons and photons are separated. A sufficient aperture and a window for the traversal through the vacuum chamber for the photons travelling to the detector is required.

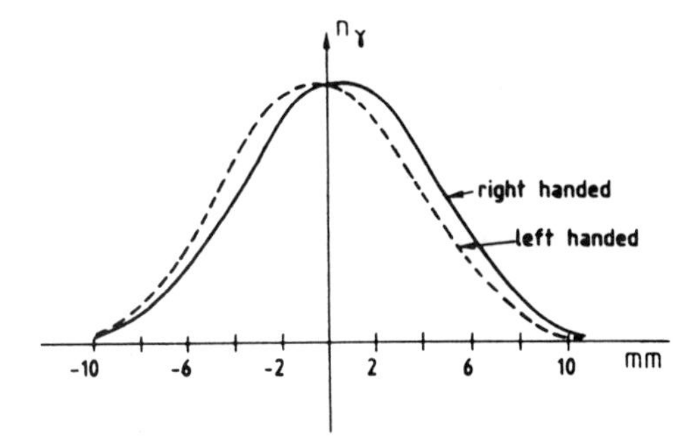

Figure 10: Measurement of the 'mean shift' in Compton polarimeter

- Detector to measure the spatial distribution of the scattered photons.
- Data acquisition and trigger system.

Single versus multiphoton technique :
In the single photon technique a continuous laser or a laser with a repetition frequency of the bunch frequency is used. The laser power is with some 10 W relatively low. Each bunch is illuminated. For one bunch the probability to scatter a photon is much less than 1 in order to avoid multiple scattered photons. Compton scattered photons cannot be distinguished from background photons from Gasbremsstrahlung. Photons in a given energy range are counted for left and right circular polarized light. The analyzing power is large, for LEP some 10 %, but might be reduced by background.

The multiphoton technique uses a laser with a low repetition frequency, but a high peak power (e.g. a Nd-YAG laser with 30 Hz repetition frequency, 10 ms long pulses of some 10 MW peak power).

Many photons per laser shot are scattered and the total energy for one shot is much larger than the beam particle energy (for LEP 100 - 1000 photons are scattered with a total energy of some TeV for each laser shot). The deposited energy is measured for left and right polarized light. An energy cut on single photons cannot be applied. The technique is insensitive to background, but the analyzing power is slightly lower than the single photon method (if the background for the single photon method is not considered).

To compare single and multiphoton polarimeter it is useful to calculate the number of scattered photons N_γ : $N_\gamma \propto P_{laser} N_{bunch} f$, with $f = max(f_{bunch}, f_{laser})$. For LEP the number of photons is 200 time larger for the multiphoton polarimeter compared to the single photon polarimeter (4 bunches are assumed, P = 30W for the single photon technique and 10 MW for the multiphoton technique).

9 The LEP polarimeter

Requirements for the LEP polarimeter :

- The expected level of polarization before the application of specific correction schemes is low : 3-20 % . With 3 % polarization the mean shift is about 15 μ. Any systematic errors should therefore be less than 5 μ in order to detect a level of 3 % with 3 sigma standard deviations.

- The degree of polarization should be measured as fast as possible. This is of particular importance for optimization of the polarization to measure as fast as possible if the polarization increases or decreases after changing a machine parameter.

- To synchronize the polarization measurement with the activities in the control room the polarimeter which is 2 km away must be remotely controllable.

Layout : The 60 mJ (compared to 90 mJ design value) light pulses from a 30 Hz Nd-YAG laser operated in the visible range at 532 nm and installed in an optical laboratory 15 m away from the LEP tunnel are guided towards the roughly evacuated beam pipe including three lenses and five multilayer dielectric mirrors (see fig.11) [24]. The light is polarized in a section with optical instuments installed on a bench at the laser output. A rotating $\lambda/2$ plate and a $\lambda/4$ plate can produce linear or circular light.

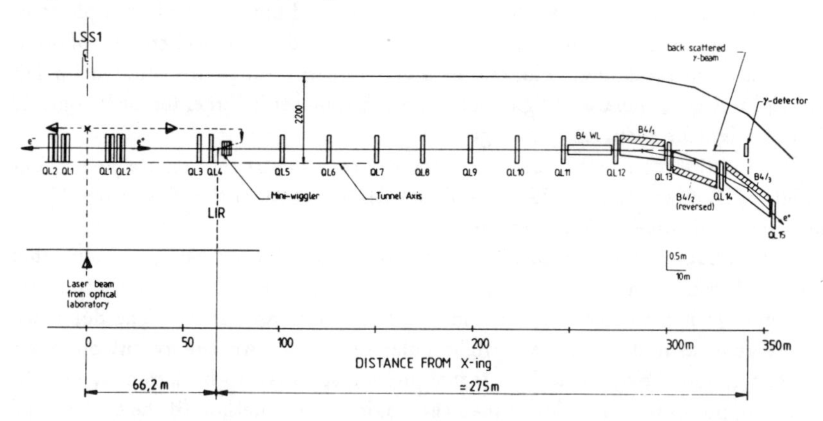

Figure 11: General layout of the LEP Compton polarimeter [24]

The final deflection onto the e- beam under an angle of 2-3 mrad is provided by (Ag + MgF_2)-coated CU mirror. Their position can be remotely adjusted. The high energy scattered γ's reach the detector 247 m downstream the interaction point through a 50 $\times$ 20 mm^2, 2 mm thick Al-window built in the modified vacuum chamber in the main dipole about 225 m from the interaction point. A silicon calorimeter is employed to measure the distribution of the recoil photons (see fig.12). Four strip detectors (S_1 - S_4) and five unsegmented pads (F_1 - F_5) are inserted between tungsten plates. The 40 mm $\times$ 40 mm strip detectors have 16 horizontal strips with 2 mm pitch. A movable lead absorber is installed in front of the silicon calorimeter to shield the detector against synchrotron radiation. An absorber thickness of 0 - 5 radiation lengths can be selected according to the intensity of the synchrotron radiation.

The asymmetry function is defined as :

$$A(y) = \frac{n_r(y) - n_l(y)}{n_r(y) + n_l(y)} \tag{5}$$

where $n_{r,l}(y)$ are the γ rates at the vertical position y for right and left circular polarized light, or horizontal and vertical polarized light. The measured asymmetries are shown in fig.13, with circular polarized light as well as with linear polarized light. The asymmetry function is derived from distributions as shown in the fig.13. For linear polarized light the rms width of the distribution changes significantly independent of any beam polarization when changing from horizontal to vertical polarization. In first order the mean of the distribution does not move and the mean shift δy is zero. With circular polarized light and beam polarization the mean of the distribution changes.

Sources for systematic and statistical errors are :

- A jitter of the e- beam orbit. This leads to displacements of the backscattered photon beam.
- If the aperture for the scattered photons is limited, the tails of the photon distribution might be lost. If the light is perfectly circular polarized, the distributions for right and left circular polarized light are identical if the beam is not polarized. If the light has some linear component the width of the distribution changes. This change together with the limited aperture can lead to a mean shift independent of polarization.
- Nonlinearities of the detector.

Although these effects are small, an optimization is needed to avoid any mean shift larger than a few μm :

- Minimization of the linear polarization component in the light.
- Precise centering of the backscattered photons in the aperture.

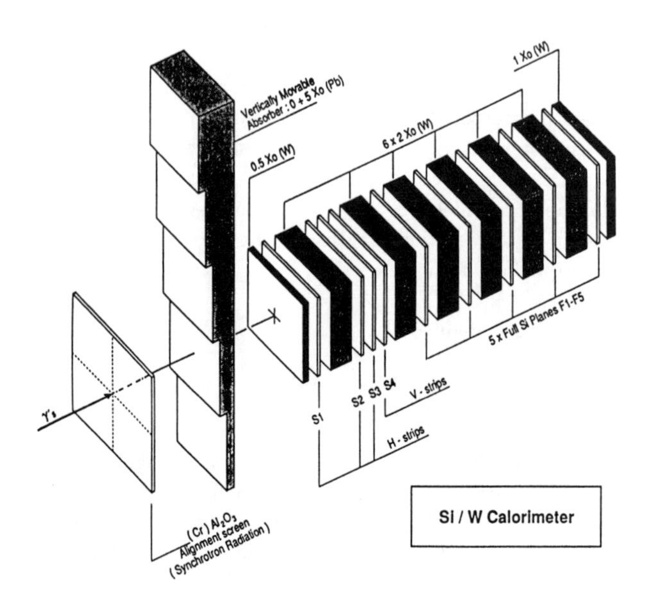

Figure 12: Silicon Strip calorimeter as a detector fo the LEP polarimeter

- All systematic errors due to electron orbit jitter are minimized by reversing the light helicity from one laser shot to the next.

10 Results from LEP

Fig.14 shows the measured mean shift as a function of time [28] [29] [31]. The initial value is about 45 μm. To ensure that the mean shift is due to beam polarization vertical closed orbit bumps are applied generating a depolarizing resonance which is expected to totally destroy the polarization. The systematic effects are then minimized and a mean shift of about zero is observed. After switching off the closed orbit bumps the polarization builds up with the time constant τ.

The polarimeter measures the mean shift δy. To calculate the degree of polarization from the mean shift two methods were employed :

- From Monte Carlo simulations the expected mean shift for a polarization level of 100 % is 500 μm for a beam momentum of 46.5 GeV/c. The polarization level is given by : P $= \delta y[\mu m]/500$.

- Another method is a three parameter fit of a function $f(t) = \delta y_{asymtotic} + (\delta y_0 - \delta y_{asymtotic}) \times exp(-t/\tau)$ to the measured data. The parameters are the initial mean shift, the mean shift which is expected asymtotically and the risetime. From the risetime and the natural polarization time constant τ_p the asymptotic polarization level can be calculated : $P = 92.4\%/(1+\tau_d/\tau_p)$, and

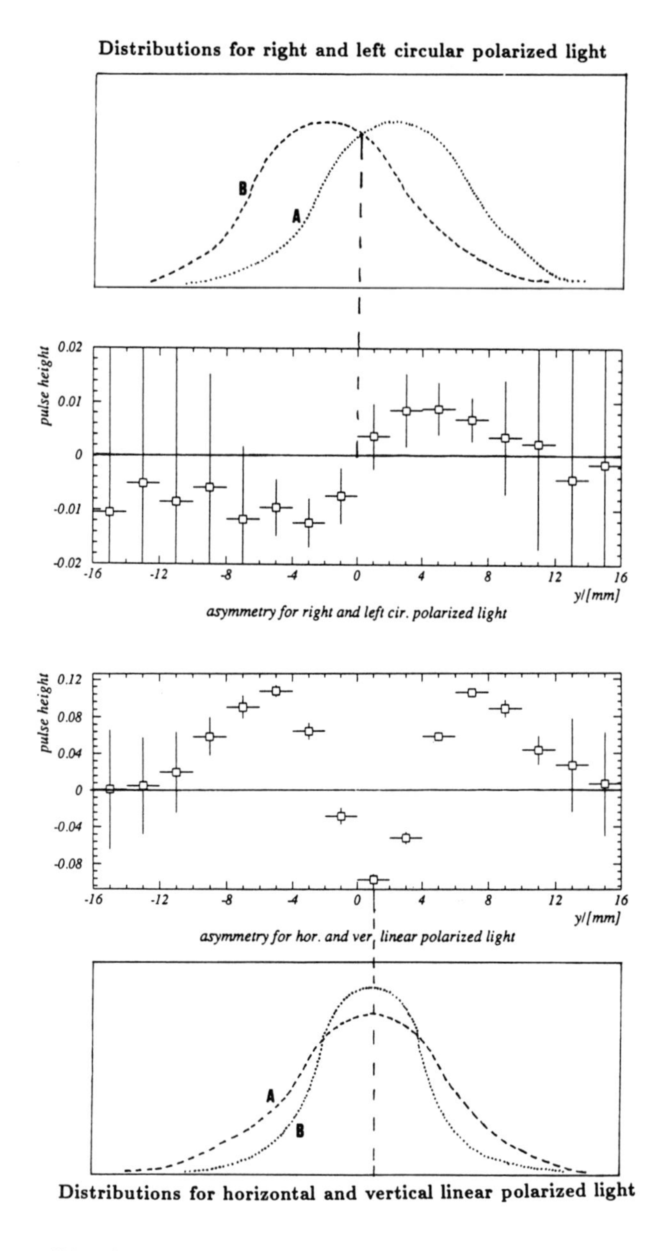

Figure 13: Distributions of scattered photons for linear and circular polarized light

$1/\tau = 1/\tau_d + 1/\tau_p$. For the measurement shown in fig.14 it has been estimated [30] : $\tau = 65 \pm 8$ min, $P_{asymtotic} = 19\% \pm 2.4\%$, $\delta y_{asymtotic} = 100 \pm 6\mu$, $P_{max} = 15.7 \pm 2\%$. In fig.14 all data points measured during an interval of 20 min do not match the curve. This observation is explained by a displacement of the distribution of the backscattered photons during this time. Due to the limited aperture the systematic effects discussed above change the mean shift, which was then compensated by changing the angle of the electron beam to recenter the distribution of the scattered photons.

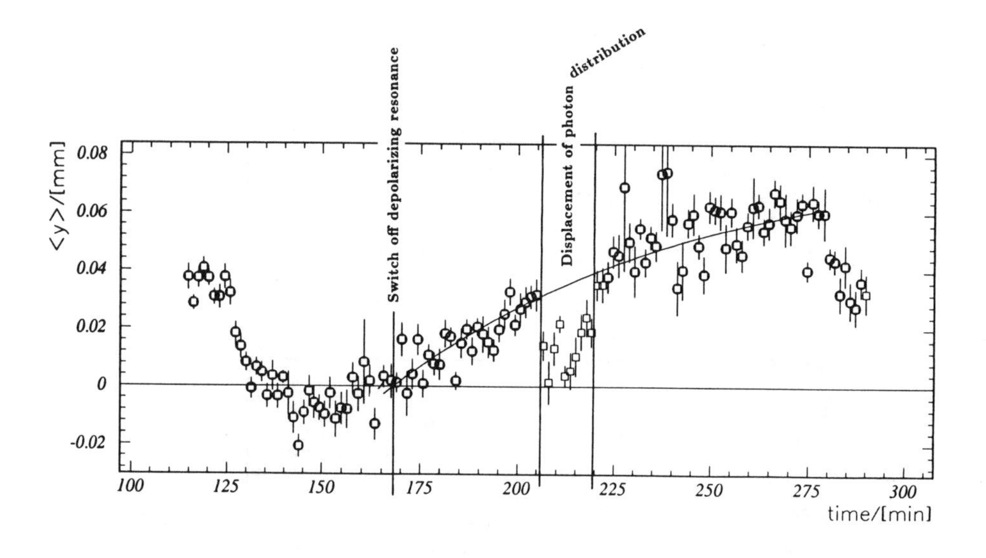

Figure 14: Polarization build up at LEP

11 Energy Calibration at LEP

To perform an energy calibration a depolarizing magnetic field in the horizontal plane with an integral strength of 4.3 Gm is applied. The frequency is slowly changed from f_1 to f_2. The frequency span is typically 112 Hz in 60-90 s corresponding to a spin tune range of 0.01 and an energy range of 4.4 MeV.

The polarization is measured at the same time and three different patterns are observed (see fig.15) [31] :

- The polarization level does not change.
- The polarization level decreases to zero or changes to the opposite direction.

- The polarization level decreases by a small amount.

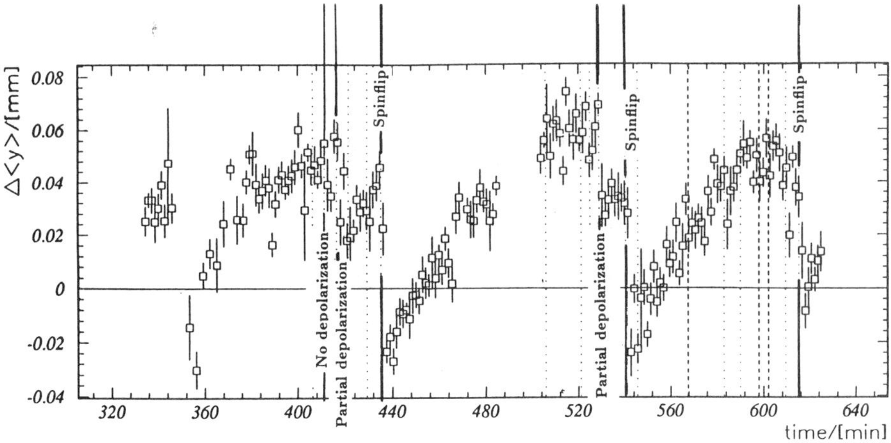

Figure 15: Resonant Depolarization for Energy Calibration at LEP : The mean shift δy is shown as a function of time

If the depolarizer frequency sweep does not cover the resonance, the polarization is not affected. If the sweep covers a frequency where the resonance condition (see eq.3) is fulfilled, the polarization is totally destroyed or the direction is reversed.

In the vicinity of the main resonance synchrotron sidebands are expected due to the modulation of the energy with the synchrotron frequency. This is the case for a frequency $f_{depolarizer}/f_{rev} = \gamma a \pm Q_s$. A sweep over these resonances causes a partial depolarization of the beam. For each resonance two synchrotron sidebands are expected and they have been observed at LEP.

12 Conclusion for LEP

The polarization is measured with a statistical error of 1 % within about 1 min. After an optimization of the polarimeter the offset is less than 1 %. The absolute level of polarization is known to an accuracy of $\delta P/P = 10\%$. The energy is calibrated with an error of about 1 MeV. An energy calibration does not require more than a polarization level of more than 5 %, although with a higher level the calibration is much easier to do.

13 Acknowledgements

I like to thank P.Castro for helping with the figures for this report and W.Herr for carefully reading the manuscript. Furthermore thanks to the collegues from the LEP polarimeter for their help, in particular B.Dehning and M.Placidi.

References

[1] B.W.Montague, Physics Reports Vol.113, No.1 (1984), 1-96

[2] V.Bargmann, L.Michel, V.L.Teledgi, Phys.Rev.Lett 2(1959), 435

[3] See various authors in : 8th.Int.Symp. High Energy Spin Physics, Minneapolis 1988, AIP-Conf.Proc 187, 1191-1237

[4] T.Niinikoski, R.Rossmanith, Nucl.Instr.Meth A 255(1987),460

[5] A.A.Sokolov, I.M.Ternov, Sov.Phys.Dokl.8 (1964) 1203 8th.Int.Symp. High Energy Spin Physics, Minneapolis 1988, AIP-Conf.Proc 187,1412-1427

[6] L.A.Ahrens, Operation of the AGS polarized Beam, in: 8th.Int.Symp. High Energy Spin Physics, Minneapolis 1988, AIP-Conf.Proc 187, 1068

[7] C.K.Sinclair, Report on e- source and e- polarimetry workshop, 8th.Int.Symp. High Energy Spin Physics, Minneapolis 1988, AIP-Conf.Proc 187,1413

[8] J.P.Koutchouk, Transverse Polarization in LEP, in: 9th.Int.Symp. on High Energy Spin Physics, Bonn 1990, Springer Verlag

[9] J.R.Johnson et.al., Nucl.Instr.Meth. 204 (1983),261

[10] A.Chao, Nucl.Instr.Meth. 180(1981), 29-36

[11] T.Roser, First Experimental Test of the Sibirian Snake Concept, 9th.Int.Symp. High Energy Spin Physics, Bonn 1990, Springer Verlag

[12] K.Nakajima et al., Preliminary Polarization Measurements of TRISTAN electron beam, EPAC 1990, Nice, June 12-16, 1990, Edition Frontieres

[13] Y.M.Shatunov, Particle Accelerators, Vol.32,139-152

[14] R.Bailay et al. LEP Energy Calibration, EPAC 1990, Nice, 12-16 June 1990

[15] Y.Derbenev et al,. Particle Accelerators 18 (1980)

[16] D.B.Barber et al., Physics Letters 135B (1984), 498

[17] V.N.Baier, Proc.Workshop Adv.Beam.Instr., KEK, Tsukuba, Japan, April 22-24, 1991 (Nat.Lab.High Energy Physics)

[18] D.G.Underwood, Summary of Polarimeter Session, 8th.Int.Symp. High Energy Spin Physics, Minneapolis 1988, AIP-Conf.Proc 187
S.Hiramatsu et al., Internal Polarimeter for the polarized proton beam at the KEK 12 GeV PS, same proceedings, p.1355
K.Imai et al.,Primakoff Polarimeter at Fermilab polarized proton / antiproton beam facility, same proceedings, p.1360
G.Pauletta et al., Tests of a Coulomb-Nuclear Polarimeter, same proceedings, p.1366

[19] W.Brefeld et al., Nucl.Instr.Meth. 228(1985) 228-235

[20] B.Wagner et al., Møller Polarimeter for cw and pulsed intermediate energy electron beams, NIM-A294 (1990) p.541-548

[21] H.Steiner, Measurement of Electron Beam Polarization at the SLC, Int.Conf.Advanced Methods for Colliding Beam Physics, March 9-13, 1987, SLAC, Stanford,CA

[22] D.B.Gustavson, Nucl.Instr.Meth 165 (1979) 177

[23] R.Schmidt, Diplomarbeit, DESY M-80/04 (1980)

[24] G.Barbagli, 9th.Int.Symp.High Energy Spin Physics, Bonn 1990, Springer Verlag, Vol.2, p.47

[25] A.S.Artamonov et al.,Phys.Lett 118B (1982), 225

[26] M.Placidi, R.Rossmanith, NIM A274 (1989) 79-94

[27] J.Badier et al.,The Commissioning of the LEP Polarimeter, Part.Acc.Conf,1991, San Francisco, 6-9 May,1991

[28] J.Badier et al., First Observation of Transverse Beam Polarization in LEP, CERN-PPE/91-125 (August 1991), to appear in Physics Letters B.

[29] J.Badier et al., Observation of transverse Polarization in LEP, Part.Acc.Conf 1991, San Francisco, 6-9 May,1991

[30] B.Dehning, private communication

[31] LEP polarization collaboration, Results from the LEP 1991 run, in preparation

INSTRUMENTATION AND DIAGNOSTICS FOR FREE ELECTRON LASERS

Todd I. Smith*
Hansen Experimental Physics Laboratories
Stanford University, Stanford,CA 94305

ABSTRACT

The operation of a Free Electron Laser (FEL) depends on the interaction of a relativistic electron beam with a beam of photons, so it is not surprising that there is a great overlap between the instrumentation needs of an FEL and those of any electron accelerator. However, the interaction requirement imposes constraints on the electron beam which sometimes lead to a different emphasis on parameters than usual, and thus on somewhat specialized diagnostics. In addition, there is the need for instrumentation of the optical part of the FEL, an area in which most accelerator workers have limited experience. Finally, the FEL itself can be used as a sensitive diagnostic for the linac.

INTRODUCTION

In a Free Electron Laser (FEL) a relativistic electron beam transfers energy to a beam of photons. In the production and control of the electron beam, the instrumentation needs are similar to those of almost any electron accelerator. However, the interaction requirement imposes constraints on the electron beam which sometimes lead to a different emphasis on parameters than usual, and thus on somewhat specialized diagnostics. In addition, in an FEL there is the need for optical diagnostics, an area in which many accelerator facilities have had limited experience. Finally, since an FEL's performance is sensitive to the parameters of the driving electron beam, the FEL itself can be used as a diagnostic for the linac.

These issues will be explored in the following sections. First will be a brief description of an FEL, an explanation of how it works and a discussion of the constraints it imposes on an electron beam. Next will be a discussion of various electron beam diagnostics used with FELs, followed by a discussion of optical diagnostics for the FEL. Finally, examples will be presented of the use of the FEL as an overall electron beam diagnostic.

THE FREE ELECTRON LASER

In this section a simple explanation of the way an FEL operates will be presented, to make the constraints which it imposes on an electron beam plausible. For detailed discussions of FELs and their requirements, please see the references[1,2,3].

* This work was supported by ONR N0014-91-C-0170

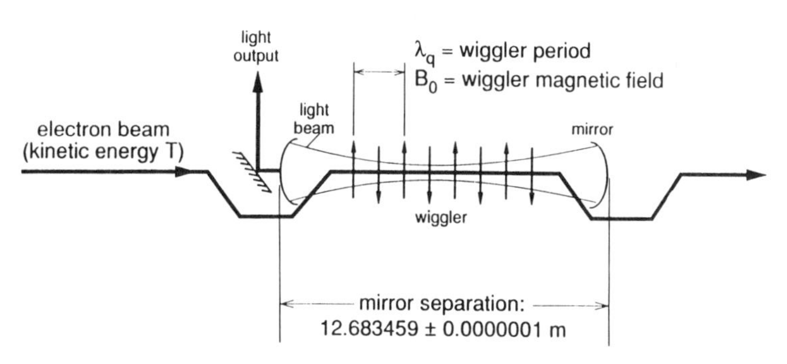

Fig 1. Basic configuration of an FEL oscillator.The mirror spacing given is appropriate for the FEL at the Stanford Picosecond FEL Center.

The basic configuration of a generic rf linac-based FEL oscillator is given in Fig 1. Fundamentally, it consists of two mirrors forming an optical resonator, and an electron beam which is co-linear with the optical beam inside a 'wiggler'. The interaction between the two beams occurs inside he wiggler, a device which produces a spatially periodic magnetic field, transverse to the initial direction of motion of the electron beam. Most commonly, the wiggler field lies in one plane and the magnetic field on the axis appears nearly sinusoidal. In this case, for small fields, the electron beam will execute simple harmonic motion in the transverse plane orthogonal to the wiggler field. Another wiggler configuration produces a helical magnetic field, with a constant field magnitude along the axis. In this case, the electron beam can spiral down the axis at constant radius, and with constant magnitude of transverse velocity. In either case, the electron beam undergoes periodic transverse acceleration with the same spatial period as the wiggler period, and thus emits spontaneous electromagnetic radiation.

As the operating wavelength of the FEL is essentially the same as that of this spontaneous radiation, is important to understand just what wavelength is produced. One of the most straightforward explanations that the observed radiation has the same wavelength as the wiggler, but Doppler shifted (as shown in Fig 2) since the emitting electrons are travelling at nearly the speed of light. The observed wavelength is given by

$$\lambda_0 = \frac{\lambda_w}{2\gamma^2}(1+a_w^2) \ , \ a_w \equiv \left\langle \frac{eB_0\lambda_w}{2\pi m_0 c} \right\rangle_{rms}$$

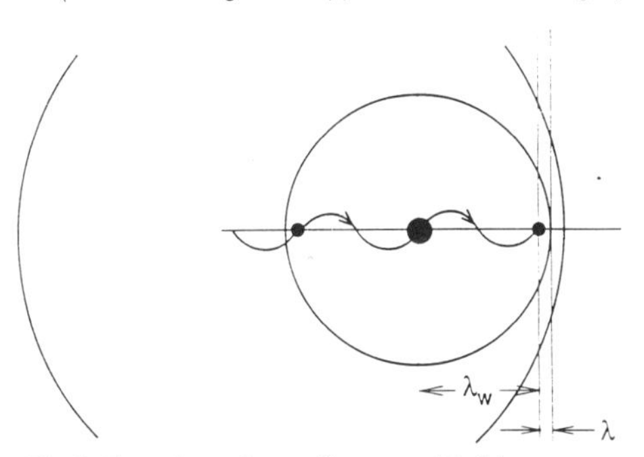

Fig 2. Snapshot of wavefronts emitted by a relativistic electron as it oscillates down a wiggler. Since the electron is travelling at nearly the speed of light, the radiation in the forward direction is Doppler shifted.

where $\gamma = (1+E/m_oc^2)$ is the usual Lorentz factor, E is the kinetic energy of the electrons, and B_o and λ_w are the amplitude and period of the wiggler field. The term containing a_w is a result of the fact that the forward velocity of the electrons decreases as the strength of the transverse field increases. In a typical wiggler, $a_w \sim 1$.

An equivalent way of looking at the emitted radiation is that it's wavelength is such that the radiation will advance by exactly one optical period over an electron during the time the electron takes to cover one wiggler period. As seen in Fig 3, this condition insures that the relative phase between a given electron's transverse velocity and the transverse electric field at that electron's position is constant. Since this phase doesn't vary as the electron moves down the wiggler, an electron can exchange significant energy with the optical field. Of course, if the electron beam's density is uniform along the axis, then over the distance of one optical wavelength just as many electrons will lose energy as will gain energy, and there will be no net energy exchange.

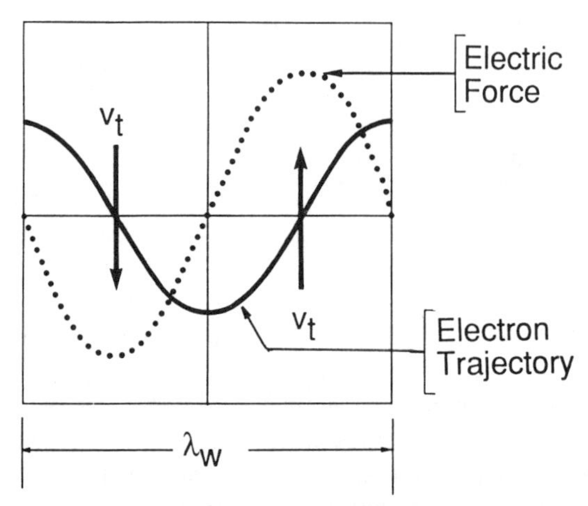

Fig 3. Path of one electron and the electric force it sees as it travels along one wiggler period.

However, just as in a traveling wave tube or klystron, the periodic energy modulation results in velocity modulation which in turn can lead to density modulation and ultimately to energy exchange. In an FEL, these three processes occur simultaneously, although their relative importance varies along the length of the wiggler. Nonetheless, all of the small signal features of the FEL can be understood in terms of an 'impulse' model in which the three processes are treated as completely independent. In this model, the FEL is treated as if it were divided into three distinct sections. In the first section only energy modulation of the uniform beam occurs. In the second section only bunching takes place. Finally, in the third section the bunched beam interacts with the optical field and yields a net energy transfer. The model and the process are depicted in Fig 4, where the FEL is shown in the optical klystron configuration, emphasizing the distinct sections.

The strength of the energy modulation is easily calculated in the model, particularly for the case where the wiggler field is assumed to be helical. In this case an electron can spiral at constant radius along the wiggler axis. The magnitude of the transverse magnetic field seen by the electron is then constant, and the transverse velocity of the electron is straightforward to calculate. The result is

$$\beta_\perp = \frac{v_\perp}{c} = \frac{1}{\gamma}\frac{eB_o\lambda_w}{2\pi m_o c} = \frac{1}{\gamma}a_w$$

where a_w is the wiggler constant defined earlier. Work done on the electron is given by the integral of the dot product between the distance it travels and the force on it. If the electron beam is assumed to overlap an optical beam of power P_o and mode area A_o, then the magnitude of the electric field is just $E_o{}^2=\eta P_o/A_o$, where η is the impedance of free space, 377 Ω. This field is perpendicular to the electron beam's longitudinal velocity, but parallel to its transverse velocity. If the length of the first portion of the wiggler is L_1, then the energy transfer between the electron and the optical field is just

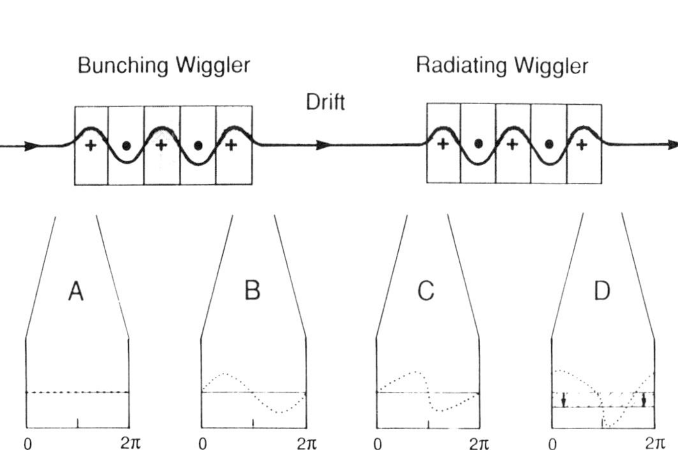

Fig. 4 'Optical klystron' FEL configuration, illustrating the impulse model used to estimate the small signal gain.

$$\Delta W = \int \bar{F}\cdot d\bar{s} = \int F_\perp v_\perp dt = eE_o\beta_\perp L_1\cos(\phi_o)$$

since ϕ_o, the relative phase between the electron's transverse velocity and the optical field, is constant. The modulated electron beam is depicted at point "B" in Fig 4.

As mentioned above, there is no net energy transfer from the electron beam at this point, since all there are as many electrons at positive phases as at negative phases. However, since higher energy electrons travel faster than lower energy electrons, even if the electrons are relativistic, the energy modulated beam is also velocity modulated. As a result of this velocity modulation, the beam will become density modulated after traveling through the middle portion of the wiggler, assumed to have a length L_2. The density modulation of the electron beam can be seen at point "C" in Fig 4. Since this modulation is periodic at the optical wavelength, there will be a component of the current with the same wavelength. The magnitude of this component can be estimated as

$$I_{optical} = I\frac{2\pi}{\lambda_o}\left[\frac{d\beta_z}{dW}\Delta W\right]L_2$$

where I is the DC (unmodulated) beam current and λ_o is the optical wavelength. Evaluating the derivative, and recalling the relation between λ_o and λ_w for an FEL, leads to an alternate expression

$$I_{optical} = I\frac{2\pi}{\lambda_o}\frac{1+a_w^2}{\gamma^3}\frac{\Delta W}{m_o c^2}L_2 = I\frac{4\pi}{\gamma}\frac{L_2}{\lambda_w}\frac{\Delta W}{m_o c^2}$$

where a_w in the center expression is calculated from the parameters of the middle section of the wiggler. (For an optical klystron, the center section of the wiggler is just a drift section, for which a_w is zero. For a uniform wiggler, a_w is the same for all sections).

Finally, in the third section of the wiggler this optical current can interact with the optical field, leading to a net energy transfer as shown at 'D' in Fig 4. The power transferred from the beam is just

$$P_b = \frac{1}{2}(I\cdot V) = \frac{1}{2}\int I_o E_o v_\perp dt = \frac{1}{2}I_o E_o \beta_\perp L_3$$

where L_3 is the length of the last section.

Combining these expressions, the gain G per pass of the FEL can be written as

$$G = 2\pi\frac{e}{\varepsilon_o m_o c^3}Ia_w^2\frac{L_1L_2L_3}{\lambda_w\gamma^3 A_o}$$

where $G=P_b/P_o-1$, and ε_o is the dielectric constant of free space. The parametric dependence of the gain equation is exactly the same as that obtained from far more detailed expressions.With the choice of L/2 for the length of each section, the absolute value of the expression is only 8% less than the exact value[4]. (Since the various processes actually occur over the length of the entire wiggler, and do not begin or stop abruptly at the model's arbitrary boundaries, the obvious choice of L/3 for the length of each section is almost certainly too small. A value of L/2 seems fairly reasonable.)

ELECTRON BEAM REQUIREMENTS

The preceding discussion implicitly assumed an ideal electron beam. All electrons were assumed to have the same initial energy, and all had exactly the same trajectory. Real electron beams have a spread in these parameters. In this section, some guidelines for acceptable tolerances on electron beams used for FELs will be discussed.

An overriding concern is obviously that the electron beam must overlap the optical beam. The optical beam in a typical FEL resonator is a TEM_{00} gaussian mode. The radius of this mode can be written as

$$\omega^2 = \omega_o^2 \left[1 + \frac{z^2 \lambda^2}{\pi^2 \omega_o^4} \right]$$

where w is the radius at distance z from the minimum, λ is the optical wavelength, and w_o is the minimum radius. If this is compared with the envelope equation for an electron beam of emittance ε in a drift region,

$$r^2 = r_o^2 \left[1 + \frac{z^2 \varepsilon^2}{\pi^2 r_o^4} \right]$$

then it is clear that spatial overlap requires that $\varepsilon <= \lambda$. This ε is the absolute, or geometric emittance, not the frequently used normalized emittance ε_n, which is Lorentz invariant. The relation between the two is $\varepsilon_n = \beta\gamma\varepsilon$. As an example, for an optical wavelength of 0.5 μ, $\varepsilon = 0.5\ 10^{-6} = 0.5$mm-mr$=0.17\ \pi$ mm-mr.

This limit on ε must be considered an upper limit. Other conditions may impose a more strict bound. For instance, the magnetic field from a real wiggler must satisfy Maxwell's equations, and therefore will have transverse gradients[5]. In fact, the transverse field will increase exponentially away from the axis, and thus electrons off axis will tend to favor FEL operation at a longer wavelength than those on axis. In addition, these same gradients cause betatron oscillations of off axis electrons, and thus transverse velocities in addition to those involved in the FEL interaction. These additional velocities result in a decrease in the off axix electrons' longitudinal velocities, and another change in the ideal operating wavelength. Both of these effects are minimized by reducing the electron beam's average radius through the wiggler, which implies the smallest possible emittance.

In addition to placing a constraint on the electron beam's emittance, the FEL interaction imposes a constraint on the energy spread of the beam. An appreciation for this can be obtained by realizing that the fractional bandwidth of the spontaneous radiation emitted by an oscillating electron as it passes through an N period wiggler is about 1/N, since the radiation will contain N cycles. Since the wavelength of radiation emitted by an electron varies as γ^{-2}, it is clear that only electrons with energies within about 1/2N of the central energy will interact. As an example, assume that radiation is to be produced with a 100 period wiggler. If the electron beam energy is 100 MeV and a_w is about 1 then $\Delta E/E \sim 1/2N = 5\ 10^{-3} = 500$keV.

Another constraint on an FEL oscillator is the simple one that the gain must exceed the system losses. If not, oscillation is clearly impossible. The gain of FELs tends to be in the percent/ampere range, so driving currents tend to be tens of amperes or more. FELs with currents less than one ampere are very difficult. It must be emphasized that the currents referred to are the instantaneous currents seen by the optical beam. They are therefore the peak currents delivered by an accelerator, and not the average. Particularly in the case of an rf linac in which only a fraction of the rf buckets is filled, the ratio of peak to average current can be substantial. As an example, the pulse structure of the superconducting linear accelerator at Stanford University is illustrated in Fig 5. The linac frequency is 1.3 GHz, and the current pulses have a FWHM of about 3 ps. However, for a variety of reasons, only one of every 110 rf buckets is filled. The repetition interval is 84.6 ns, and the ratio of peak to average current is almost 30,000.

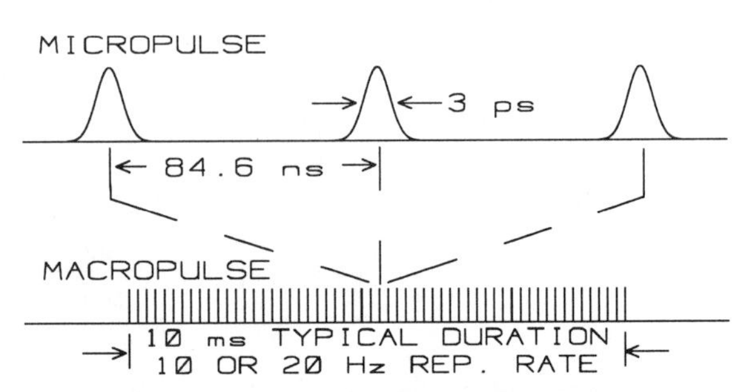

Fig. 5 Pulse format of the Stanford Picosecond FEL.

A consequence of the fact that electrons must overlap with photons in order to interact is that the length of the optical resonator must be carefully selected and adjusted in pulsed machines so that the round trip time of the photons match some multiple of the pulse repetition rate. This requirement is illustrated in Fig 6. The FEL resonator at the Stanford Picosecond FEL Center is 12.7 meters long, chosen to just match the interpulse spacing. When operating the FEL at 3μ, changes in the resonator length of 1μ are observable. If the absolute length is off by more than 20μ the FEL won't oscillate. These constraints are equivalent to 3 fs and 70 fs respectively.

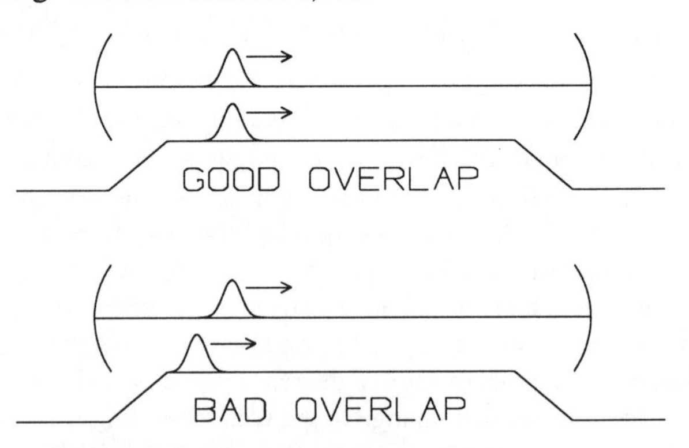

Fig. 6 If successive photon and electron pulses do not overlap adequately, oscillation is impossible.

ELECTRON BEAM DIAGNOSTICS

Virtually all accelerators have an energy spectrometer in order to measure the beam energy and energy spread. Accelerators also require a beam dump of some kind, and this is frequently incorporated into the spectrometer. Designing a combination spectrometer and beam dump for the beam used to power an FEL presents some special challenges. When the FEL is not operating, the electron energy spread will be fairly small, and for diagnostic purposes the dispersion of the system will have to be reasonably large in order to have acceptable energy resolution. On the other hand, when the FEL is operating, its interaction with the electron beam will result in a large energy spread, typically several percent, incompatible with a high dispersion spectrometer/dump.

The Stanford Center has designed and incorporated a variable dispersion spectrometer/dump[6] which can be set up to resolve energy spreads as low as 0.01% at a dispersion of about 1 mm/%, or adjusted for any lower dispersion (including zero) to function as a beam dump. The energy acceptance in the dump mode is a bit less than 10%. The system is shown schematically in Fig 7. The two quadrupoles upstream of the bending magnet are used to adjust the system dispersion, while the two downstream control the image size at the detection plane. A Panofsky quadrupole is used immediately after the bending magnet because of the highly stigmatic beam at that point.

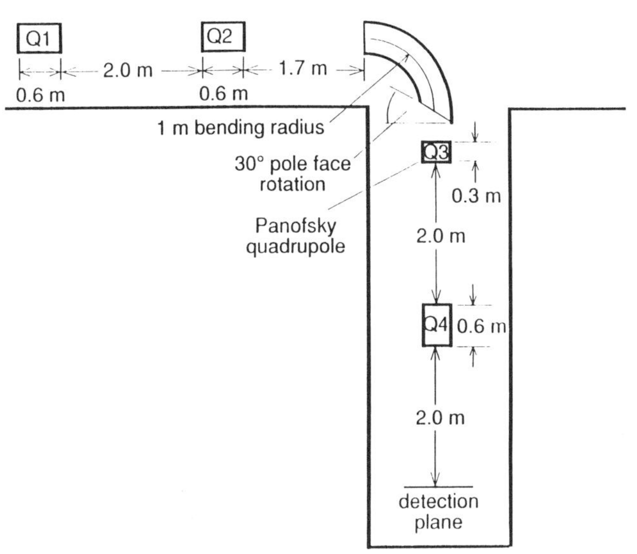

Fig. 7 Schematic view of the variable dispersion spectrometer/beam dump for the Stanford Center.

There are two detectors located at the image plane. One is a fluorescent screen made of Cerium doped material with a short fluorescence lifetime to reduce saturation effects. The other is a wire array. The screen provides high spatial resolution and poor temporal resolution while the wires provide good temporal resolution (better than 30 μs) and limited spatial resolution. Fig 8 shows sample data taken with the screen. The narrow trace was taken while the FEL was prevented from oscillating by means of an absorber inserted in front of one of the mirrors. When the absorber was removed, the broad trace appeared. The centroid shift represents the energy removed from the electron beam. (Note that a small part of the beam was

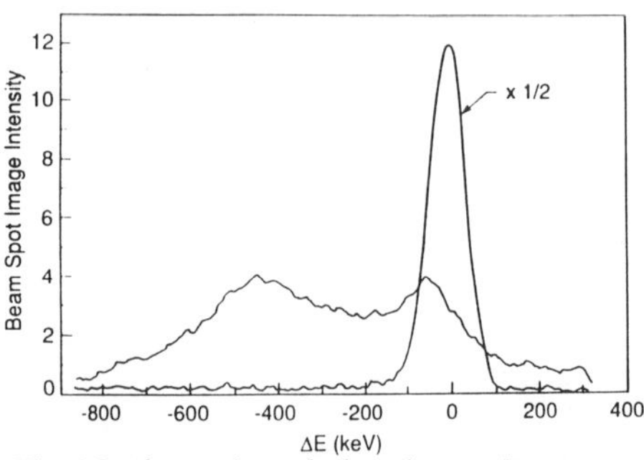

Fig. 8 Lasing and non-lasing electron beam energy spectra taken with the fluorescent screen on the variable dispersion spectrometer.

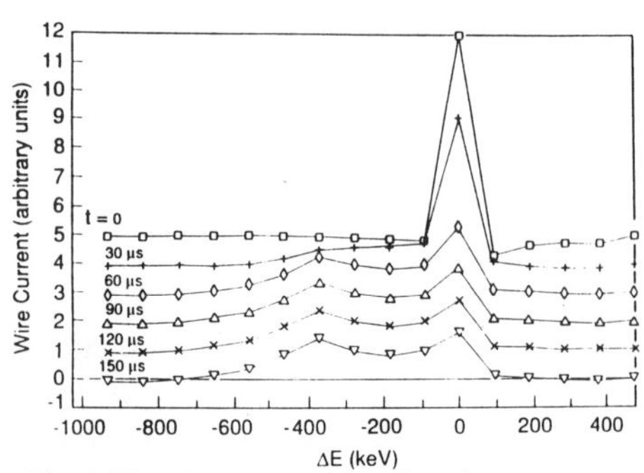

Fig. 9 Electron spectra from the wire array, taken at 30 μs intervals as the FEL turns on.

accelerated by a few hundred keV!) Fig 9 shows a sample of wire array data. The figure shows the single micropulse energy spectrum at 30 μs intervals as the FEL fields build up at the start of the macropulse. The evolution of the energy spectrum from a single narrow spike to a broad double peaked spectrum is evident.

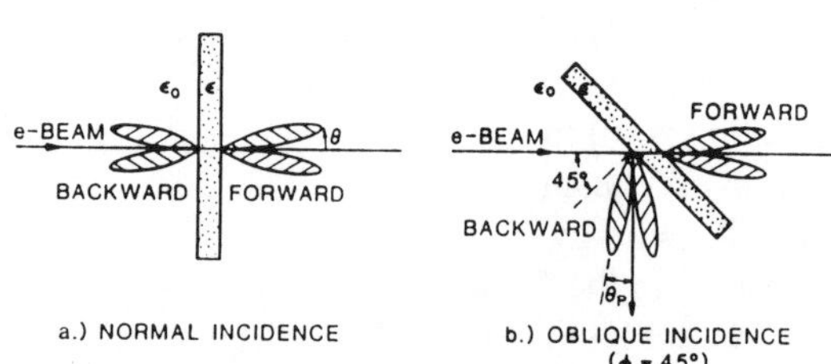

Fig. 10. Optical transition radiation patterns emitted by electrons incident at (a) 90° and (b) 45° to the surface.

The electron beam position and shape must be accurately known in the FEL, in order that the overlap with the optical beam can be optimized. The same devices normally used in linacs are used in FELs, the most commonly used being TV cameras and screens. At present the most popular screens are fluorescent, but for various reasons transition radiation[7,8] screens are gaining in popularity. Among the reasons are: thinner foils resulting in less bremstrahlung, linear response, no burning, and the possibility of obtaining angular and emittance information at the same time as position information[9]. Transition radiation

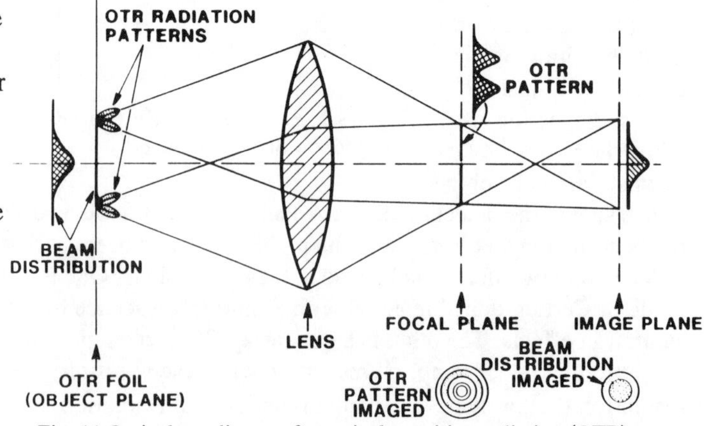

Fig. 11 Optical ray diagram for optical transition radiation (OTR) imaging. Note that angular information is available at the focal plane, and spatial information is available at the image plane.

occurs when an electron beam passes from one medium to another with a different index of refraction. The radiation is quite broadband, and as shown in Fig.10, is

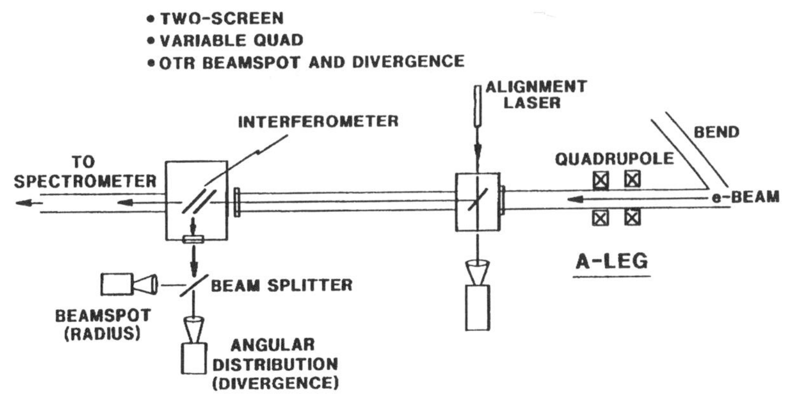

Fig. 12 Schematic of OTR interferometer setup on Boeing FEL for emittance technique comparison.

emitted in both the forward and backward directions. The backward radiation is emitted as if reflected from the surface, while the forward radiation is emitted along the electron beam direction. In both cases, the radiation is emitted as a cone of light with opening angle ~$1/\gamma$, and width also $1/\gamma$. As the direction of emission of the cone tells the direction of the emitting electron, and the apex of the cone tells the position of the electron, imaging the angular information from the foil provides information about the beam direction, and imaging the foil itself provides the conventional beam shape. Figure 11 shows these two cases. In fact, much more accurate angular information can be obtained by placing two transition radiation screens so that the forward radiation from one interferes coherently with the backward radiation from

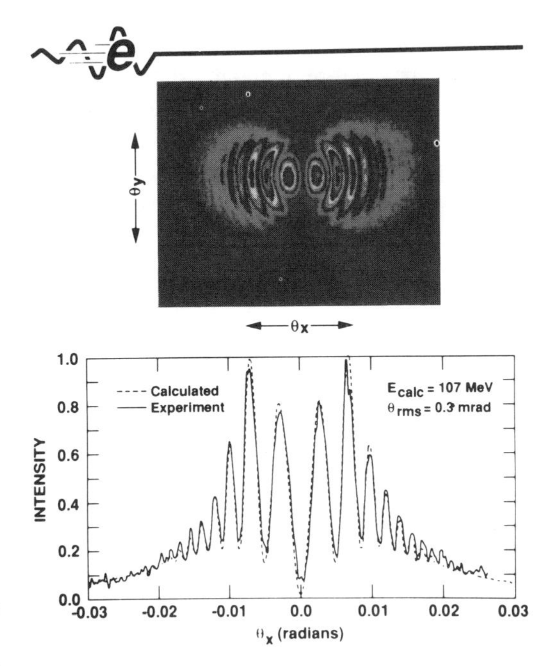

Fig. 13 OTR interferometer image and data taken with the setup of Fig 12. Electron beam energy and angular spread are obtained from the data.

the other. Such an arrangement is shown in Fig 12, as it has been used on the Boeing FEL experiment. Data taken from this setup is shown in Fig 13. The photograph is a picture of the interference pattern observed, and the curves compare calculations with data. The calculated curve is fit to the data by choosing values of θ_{rms} and beam energy. Table I shows a comparison between three experimental methods of determinations the emittance, and PARMELA simulations. All three methods are in good agreement, and the transition radiation method also gives a measure of the beam energy. There is some question about the minimum angles which can be measured with transition radiation. It seems that at present, 0.1 mr is about the limit. Other methods of using the radiation, particularly taking advantage of the wide range of wavelengths emitted, may improve this number.

Table I
Comparison of emittance measurements on the Boeing FEL

	ε_{nx}(edge) (mm-mr)	x_{rms} (mm)	θ_{rms} (mrad)	E (MeV)
Single Screen Quad Scan	158±24			
Two Screen Method	167±25	0.57	0.35	
OTR Interferometer	146±25	0.57	0.3±.05	106.8
Parmela Simulation	143	0.81	0.21	106.9

Temporal information about the electron beam is necessary, both in order to determine the peak current, and to establish limits on timing jitter between pulses. The current averaged over a macropulse is straightforward to measure, but knowledge of the peak current requires information about the micropulse length. Timing jitter information is necessary to verify that consistent overlap between the optical pulses in the resonator and the micropulses is possible.

One way of obtaining this information is through the use of an rf deflection cavity placed on the beam line, and phased so that the time rate of angular deflection imparted to electrons passing through the cavity is a maximum[10]. Then, the duration of the beam is converted to transverse information which can easily be read out on a view screen or by a scanning wire. Such a system is indicated schematically in Fig 14. The system can be calibrated by changing the phase of the rf in the deflecting cavity by a known amount and observing the resulting displacement on the screen or wire. Data taken by such a system on the Stanford Center's FEL is shown in Fig 15. The three families of curves

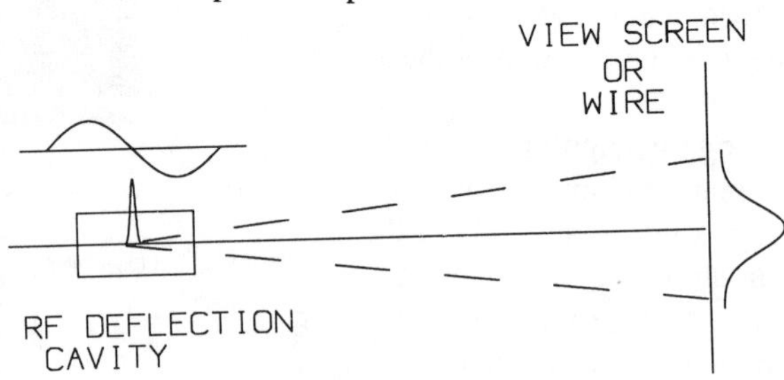

Fig 14. Schematic of an rf deflection cavity used to convert micropulse duration into transverse extent.

were taken with different settings of the phase of one of the accelerating structures in the low energy end of the linac, where bunching is still taking place. Notice the narrowing of the micropulse width from 7.7 ps to 2.9 ps. The charge per bunch remained constant, so that the peak current varied by a factor of almost 2.7. As each curve required several seconds to acquire, any jitter on the beam timing is included in the width.

The system just described is unable to provide information about changes in the micropulse structure within a macropulse. However, such information can be obtained by introducing a relatively slow sweep in the plane transverse to that of the deflection cavity[11]. Specifically, if the rf cavity deflects the electron beam horizontally, and the slow sweep deflects the beam vertically, then not only will temporal information about the micropulses be displayed horizontally on a screen, but the information from individual micropulses will be separated vertically.

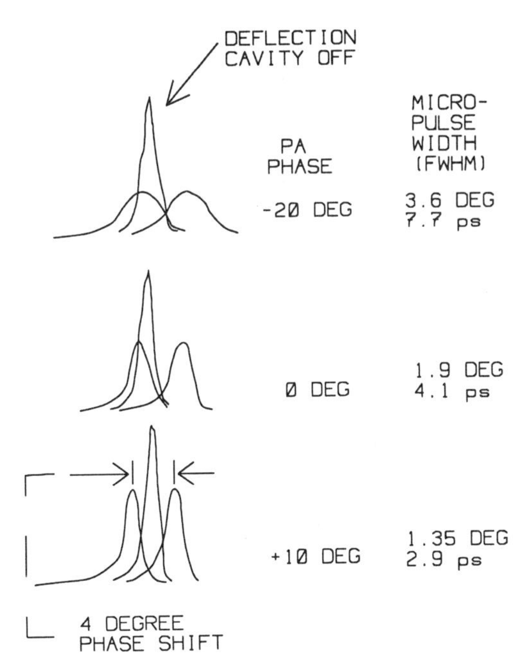

Fig. 15. Data taken with an rf deflection system such as that in Fig. 14, on the Stanford Center's electron beam. For each family of curves, a different setting of the phase of an rf cavity in the injector was used. The tall peak in each family shows the beam width with the deflection cavity off. The other two traces were taken with the cavity on, but shifted by 4° relative to one another.

A streak camera can also be used to obtain information not only about the micropulse structure of the electron beam, but of the FEL optical beam as well. To date the temporal resolution and timing stability (jitter) available from streak cameras is not as good as might be desired, but as suggested in Table II, performance is improving[12]. The best temporal resolution is available from a single shot fast sweep system, but the FWHM timing jitter for this system is not better than 10 to 15 ps. The relatively recent 'synchroscan' systems provide much better timing stability at the expense of some temporal resolution. As seen in the Table, FWHM jitter of about 2 ps can be achieved with a resolution of 3 to 4 ps. A schematic of the dual sweep technique is shown on Fig 16. A high speed oscillating signal, phased locked to some reference source is used to move the beam in the vertical direction, while a much lower speed ramp deflects the beam horizontally. The phase locking system generally has a fairly narrow bandwidth, and needs to be optimized for each specific use. The data in Table I, and the streak data to be presented later were taken with a system optimized for use on the Boeing FEL, and are locked to a signal at 108 Mhz. The similarity between the motion of the signal on the screen in Fig 16, and that of the electron beam on the screen in the rf deflection cavity system emphasizes a strong parallel between the two diagnostic systems. The rf cavity system can be thought of as a means for performing some of the functions of a streak camera by manipulating the linac electron beam directly. The streak camera typically uses the light from the FEL wiggler (either spontaneous emission or lasing) to obtain its signal. Which diagnostic is better suited to a given application depends on many factors.

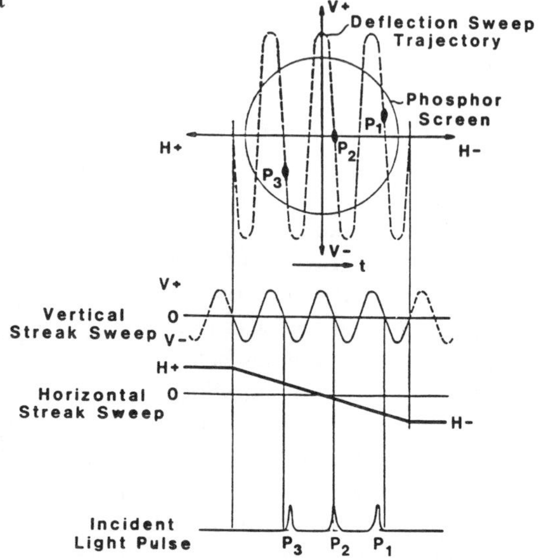

Fig. 16. Schematic of the dual sweep streak camera technique.

Table II
Summary of Streak Camera Module Properties

Module*	Temporal Resolution (FWHM, ps)	Phase Jitter (FWHM, ps)	Phase Jitter (rms, ps)
Fast Sweep (single)	<2	10-15	4-6
Synchroscan (M 1954)	6-8	4	1.7
Synchroscan (M1954-10)	~3-4	2	~0.8

*Based on Hamamatsu C1587 streak camera

Figs 17 and 18 are examples[7,13] of dual sweep synchroscan streak camera images obtained from spontaneous radiation emitted by the Boeing FEL wiggler. Fig 17 shows the streak camera's ability to provide simultaneous micropulse and macropulse information. Both halves of the figure show that the micropulse is substantially constant throughout the 100 μs macropulse. However, the FWHM of the train of micropulses in the left half is only 11 ps, while that of those in the right half has increased to 17 ps due to deliberate misadjustment of the subharmonic buncher cavity. Figure 18 shows similar data taken at a different time, and illustrates the streak camera's ability to help resolve a potentially difficult problem. In the upper picture, it appears as though the micropulse length is decreasing throughout the macropulse. However, from the picture it can be seen that the micropulse intensity decreases throughout the macropulse. These effects were traced to beam scraping on energy slits in the beam transport system prior to the wiggler. The lower picture shows the streak trace taken with the energy slits removed. Note that the micropulse is now constant throughout the macropulse.

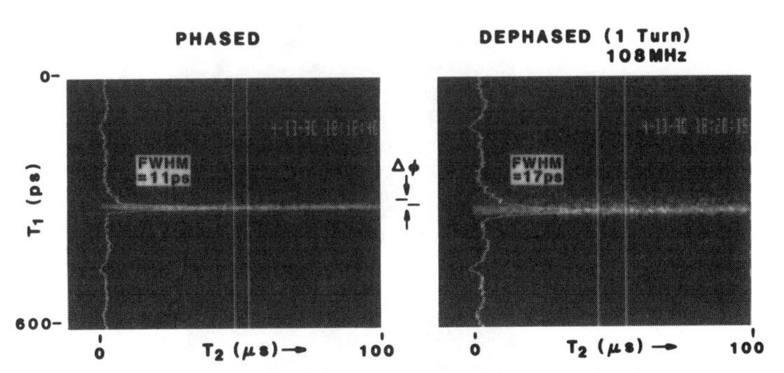

Fig. 17. Examples of dual sweep streak images. Note the simultaneous presentation of data on both the μs and ps time scale.

Fig. 19 is an example of FEL data taken with the dual sweep synchroscan technique. The upper picture in Fig 19 shows the intensity of the FEL output on both the micropulse and macropulse time scales, and explicitly displays the intensity throughout the macropulse. The lower picture is of a similar, but different macropulse, and shows the micropulse intensity. The FWHM of the optical micropulse is seen to be 8 ps.

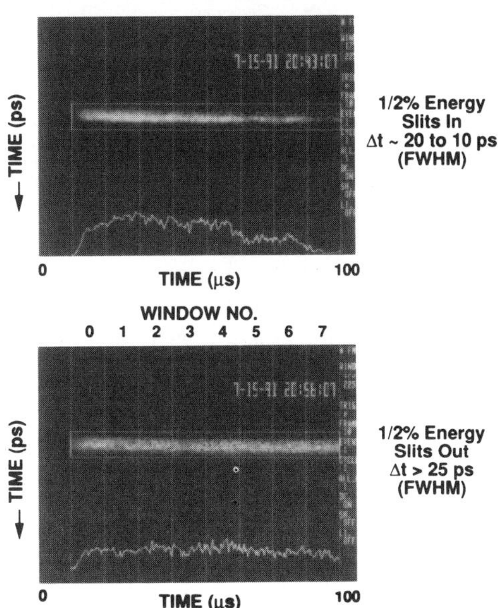

Fig. 18. Dual sweep streak camera data which helped resolve an energy vs time slew in the Boeing linac macropulse.

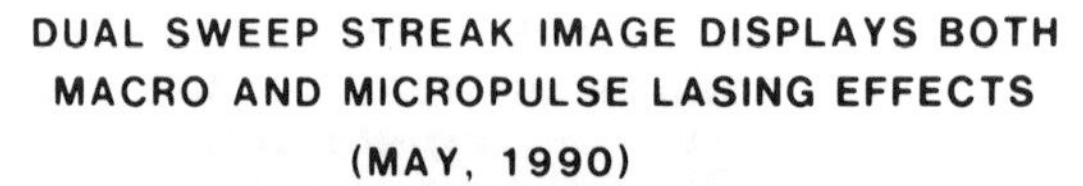

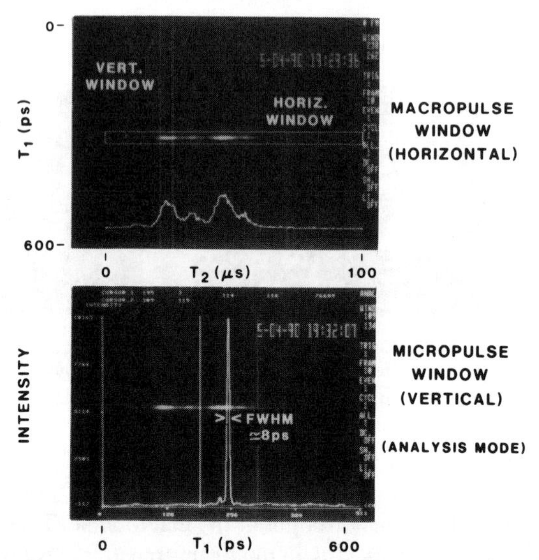

Fig. 19. Dual sweep image recording FEL light directly, and showing both micropulse and macropulse information. Note the direct measurement of a single micropulse width of 8 ps.

FEL DIAGNOSTICS

To be as useful as possible, an FEL must provide light with carefully controlled properties[14]. This means that any parameter deemed important by users must be monitored so that deviations from desired conditions can be detected and corrected. Important parameters include the center wavelength, the bandwidth, the macropulse and micropulse temporal structure, the macropulse and micropulse power and energy, and positional stability. The requirement that virtually all aspects of the FEL's output be continually monitored leads to the conclusion that a comprehensive set of optical diagnostics must be continuously on line.

As an example of an FEL diagnostic system, Fig. 20 is a block diagram of the system recently used on the Stanford Center's FEL. The power detectors at each end of the FEL are relatively slow devices, and are used to monitor the optical power on a macropulse time scale. The fast detectors shown on the right side of the figure are used to obtain micropulse data. Together with the monochromator, the fast array detector provides micropulse measurements of the spectral output of the FEL. The autocorrelator is used to measure the optical pulse length with sub-picosecond resolution. The wire array shown on the left is used to measure the electron beam energy spectrum, as mentioned in the earlier section on electron beam diagnostics. The interferometer is used to measure and control the optical cavity length of about

12.68 m to sub-μ resolution. The data is collected by an AT class computer and made available for presentation in the control room and user labs.

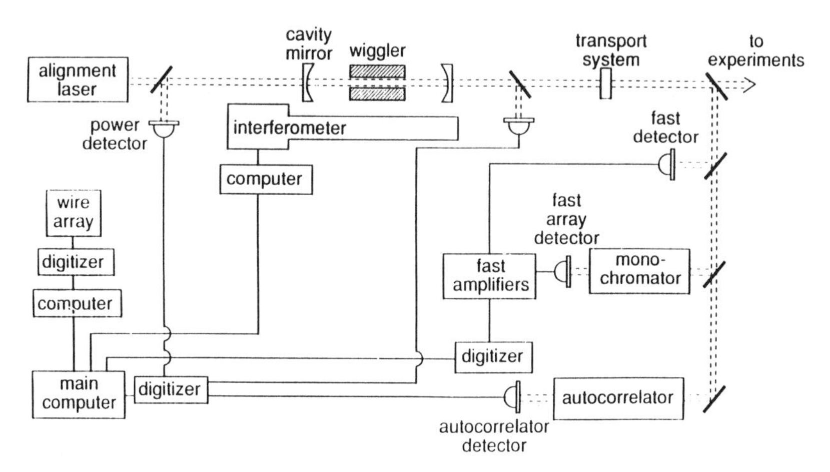

Fig. 20 Block diagram of FEL data acquisition system employed at the Stanford Center.

Some of the subtleties of the FEL interaction are illustrated by data taken while Stanford's FEL was operating near 3.5 μ, and presented in Figs 21 and 22. Fig. 21 shows the time averaged spectrum. For relatively short cavities the spectrum is narrow, but as the cavity is lengthened, the spectrum develops sidebands and ultimately becomes chaotic. Fig. 22 a) shows the total optical power as a function of time within a macropulse, and Fig. 22 b) shows the power averaged over several macropulses. For short cavities the power is constant throughout the macropulse, and increases smoothly as the cavity is lengthened. However, as the length approaches that corresponding

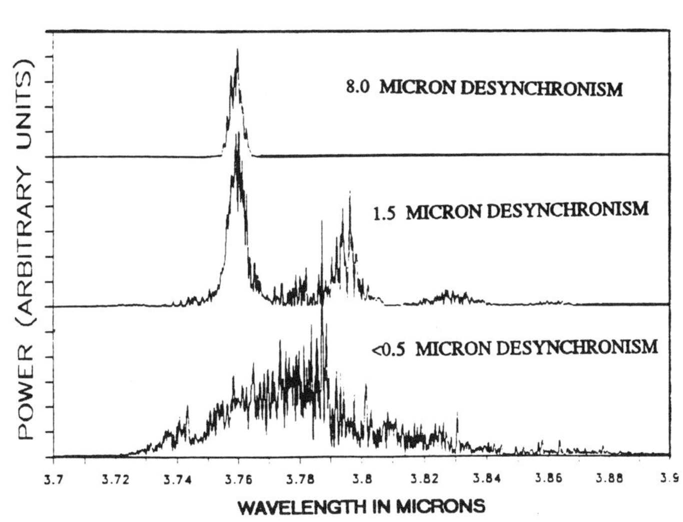

Fig 21. Measured optical spectra at 3.7 m for the Stanford FEL at various optical cavity lengths.

to maximum average output the power level begins to fluctuate substantially throughout the macropulse. As is evident from Fig. 22 b) the time averaged power of the noisy state is two to three times that in the quiet state. However, from Fig 21, it is clear that the noisy power trace is a symptom of FEL sideband instabilities and large bandwidth operation.

The ability to take spectral data at the micropulse level allows several ways of verifying stable, narrow band operation of the FEL. One of these is shown in the upper part of Fig 23, in which the spectrum of a single micropulse of operation at 1.65 μ is compared with the average spectrum of all 25,000 micropulses in a 2 ms macropulse. The FWHM of the micropulse (0.08%) is only slightly narrower than that of the macropulse average (0.09%), indicating that pulse to pulse fluctuations must be small. The autocorrelation signal for these micropulses was consistent with a gaussian temporal profile of 3 ps FWHM, within 10% of the Fourier transform limit. Another way of verifying stable operation is uses a contour plot of wavelength vs time throughout a macropulse, with contours indicating profiles of constant intensity. The left side of Fig. 24 shows such a display recorded during operation at the length for maximum power. The right side shows one taken during normal operation.

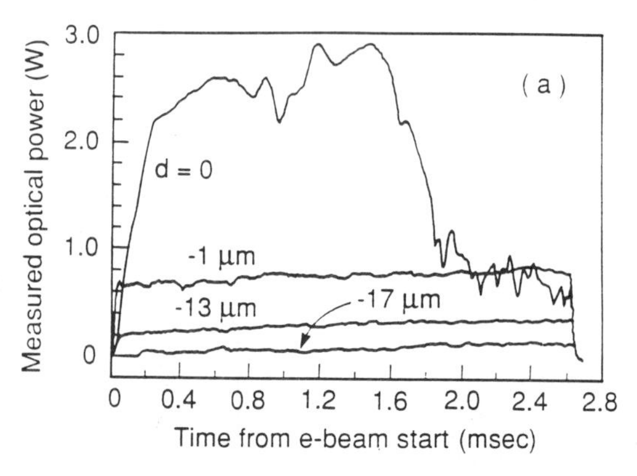

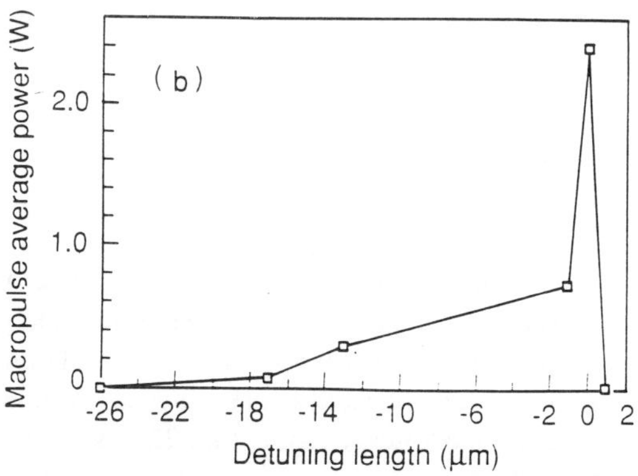

Fig. 22. a) Total power during the macropulse from the Stanford FEL as a function of cavity length. The center wavelength is near 3.7 μ.

b) Total power averaged over many seconds.

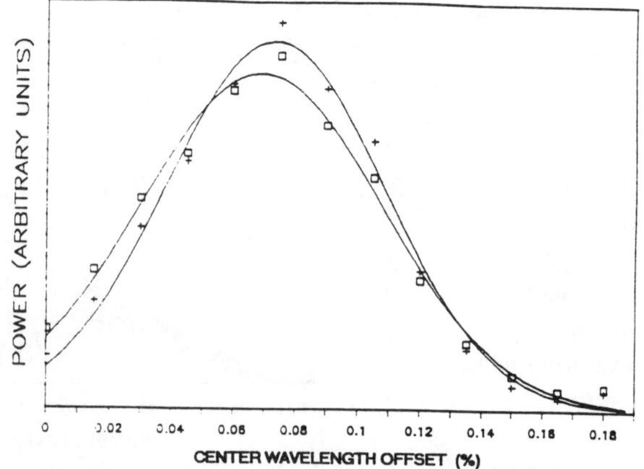

Fig 23. Comparison of the optical spectrum of a single micropulse (FWHM=0.082%) with that of the entire macropulse (FWHM=0.091%). The macropulse consists of some 25000 micropulses. The center wavelength is 1.65 μ.

The contrast between the chaotic behavior in the first case and the stable behavior in the second is striking.

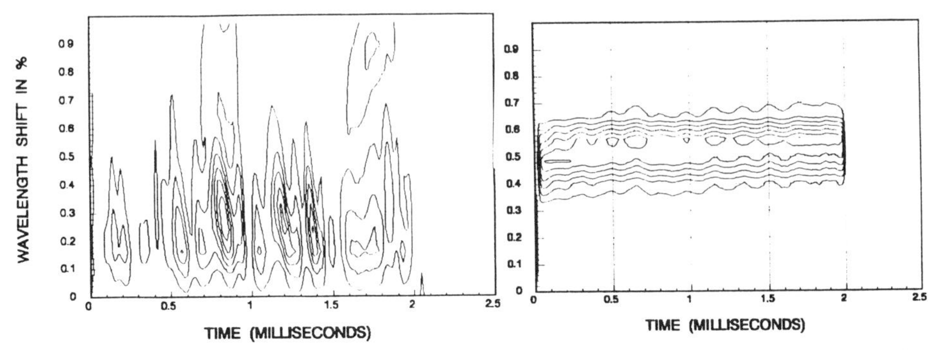

Fig. 24. left) Time resolved spectrum of the Stanford FEL 1.5 μ beam with the optical cavity length adjusted for maximum average power. The chaotic nature of the output is evident.
right) Same as above except cavity length adjusted by a few μ for stable operation.
The contour intervals are linear.

THE FEL AS A LINAC DIAGNOSTIC

Since the properties of the FEL depend critically on the properties of the electron beam driving it, it is reasonable to assume that the FEL can be used as a sensitive monitor of the electron beam's quality. In some cases, it will be straightforward to infer the specific electron beam parameter affecting the FEL. In other cases, the FEL data may serve to imply strongly that the linac is not operating ideally, but it will not point directly to the problem.

Since the FEL's output wavelength is a strong function of the energy of the driving electron beam, wavelength fluctuations will probably be due to energy fluctuations. (Steering and focussing can also have small effects on the operating wavelength). Fig. 25 shows the deviation of the center wavelength of the FEL during a single macropulse. The deviation extremes are about ±0.02%, implying an electron beam energy stability of ±0.01%. The rms value of the energy fluctuations is of course substantially less than 0.01%. However, of perhaps more interest is the

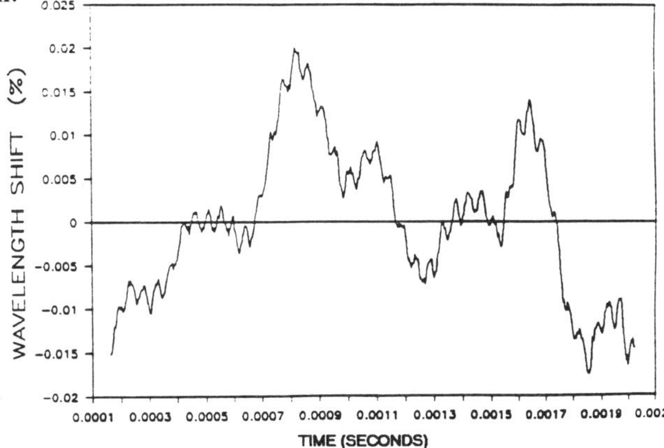

Fig. 25. The fractional deviation of the center wavelength of the 1.5 m Stanford FEL beam as a function of time within a macropulse.

unexpected presence of an oscillation at 10 or 20 KHz. The source of this oscillation has not yet been identified.

The upper part of Fig. 26 shows the output power of the FEL vs time during a macropulse while operating at about 1.5 μ. The lower part of the figure shows electron beam current measured on the electron beam dump during a macropulse. The FEL power clearly has a high frequency modulation, which is not present on the electron beam current. (The current data was not taken on the same macropulse as the lasing data, so the comparison isn't entirely legitimate. However, the current trace is quite typical, and has never shown high frequency modulation.) The modulation of the FEL power is not understood. It may be correlated with the 'energy' oscillations discussed in the previous paragraph, although it doesn't appear that this is the case. Even if the two fluctuations were due to the same source, its not clear why such a small energy change should be connected with such a relatively large power change.

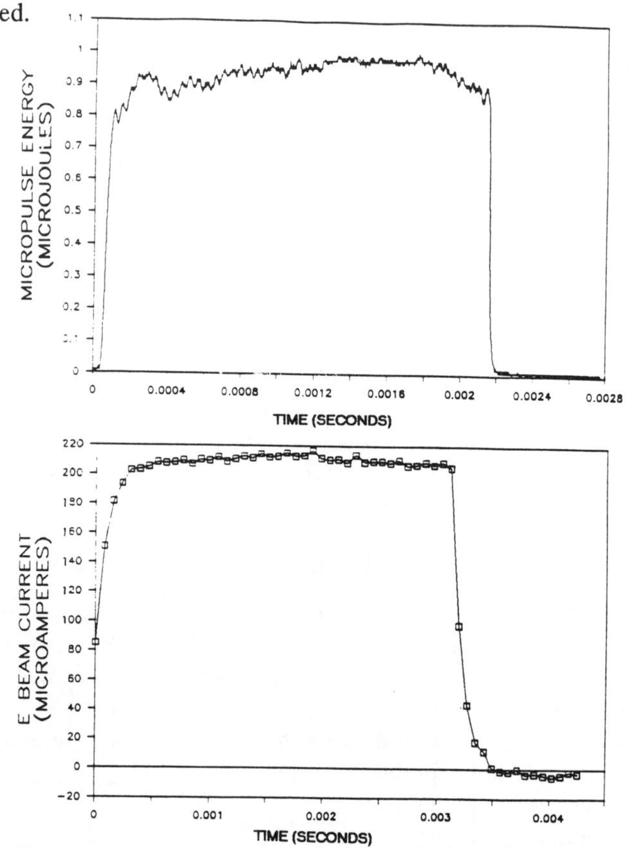

Fig. 26 upper) Output power at 1.5 μ from the Stanford FEL as a function of time within the macropulse.
lower) Electron beam current vs time.

CONCLUSIONS

An introduction to the principles of operation of an FEL has been presented, along with a discussion of the constraints imposed on the parameters of an electron beam intended as an FEL driver. This was followed by examples of electron beam diagnostics particularly appropriate for FELs, and by examples of optical diagnostics for FELs. Finally, there was a brief discussion of the use of an FEL as a diagnostic for the linac providing the electron beam.

ACKNOWLEDGEMENTS

Much of the data presented was collected using the facilities of the Stanford Picosecond FEL Center. Without the contributions to its operation and management by H. Alan Schwettman and by Richard L. Swent, and without the efforts of Josef C. Frisch(now at SLAC) on the FEL optics, none of the data would have existed. Alex Lumpkin of the Los Alamos National Laboratory provided the figures and the data regarding streak cameras, and Ralph Fiorito of the Naval Surface Warfare Center provided the material and figures on optical transition radiation.

REFERENCES

1. W. B. Colson, C. Pellegrini, A.Renieri, Laser Handbook: Free Electron Lasers, Vol 6. (North Holland, 1990)
2. C. A. Brau, Free Electron Lasers, (Academic Press 1990)
3. C. W. Roberson and P. Sprangle, "A Review of Free Electron Lasers", Physics of Fluids, **1**, 3 (1989).
4. W. B. Colson, Physics Letters **64**A, 190 (1977).
5. T. I. Smith and J. M. J. Madey, "Realizable Free Electron Lasers", Applied Physics B 27, 195-199 (1982).
6. R. L. Swent, J. C. Frisch, T. I. Smith, "The Variable Dispersion Spectrometer at the SCA/FEL", Nucl. Inst. and Meth. A296 736-738 (1990).
7. A. H. Lumpkin, "Advanced, Time-Resolved Imaging Techniques for Electron Beam Characterizations", AIP Conference Proceedings No. 229, Accelerator Instrumentation. Particle and Fields Series 44. pp 151-179, (E. S. McCrory ed.) AIP, New York, 1991; see other OTR references therein.
8. D. W. Rule and R. B. Fiorito, "Imaging Micron-Sized Beams Using Optical Transition Radiation", Ibid, pp 315-321.
9. R. Fiorito et. al., "Emittance Measurements of FEL Accelerators Using Optical Transition Radiation Methods", Proc of the IEEE Particle Accelerator Conference, May 6-9, 1991, San Francisco, CA. (in press)
10. T. I. Smith, "The Stanford Superconducting Linear Accelerator", Physics of Quantum Electronics, Vol 8: Free Electron Generators of Coherent Radiation, pp77-87, (S.F. Jacobs et. al. eds.) Addison-Wesley 1982.
11. A. H. Lumpkin and D. W. Feldman, "Diagnostics of the Los Alamos Free Electron Laser Using Streak Systems", Nucl. Inst. and Meth. A259, 13-18 (1987).
12. A. H. Lumpkin, "Advanced Diagnostic Concepts for Emerging RF-FEL Designs", Proc. 13th International FEL Conf., Santa Fe, New Mexico, Aug. 25-30, 1991, (to be published-Nucl. Inst. and Meth.).
13. A. H. Lumpkin, "The Next Generation of RF FEL Diagnostics:Synchroscan and Dual-Sweep Streak Camera Techniques", Nucl. Inst. and Meth. A304, 31-36 (1991).
14. T. I. Smith and H. A. Schwettman, "Facilities for using the FEL as a Research Tool", Nucl. Inst. and Meth. A304, 812-821 (1991).

DEVELOPMENT OF A MODEL FOR RAMPING IN A STORAGE RING

Kevin J. Cassidy
Tektronix Federal Systems, Inc., Irvine CA 92714

Sam Howry
Stanford Linear Accelerator Center, Stanford, CA 94305

ABSTRACT

This paper describes the process of ramping in a storage ring such as the SPEAR Synchrotron Radiation Ring at SLAC/SSRL. A definition of ramping is presented first. Then an "ideal" ramp that includes the necessary calibrations is presented. This is refined to account for nonzero response times that may occur in an "actual"ramp. Parameters are identified which cause the "actual" model to deviate from the "ideal." A process to estimate these parameters is described that depends on rapid measurement of the ring's tunes. Finally, the paper describes a digital signal processor (DSP) that was used at SPEAR to measure the tunes during a ramp.

INTRODUCTION

This paper describes the process used to ramp electrons in the SPEAR Synchrotron Radiation Ring. Although ramping in a storage ring is certainly not new[1], it is detailed here, so that future ring designers can contemplate the problem. The ramping software is part of the PEP/SPEAR control system.[2] The ramping model is defined first; then values of certain control parameters are refined with the help of a digital signal processor (DSP) that can measure actual ramp tunes on-line.

SECTION 1: DEFINITION OF THE RAMP

Lattice designs for storage rings[3] work with a set of strengths S of controllable elements (such as quadrupoles, sextupoles, rf voltages) and with a set T of Twiss functions, which describe the focussing characteristics of the beam at any geometric position s along the beamline. In a storage ring, Twiss functions $\beta(s)$, $\alpha(s)$, $\eta(s)$, and $\eta'(s)$ are periodic (period = 1 ring revolution). The monotonic tune advance function Δ**nu**(s, s + Δs) increases by a fixed constant **nu**, called the tune, whenever $\Delta s = 1$ period. There are horizontal, vertical, and longitudinal components of the Twiss functions, although in most systems some components are identically zero. The fractional part of the three tune constants: nux, nuy, and nus can be directly measured when there is a stored beam in the ring.

Typically, the output from a lattice design program is a set of strengths S that produces a specified set of Twiss functions T. Often it is desirable to inject particles into a ring configured at energy E_0 and strengths S_0 (corresponding to Twiss functions T_0) and then to slowly change the ring to a final configuration of energy E_{final} and strengths S_{final}, while keeping the stored particles in the ring. This operation is called ramping. In a pure energy ramp, $S_0 = S_{final}$; only the energy is changed. Ramping allows a storage ring to operate with a relatively inexpensive, lower energy injector.

SECTION 2: THE "IDEAL" RAMP MODEL

The energy of a storage ring is determined by $\int B\, d\ell$ of all of its dipole (bending) magnets. The dipoles are often all on one series circuit. This circuit may also supply part of the current that determines some of the strengths, with the exact values of these strengths being controlled by individual shunt or booster supplies.

The ramping model must convert energy and strengths (E,S) into integers for actual power supply controllers, and vice versa. This is the one to one, invertible mapping:

$$\Phi(E,S) \Rightarrow \text{DAC integers} \tag{1}$$

for the dipole controller(s) and for controllers of all of the focussing elements (see figure 1.).

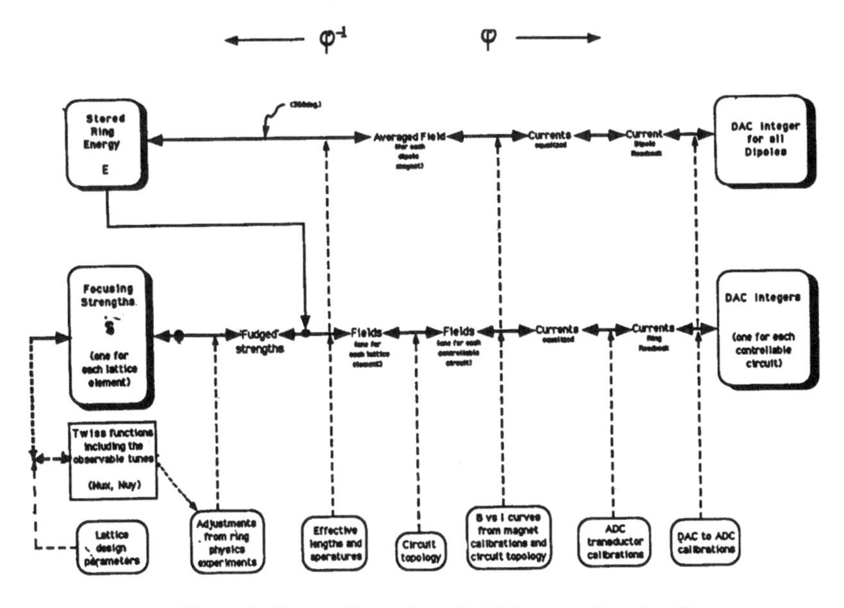

Figure 1. Energy/Strengths to DAC Integers Mapping Φ

The map is roughly linear for most systems, but it is by no means linear enough for ramping. Before the ramp begins, DAC setpoints are calculated from the map and stored in a table for each independent control and for each ramp step. The number N of ramp steps and the number Δt of milliseconds between ramp steps are parameters that can be changed by the operator. The map must include considerations such as:

- Circuit topology. Which magnets are connected to which controllers; series connections, shunts and boosters must be taken into account.

- Field/current calibrations of the magnet elements. Field/current curves have been estimated for each circuit from the magnet calibration data. Note that unavoidable error is introduced here if several magnets are on one circuit (with one controller). Effective lengths and apertures of the magnets are also required by the map.

- Electromagnetic hysteresis. Hysteresis effects are avoided by executing a standardizing "degauss" cycle for all controllable elements before injection, and by keeping the ramp monotonic (i.e. no overshoot). Under these conditions the map follows the lower paths of the field/current curves, when going to a higher current.

- Strength fudges. These are empirically determined corrections to strengths as calculated by lattice design programs.
- ADC transducer calibrations. This calibration must be done so that all control readbacks can be equivalenced relative to each other.
- DAC to ADC calibrations. This calibration insures that each individual control setpoint can be equivalenced to its corresponding readback.

Our model assumes only one dipole controller. All dipole magnets are connected to this circuit in series. Successful ramping only requires that each dipole/focussing controller value be consistent relative to the others. Hence one controller ramp function is arbitrary, and the others must match it. Our model selects the dipole controller to be a straight linear DAC ramp from the initial to the final energy. That is, for a ramp of N steps:

$$\mathrm{DAC}_{dipole}[0] = \mathbf{\Phi}(\mathrm{E}[0],\mathrm{S}); \text{ the initial dipole controller setting} \tag{2}$$

$$\mathrm{DAC}_{dipole}[\mathrm{N}] = \mathbf{\Phi}(\mathrm{E}[\mathrm{N}],\mathrm{S}); \text{ the final dipole controller setting} \tag{3}$$

$$\mathrm{DAC}_{dipole}[\mathrm{t}] = \mathrm{DAC}_{dipole}[0] + \mathrm{t} * \Delta\mathrm{DAC}_{dipole}; \quad \mathrm{t}=0, 1, \ldots, \mathrm{N} \tag{4}$$

where,

$$\Delta\mathrm{DAC}_{dipole} = \left(\mathrm{DAC}_{dipole}[\mathrm{N}] - \mathrm{DAC}_{dipole}[0]\right) / \mathrm{N} \tag{5}$$

At any step t, the focussing controllers must match the actual energy produced by the integer $\mathrm{DAC}_{dipole}[\mathrm{t}]$. If the dipole circuit had an instantaneous response, this actual energy would simply be the applied energy $\mathbf{\Phi}^{-1}(\mathrm{DAC}_{dipole}[\mathrm{t}])$.

SECTION 3: THE "ACTUAL" RAMP MODEL

In a real system the actual energy is not the same as the applied energy during a ramp even if the DAC values are simultaneously loaded into the controllers (within microseconds). Nonzero response times in the control circuits and in the individual magnets will cause deviations from the ideal map $\mathbf{\Phi}$, particularly at the start and at the end of the ramp. This becomes more severe as operation parameters N and Δt are adjusted to get faster ramps. In our model, the combined nonzero response times of the dipole electrical circuit and of the domains in the dipole magnets are lumped into a single response time $\tau_{dipole} > 0$. Then at each step t, the actual energy change differs from the applied energy change by a constant factor:

$$\Delta\mathrm{E}_{actual}[\mathrm{t}] = \Delta\mathrm{E}_{applied}[\mathrm{t}] * \mathcal{F}_{dipole} \tag{6}$$

where,

$$\mathcal{F}_{dipole} = \left(1 - e^{-(\Delta t/\tau_{dipole})}\right) \tag{7}$$

From the increments of equation (6), the actual energy, $\mathrm{E}_{actual}[\mathrm{t}]$, can be accumulated for each step t. Now, the focussing strengths required to match this are:

$$\mathrm{S}[\mathrm{t}] = d\mathrm{S}/d\mathrm{E} * \left(\mathrm{E}_{actual}[\mathrm{t}] - \mathrm{E}[0]\right) + \mathrm{S}[0] \tag{8}$$

where the vector $d\mathbf{S}/d\mathrm{E}$ is a constant defined by:

$$d\mathbf{S}/d\mathrm{E} = \left(\mathbf{S}_{final} - \mathbf{S}[0]\right) / \left(\mathrm{E}_{final} - \mathrm{E}[0]\right) \tag{9}$$

Note that $d\mathbf{S}/d\mathrm{E}=0$ for a pure energy ramp, and the focussing strengths do not change during the ramp. However, the controller integers that correspond to these focussing strengths always depend on the actual energy at step t:

$$\mathbf{DAC}_{actual}[t] = \mathbf{\Phi}(\ \mathrm{E}_{actual}[t],\ \mathbf{S}[t]\ ) \tag{10}$$

Finally, each of the focussing circuits k also has a lumped parameter to represent a nonzero response time. So each component k of the vector for the actual controller integer increments differ from the corresponding component of the applied controller integer increments vector, by a constant factor:

$$\Delta\mathrm{DAC}[k]_{actual}[t] = \Delta\mathrm{DAC}[k]_{applied}[t] * \mathcal{F}[k] \tag{11}$$

where,

$$\mathcal{F}[k] = \left(1 - e^{-(\Delta t/\tau[k])}\right) \tag{12}$$

or,

$$\Delta\mathrm{DAC}[k]_{applied}[t] = \Delta\mathrm{DAC}[k]_{actual}[t] \,/\, \mathcal{F}[k] \tag{13}$$

From the increments of equation (13), the integers $\mathrm{DAC}[k]_{applied}[t]$ can be accumulated for each controller k of a focussing element and for each step t. All of the integers:

0	$\mathrm{DAC}_{dipole}[0]$	$\mathrm{DAC}[1]_{applied}[0]$	$\mathrm{DAC}[2]_{applied}[0]$	...	$\mathrm{DAC}[k]_{applied}[0]$	...
1	$\mathrm{DAC}_{dipole}[1]$	$\mathrm{DAC}[1]_{applied}[1]$	$\mathrm{DAC}[2]_{applied}[1]$	...	$\mathrm{DAC}[k]_{applied}[1]$	...
2	$\mathrm{DAC}_{dipole}[2]$	$\mathrm{DAC}[1]_{applied}[2]$	$\mathrm{DAC}[2]_{applied}[2]$	...	$\mathrm{DAC}[k]_{applied}[2]$	...
⋮	⋮	⋮	⋮		⋮	
t	$\mathrm{DAC}_{dipole}[t]$	$\mathrm{DAC}[1]_{applied}[t]$	$\mathrm{DAC}[2]_{applied}[t]$	...	$\mathrm{DAC}[k]_{applied}[t]$	...
⋮	⋮	⋮	⋮		⋮	

are computed as described and arranged in a table before the ramp begins. The râmp model asserts that each row of the table represents a consistent set of setpoints. When the ramp starts, each row is emitted to the controllers at timed intervals that are Δt milliseconds apart.

Note that step N usually is not be the last step of the ramp, even though all of the dipole steps have been implemented. The ramp continues until all of the setpoints for each focussing controller have been implemented. This occurs when E_{actual} finally catches up and becomes essentially equal to $\mathrm{E}_{applied}$ (which is E_{final}).

SECTION 4: ESTIMATION OF THE MODEL PARAMETERS

The ramp mapping equations have parameters, such as the response times τ[k], that must be determined empirically. This determination is done by performing repeated ramps during a "ring physics" experiment. For a pure energy ramp, the tune constants nux[t] and nuy[t] should not change at any time during the ramp. The tunes are observed and recorded as very precise functions of time and frequency while ramping. As the time dependencies of nux[t] and nuy[t] are observed, parameter values such as $\tau[k]$ are systematically assigned until the tunes are constant during pure energy ramps.

To measure the tunes, a pickup is connected to an electrode of one of the storage ring's beam position monitors. This signal, when converted to a frequency spectrum, shows resonances at the tune values and their harmonics. The performance of the ramp can then be observed by processing this signal many times during the ramp, at a rate higher than once per ramp step interval Δt. Tune transients will be displayed as "blips" near the first and last steps of the ramp as shown in figure 2. Errors in the map $\mathbf{\Phi}$ appear as a bow or wiggle in the middle of the ramp. Failure to compensate for these effects can result in the beam blowing up, especially in vacuum chambers that have small stay clear specifications. Once the mapping equations' parameters have been determined, they are valid for ramps, where nux and nuy would linearly move from their values corresponding to (E_0,S_0) to those for (E_{final},S_{final}). However, the parameters may be different for other values of the step interval Δt.

SECTION 5: MEASUREMENT EQUIPMENT

Measuring the tune values requires equipment with good frequency resolution (1 kHz or less) and good time resolution (100 milliseconds or less), wide bandwidth (100's of kHz or more), and a enough memory to capture several minutes of frequency spectra for ramp display. Most off-the-shelf test equipment is inadequate to properly measure the tunes during an energy ramp. Analog spectrum analyzers and digitizing oscilloscopes have been tried at SPEAR with little success.

An analog spectrum analyzer measures the frequency of signals by down converting the input signal into an intermediate frequency where it is measured by a resolution filter. It can also be visualized as a band pass filter "sweeping" through the frequencies of the signal. Such devices fail to observe the tune's transient behavior, because the sweep time is usually large compared to the ramp step interval Δt.

A digitizing oscilloscope might be used to measure the tunes. However, "the wide bandwidth given by the [beam's] short bunches requires very high sampling rates and the fast growth rates imply fast data rates. Fast data acquisition systems or digital oscilloscopes with sampling rates of many Giga samples per second have restricted triggering speeds and are usually limited in data storage."[4] In general, a digitizing oscilloscope that is setup to sample fast in order to capture tune transients has insufficient memory to capture an entire ramp. If the oscilloscope is set up to conserve memory, the sample rate is so slow that samples of the tunes will exhibit aliasing.

With modern DSP hardware, a parallel, filter bank analyzer has been implemented to cover wide bandwidths with narrow frequency resolution and good time resolution. A digital, parallel, filter-bank analyzer is similar to a spectrum analyzer in that it measures a signal's frequency content or spectrum. However, "the digital filter bank analyzes an entire span of frequencies simultaneously, rather than by sweeping, and thus has an inherent speed advantage. Its frequency span is divided into side-by-side, stationary resolution bands that are slightly overlapped. The filter bank acquires signal data in a relatively short span of time and performs parallel computation of signal amplitudes in all of its resolution bands from that data. In this regard, the filter bank is similar to a conventional FFT analyzer, of which it is an outgrowth."[5] In addition, the output from a parallel, filter-bank can be captured in memory for spectral analysis of long periods of time. Because it exhibits a long memory length, narrow frequency resolution, short time resolution, and wide analysis bandwidth, the digital, parallel, filter-bank analyzer is an ideal instrument for measuring the tunes on-line during an energy ramp.

The Tektronix 3052 DSP System is an implementation of the digital, parallel, filter-bank analyzer described above, tunable from 0 to 10 MHz. The filter bank is implemented using DSP hardware on VMEbus boards. These boards process spectrums and pass data to a microcomputer on the VMEbus. The VMEbus has been modified in a fashion allowed by the VMEbus standard to create a processing "pipeline." A RAM board at the end of the processing pipeline is large enough to hold 500 frequency spectrums.

The 3052 is being used at SPEAR to observe tunes during energy ramps. In addition to fast spectral processing, the 3052 has several data display modes. First of all, it can display the spectral data in a traditional spectrum analyzer amplitude vs. frequency display. The 3052 can also display phase vs. frequency, an amplitude vs frequency waterfall, and finally a "Color Spectrogram" (see figure 2). The Color Spectrogram is most useful for observations of tunes. It is a three dimensional time versus frequency versus power plot that uses color as a third axis. The X axis depicts frequency, the Y axis time, and log power is represented by multi-color scaling. The time axis scrolls continuously upwards as information enters from the bottom of the display. In addition, software was written on the 3052 to display the tune fractions on the screen while the digital, parallel filter-bank is processing spectrums.

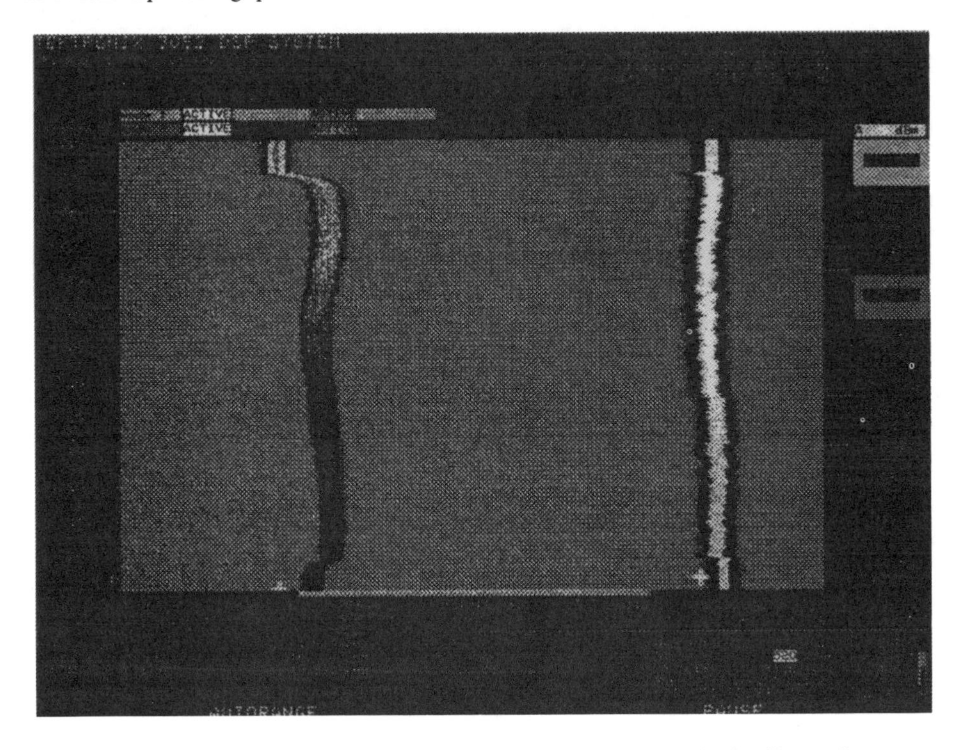

Figure 2. Time vs. Frequency vs. Power Display (Color Spectrogram) of an Energy Ramp

SECTION 6: REFINEMENTS AND EXTENSIONS

At SPEAR, all strengths save one have independently controlled supplies. One quadrupole string has a booster supply that modifies current supplied by the dipole circuit. It was found that an additional parameter may be needed to model the combined electrical response of this circuit.

Given enough ring physics experiment time, longer ramps could be developed that would avoid possible tune resonances. These would use non-constant strength ramp functions dS/dE so that $(nux[t], nuy[t])$ follows a prescribed path to avoid resonance points in "tune" space. Another application would be to "top off" stored beams. However, since this involves a ramp down, then a ramp up after the fill, repeated hysteresis effects would have to be taken into account in the ramp model.

CONCLUSION

The model described in this paper can be used to develop a ramping process in a storage ring. Values for some of the model parameters can be obtained by on-line experiments if tunes can be measured with precision, and with frequency greater than the ramp stepping rate.

Acknowledgement: The authors would like to thank Martin Donald of SLAC for his help on the equations of section 3.

---

[1]A. Boyarski, et al, "Automatic Control Program for SPEAR," IEEE Transactions on Nuclear Science NS-20 No. 3, (1973).
[2]S. Howry, T. Gromme, A. King, M. Sullenberger, "A Portable Database Driven Control System for SPEAR," SLAC PUB 3618, (1985).
[3]M. Sands, The Physics of Electron Storage Rings An Introduction, (1970)
[4]D. Boussard, G.Lambert, R. Lauckner, T. Linnecar, W. Wingerter, "Real Time Fourier Analysis for the Observation of Fast e+ e- Instabilities in the CERN SPS," Proceedings of the Second European Particle Accelerator Conference, (1990).
[5] K. Cassidy, J. Snell, "Fast, Wideband Search for Spurious Responses," IEEE AUTOTESTCON '91 Conference Record, (1991), pp. 383-391.

The Programmable Controller-based CEBAF Personnel Safety System

R. Bork, J. Heefner, H. Robertson, R. Rossmanith
CEBAF,* Newport News, VA 23606

1.0 Introduction

The Continuous Electron Beam Accelerator Facility (CEBAF), presently under construction, is a five pass recirculating electron accelerator with an energy gain of 800 MeV per pass.[1] The accelerator itself consists of a pair of 400 MeV linacs together with two sets of five recirculating arcs stacked on top of each other. The accelerator is capable of producing a 200 μA CW electron beam at an energy of 4 GeV. As indicated in figure 1, the entire beam or any fraction of the beam can be directed into each of three end stations. The project itself will be tested and commissioned in stages and requires that the personnel safety system be modular in construction and capable of operation in the various configurations necessary during these phases. For obvious reasons the system must also be fail-safe and have a high availability. The system must also be user friendly since the personnel that operate the system during the construction phase will not be full time operators intimately familiar with the interlock logic. The benefit of high availability is that few maintenance personnel are required. This is an important feature when the maintenance personnel are also required for the construction tasks being performed in parallel with the accelerator testing. The Personnel Safety System designed and currently in use at CEBAF uses commercially available programmable controllers in a redundant configuration to perform the interlock logic required to operate the accelerator in a manner that is safe to personnel.

The CEBAF Personnel Safety system uses redundancy, isolation, system self-test, strict test and operating procedures, reliability studies, and many published guidelines and standards[2–9] to ensure that the system is not only operated in a manner that is safe, but also that the system is fully operational at all times.

All critical components, such as door switches, hatch switches, and crash buttons, are monitored by independent programmable controllers. The use of redundant hardware helps to minimize the chance of a common-mode hardware failure, but does nothing to ensure that the logic used by the interlock system to determine if the machine is operating in a safe manner is actually correct. Therefore, the logic used by each of the programmable controllers was developed by different individuals working to a common system specification. The use of different logic helps to eliminate any common-mode software failures that might occur.[9]

In addition to redundancy, the Personnel Safety System uses isolation to increase the reliability and integrity of the system. Each of the PLC systems is electrically and physically separated from the other and from all other accelerator systems. All cables for the system are run in dedicated conduits or box duct. All components are housed in locked racks and enclosures, with keys controlled by the operations crew chief.

2.0 Why Programmable Controllers?

There are many reasons for considering Programmable Controller (PLC) based interlock systems. The following is a list of some of the points that were considered by the Safety System engineering staff prior to selection of a PLC based system for CEBAF.

Complexity – It was estimated that the interlock logic for the accelerator alone would consist of approximately 2000 I/O points. Each of these I/O points would in turn have a certain

* Supported by D.O.E. contract #DE-AC05-84ER40150

amount of logic associated with it. This would have required that a relay-based system be extremely large and therefore quite complex to design, test, and ultimately maintain. The size of the system could have been reduced by monitoring several interlock points in series, but this would have made system maintenance more difficult. A PLC based system makes it reasonable to monitor every input device individually. The increase in monitoring capability is possible with virtually no increase in system complexity. This decreases the amount of time necessary to locate faulty devices and therefore increases the availability of the system.

Modularity – As was stated above, CEBAF will be commissioned and tested in several stages. During each of the testing stages the interlock system will need to be expanded and re-configured. A PLC based interlock system, because of its inherent modularity, would allow logic and interlocks for new sections of the system to be added without destroying what has already been installed (i.e. more remote I/O drops would be added and blocks of logic would be incorporated into the existing system).

Documentation – A PLC based interlock system is essentially self-documenting. The software package that controls the ladder logic programming of the CPUs also produces ladder logic diagrams of the program that is being run by the CPU. The only drawings that need to be produced are those that show the actual device connections to the controller; these tend to be extremely simple since one device is connected to only one input or output.

Resource limitations – Rack space and manpower are extremely limited resources at CEBAF. The high density nature of PLC systems allows the interlock system to be housed in a fraction of the space that a relay-based system would require. This saves rack space and also makes system maintenance easier. The software logic eliminates all of the wiring and interconnections used by relay-based systems to perform interlock logic, thus saving installation manpower. The self-documentation features reduce the manpower required to generate and check the system documentation.

Cost and Schedule – CEBAF is no different from any other research institution; therefore, cost and schedule impacts must always be taken into consideration when any decision is being made. It was estimated that a PLC based interlock system would initially cost the same as a relay based system, but that the installation, test, and maintenance would be faster and less expensive.

Reliability – A Failure Modes and Effects Analysis (FMEA) of the CEBAF PLC based system shows that the reliability of the system would be comparable to a relay based system.[10]

Familiarity – Several of the personnel involved in the design and installation of the system had previous experience with PLC based systems and felt comfortable with the hardware. They had used both PLC and relay based systems in the past and were familiar with the advantages and disadvantages of both.

Industry input – CEBAF personnel spent a great deal of time interfacing with different vendors and customers. The purpose of this interfacing was to become familiar with each vendor's equipment so that an informed choice between different types of systems could be made. Books and articles by industry experts were reviewed and referenced for pertinent information and opinions on safety related subjects.[11–13]

Training – All PLC manufacturers offer intense and specific training on how to operate their equipment. All CEBAF Safety System Personnel were trained by the PLC vendor. This eliminated the problems associated with having a limited number of system experts.

3.0 CEBAF'S PLC System

As indicated in figure 2, the CEBAF Personnel Safety System consists of two independent and redundant PLC systems. Each system consists of one master rack located in the Machine Control Center (MCC) and six remote I/O racks located in equipment racks around the accelerator. The location of each of the remote racks is indicated in figure 3. All PLC hardware is commercially available hardware purchased from Modicon.

The master rack of each system contains a 984-685 processor module, an S908 remote communications module and various input and output modules. The 984-685 module performs all logic operations and communicates with the CEBAF supervisory control system. The S908 remote communications module handles the communications with each of the remote racks on a 1 Mbaud proprietary bus.

The remote rack of each system contains a P890 communications processor, B827-032 and B825-016 input modules, and B824-016 and B838-032 output modules. All of the input and output modules are 24 VDC true-high devices.

The logic for the CEBAF Personnel Safety System is developed off line on a test stand that is similar in configuration to each of the two PLC systems used for the accelerator. The software package used to develop logic and program the PLC's is a commercially available package available from Taylor Industrial Software.[14] The package runs on a PC-compatible computer and communicates with the PLC via the Modbus (serial) port on the 984-685 processor module. Programs are written in ladder logic similar to the relay diagrams and schematics that are used to document relay based interlock systems. The main advantage of using ladder logic instead of a more complex programming language is that maintenance personnel familiar with relay diagrams can immediately become familiar and feel comfortable with the system logic documentation. Some of the features inherent with the Taylor programming package are on-line and off-line programming and documentation, on-line PLC diagnostics, and block move and copy functions that assist the logic designer when the logic is repetitive. Search and cross referencing functions are also included to help the designer and maintenance personnel locate various inputs and outputs in large logic arrays.

4.0 Control System Monitoring of Interlocks

The CEBAF Personnel Safety System PLCs are monitored by the CEBAF control system computers. The CEBAF control computers are Hewlett-Packard series 300, 500, and 800 computers running a logic-based application software package, Thaumaturgic Automated Control Logic (TACL), developed at CEBAF.[15] The use of TACL provides a user-friendly interface to the computer system and is used to graphically display the status of the various interlocks monitored by the PLCs. The status of each of the interlocks is read by the HP computer via an RS232 (9600 baud) link to the Modbus port of each of the PLCs. This status information is then conveyed to the operator through a series of displays on the HP computer. These displays range from overall views of the entire interlock system to close-up views of each zone or subsystem. In addition to the present status of each of the interlocks in the system, first fault status of each interlock can also be displayed. At the time of a system fault, the status of each interlock being monitored by the PLC is stored in a group of registers that can be read by the HP computer. This information is then used by the operator to determine which device was responsible for disarming the interlock system. This feature has proved to be invaluable and would be virtually impossible to implement in a relay-based interlock system.

The HP computer also provides an additional back-up to the redundant PLCs. The TACL logic constantly compares the status of each device being monitored by the interlock system. If a discrepancy is detected, an alarm is activated on the monitor screen. This information is used by the operators to detect faulty switches, inputs, etc.

5.0 Interfaces to Other Systems

The Personnel Safety System must interface with other systems within the accelerator. These include the RF control system, the Magnet control system, and the Injector control system. These systems provide each of the PLCs with contacts that reflect their operational status, such as "off/safe" and "ready for beam". The PLCs in turn provide each system with contacts that indicate permission for energization, emergency off, and whether or not personnel are present in areas that are controlled by the safety system. The exchange of information between the safety system and other systems takes place through special interface chassis that have been developed at CEBAF. These chassis are located in the remote system racks and use relay contacts to pass the information. Relay contacts are used for several reasons. Most power supplies and control circuits use relays to perform their interlock logic and the safety system contacts can then be placed directly in the turn-on chain for the particular device being controlled. The relay contacts also provide electrical isolation between the safety system and all external systems. It should be noted that the chassis used to interface to virtually all external systems at CEBAF are identical in design, allowing them to be mass produced. Uniformity of design also eases maintenance and documentation requirements.

There are a few exceptions to the rule when it comes to interfacing systems to the safety system. The Oxygen Deficiency Monitoring system and the Controlled Area Radiation Monitors are examples of these. These systems are designed to provide alarm, warn, and system failure contacts directly to PLC input modules. They also continuously communicate analog level information directly to the CEBAF Supervisory Control System via CAMAC.

6.0 Procedures and Change Control

The interlock system integrity and performance are protected through a series of procedures that outline specific steps that must be followed whenever a system change or maintenance is required. These procedures cover daily operations, system maintenance, system validation, operator training, and logic control and documentation. Approvals and audits are controlled by the accelerator operations crew chief and the Accelerator Safety Department.

7.0 System Reliability and Performance

A Failure Modes and Effects Analysis was performed on the CEBAF PLC systems by engineers at Modicon.[10] For the study, a "catastrophic failure" was defined to be any failure of the system in which both PLCs failed in exactly the same manner within one hour of each other. The mean time to catastrophic failure for the CEBAF PLCs was calculated to be 320,000 years. The mean time to any single failure in the PLCs was calculated to be greater than 10,000 hours. These figures are for the PLCs only and do not include any other parts of the system. A separate mean time to failure calculation was performed by the CEBAF staff; it was determined that the mean time to any single failure in the entire system was approximately three months.

The CEBAF Personnel Safety System was first commissioned on 3 January 1991. The failures and problems that have been observed in the first six months are as follows.

1. A faulty switch was discovered during the initial commissioning of the system. The switch was replaced and returned to the manufacturer. The resulting downtime was approximately one hour.
2. A logic coordination problem between the two PLCs was discovered while operating the system in a mode that was not originally foreseen. The problem was the result of having different logic running on the two systems and did not present a problem from the standpoint of system security. The resulting downtime was one hour.
3. Several of the display symbols on the monitoring computer were discovered to be in error. These symbols were corrected while the system was running and did not result in any system downtime.

4. The Uninterruptable Power Supply (UPS) in one of the remote racks had to be reset several times following prolonged site power outages. Total downtime was approximately one hour.

Using the failures and downtimes listed above the availability of the system was calculated to be greater than 99.9% for the first six months of operation. It should be noted that only the logic coordination problem described above actually resulted in a loss in of system availability during accelerator operations, therefore the real availability number is much greater than 99.9%.

8.0 Upgrades and Future Additions

As the construction and testing of the accelerator progresses the Personnel Safety System will be expanded to include the new areas. These expansions are as follows:

1. The entire north linac of the accelerator will be tested in early 1992. A temporary wall will be installed at the end of the linac ahead of the first arc. The area of coverage for the safety system will be extended to this temporary wall.
2. A temporary wall will be installed at the end of the first arc following the north linac tests. The safety system will be expanded to accommodate these tests.
3. The safety system will then be expanded to cover the entire accelerator enclosure. Single and multiple pass operation of the accelerator will be tested.

It should be noted that the entire safety system is actually being installed all at once and these future expansions of the system will only involve the addition of new blocks of logic and a few temporary switch connections to the PLCs.

CEBAF is also planning to use PLCs to perform the interlock logic required for the three end stations. These PLCs will be separate from the accelerator PLCs and will communicate with these via a peer-to-peer communications link. This will allow the end station logic and systems to be installed independently of the accelerator system. In this way, any changes or additions to either the accelerator or end station systems will not require that the actual logic for the other system be modified.

9.0 Conclusions

The use of commercially available programmable controllers has proven to be a labor- and time-saving way to implement the CEBAF Personnel Safety interlock system. The system availability over the first six months of operation has been greater than 99.9%. It is felt by the CEBAF staff that with the severe schedule and manpower constraints of the CEBAF construction project an interlock system based entirely on relay implemented logic would have been impossible to implement.

Other facilities wishing to implement a PLC based interlock system should follow all existing standards and guidelines, several of which are listed in the references below. It is also recommended that the staff of the facility thoroughly research and become familiar with several PLC vendors and distributors. A good working relationship with these individuals has proved to be invaluable during the design and construction of the CEBAF system.

10.0 References

1. B. Hartline, "CEBAF Progress Report", Proceedings of the 1990 Linear Accelerator Conference, Albuquerque NM, September 1990.
2. CEBAF Design Handbook, January 14, 1991.
3. R. C. McCall, W. R. Casey, L. V. Coulson, J. B. McCaslin, A. J. Miller, K. F. Crook, T. N. Simmons, "Health Physics Manual of Good Practices for Accelerator Facilities", SLAC-Report-327, April 1988.
4. CEBAF Safety Manual, March 30, 1989.
5. CEBAF Electrical Standards Manual, November 1988.
6. NCRP Report No. 88, "Radiation Alarms and Access Control Systems".
7. American National Standard N43.1, "Radiological Safety in the Design and Operation of Particle Accelerators".
8. CEBAF TN-90-233, "Workshop on Personnel Safety Interlocks", March 1990.
9. Health and Safety Executive, Programmable and Electronic Systems in Safety Related Applications, 1987, ISBN 0 11 883906 3.
10. MODICON/AEG Quality Assurance, "Failure Modes and Effects Analysis", Revision A, Prepared for CEBAF July 6, 1990.
11. ISA Transactions, "Programmable Controllers", Volume 29, Number 2, 1990, Instrument Society of America.
12. L. A. Bryan, E. A. Bryan, "Programmable Controllers Theory and Implementation", Industrial Text Co., 1988.
13. R. C. Waterbury, "Fault-Tolerant/ Fail-Safe Systems are Fundamental", INTECH Volume 38, Number 4, April, 1991, ISA Publications.
14. Taylor Industrial Software, "Ladder Logic Development Series for Programmable Controllers", Modicon 984/584, Version 1.24, August 1989.
15. R. Bork, C. Grubb, G. Lahti, E. Navarro, J. Sage, T. Moore, "CEBAF Control System", CEBAF PR-89-013.

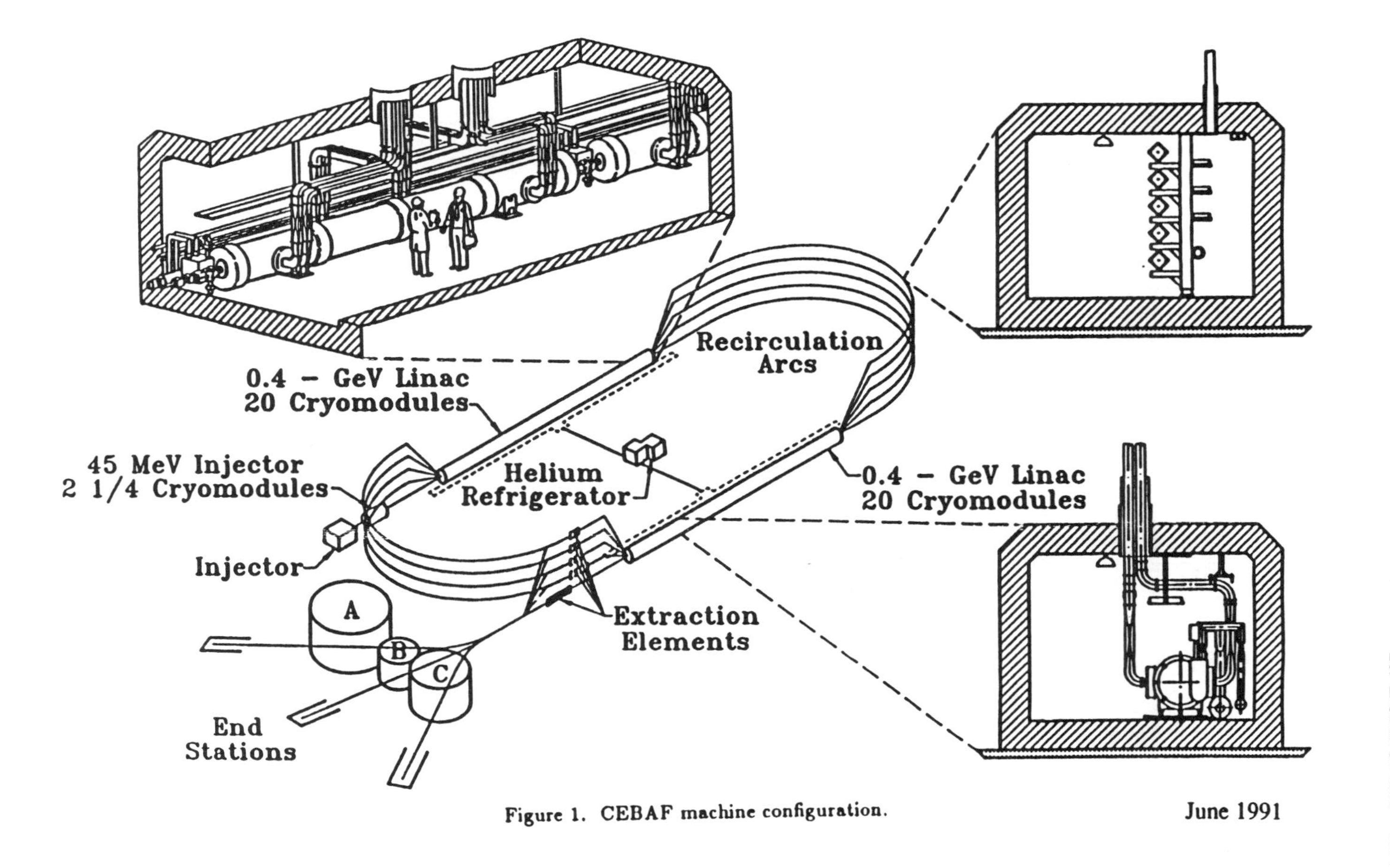

Figure 1. CEBAF machine configuration.

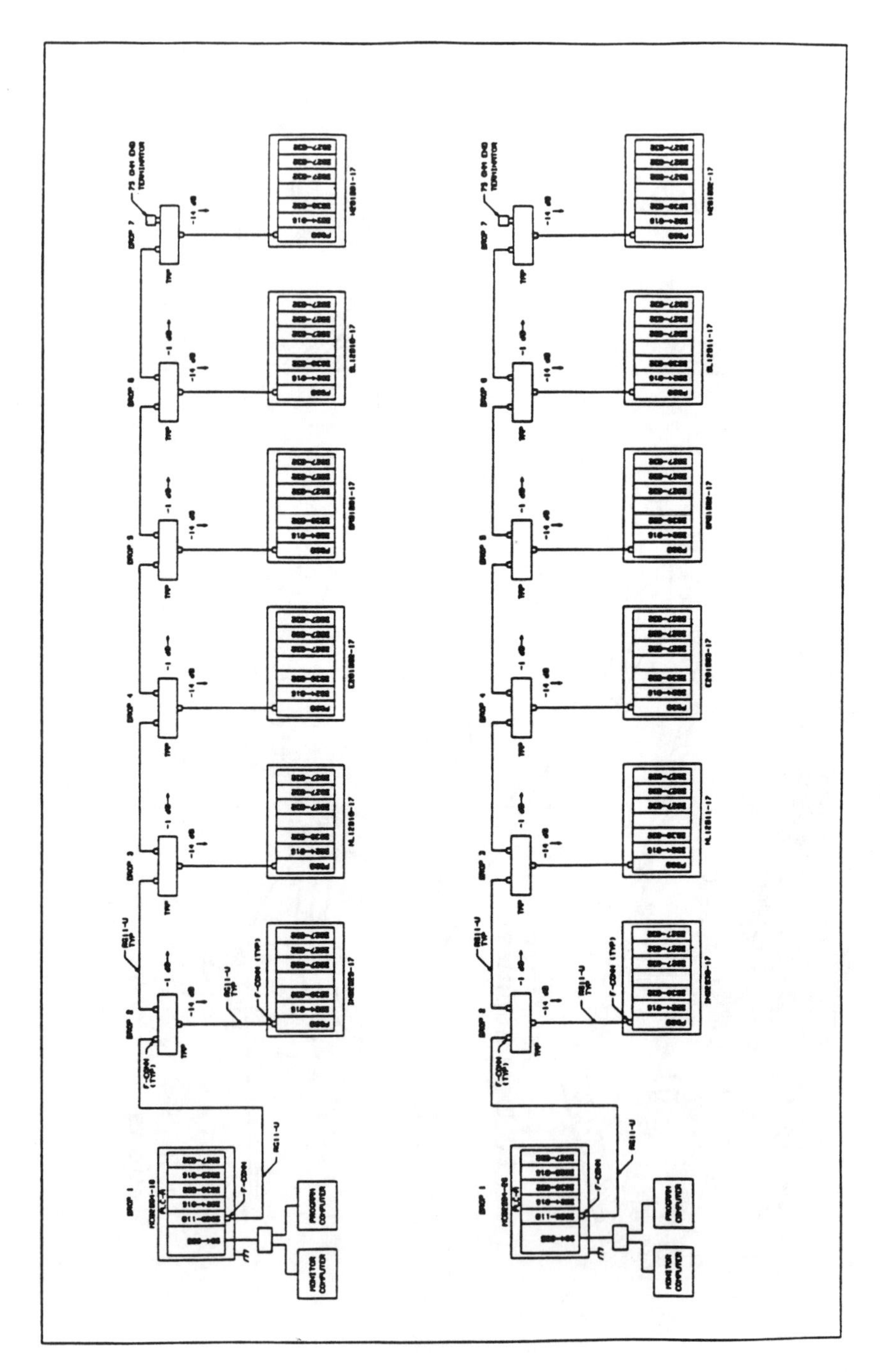

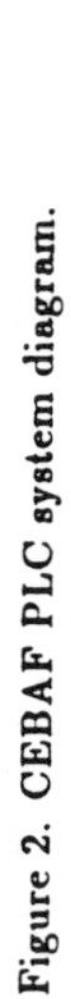

Figure 2. CEBAF PLC system diagram.

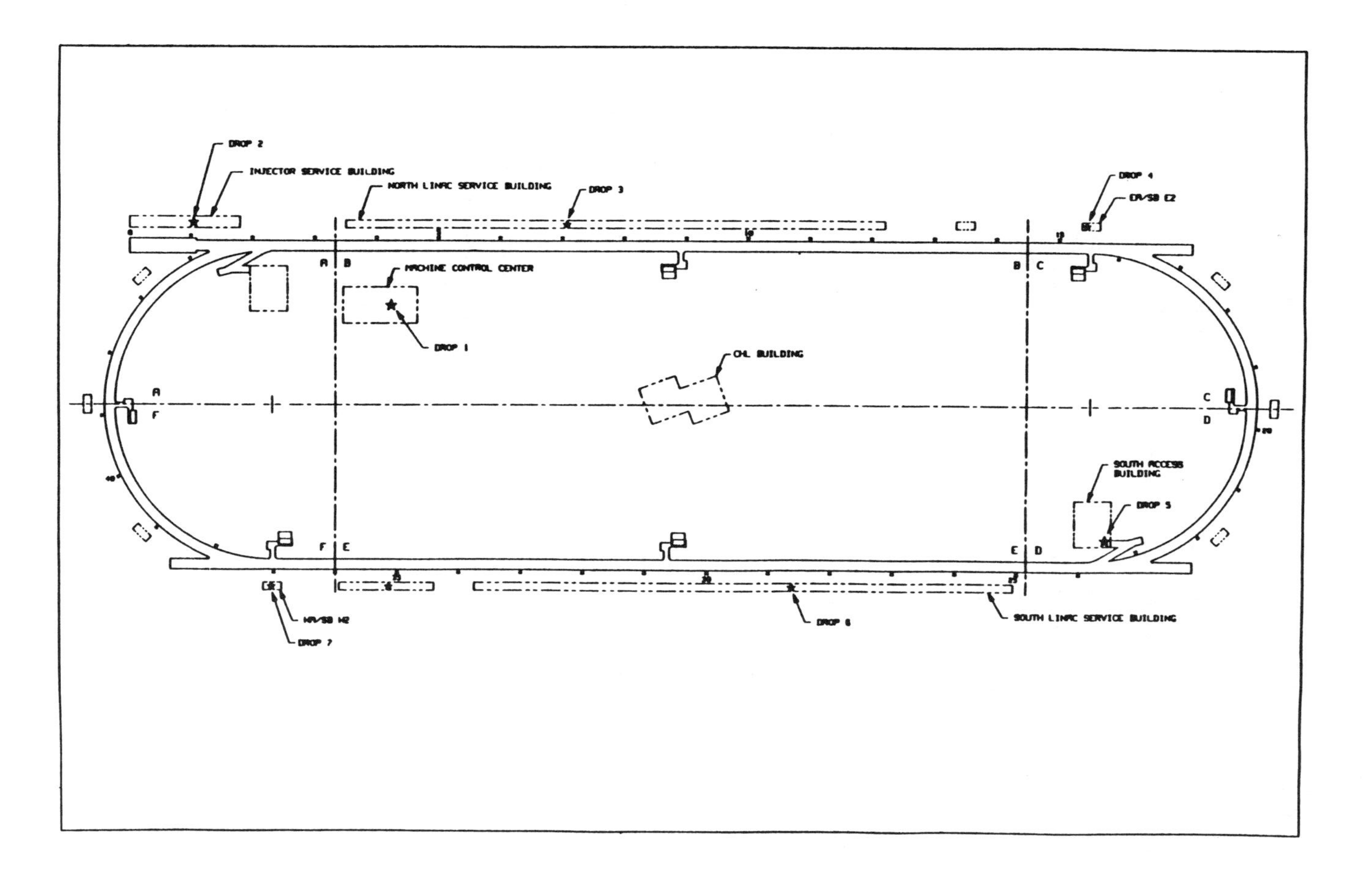

Figure 3. CEBAF PLC drop locations.

The Crawling Wire Method for Transverse Beam Diagnostics

C. Johnstone, J. Lackey, and R. Tomlin

Fermi National Accelerator Laboratory, P.O. Box 500, Batavia, IL 60510

An automated profile measuring system using single wire scanners has been implemented in the Fermilab Booster Synchrotron which allows the beam profile to be measured on a turn-by-turn basis. More than 30 profiles after injection can be accumulated before the wire degrades the beam significantly. An analysis of the crawling wire profiles yields quantitative information on the various transverse modes of the circulating beam. A complex injection pattern emerges from the analysis which includes a direct correlation between injection efficiency and the quadrupole breathing modes of the Booster beam.

I. INTRODUCTION

In a circular accelerator, a single wire scanner proves to be an effective diagnostic tool from which profiles of the injected and circulating beam can be extracted on a turn-by-turn basis. The profile is obtained by reading the current on the wire at selected time intervals within the beam cycle and then advancing the wire to different positions and recording the same information on subsequent beam pulses. For thin wires the losses on a turn-by-turn basis are small and readily calculable. For wire thicknesses of a few mils, effective beam profiles can be generated for up to 60 turns before wire distortions significantly alter the beam characteristics. In this way a detailed history of the injection process and post-injection beam behavior can be recorded using the crawling wire diagnostic. Examples of wire profiles taken upstream of the injection septum and at injection are shown in Figure 1.

II. THE CRAWLING WIRE EXPERIMENT

The Fermilab Booster utilizes a multiturn injection scheme where the incoming H^- beam from the Linac is injected atop the circulating proton beam. Four pulsed magnets called ORBUMPs (for they bump the closed orbit) perform injection into the Booster. The H^- beam is injected parallel to the Booster closed orbit beam, but offset by 6 cm. The first pair of ORBUMPs creates a dogleg which shifts the H^- beam by 3 cm onto a 200μgm/cm^2 carbon foil used for stripping. This foil is located in a half-meter drift section between ORBUMP pairs. Once stripped, the beam, now completely protons, undergoes an identical dogleg in the two remaining ORBUMPs for the additional 3 cm displacement needed to put it on the Booster closed orbit. The proton beam, after circulating around the Booster and back to the injection point, undergoes a dogleg in the opposite direction to the incoming H^-. The result is complete overlap of the two beams immediately downstream of the first two ORBUMPs. The advantage to this injection scheme is that H^- can be injected into the same phase space as the circulating proton beam with the result that many turns of beam can be injected into the Booster before filling the available aperture. After H^- injection is complete, the ORBUMPs require about 25 microseconds to decay so that the circulating beam makes 6-10 revolutions before the trajectory settles into its closed orbit position. For about 6 of the revolutions, the proton beam still intercepts the stripping foil. (The effect of additional stripping foil passes on the beam has proved to be insignificant. Losses are less than .5% per pass.) Beam profiles measured using a wire just upstream of the stripping foil are shown in Figure 1. The profiles clearly show the orbit displacment due to the ORBUMPs during injection.

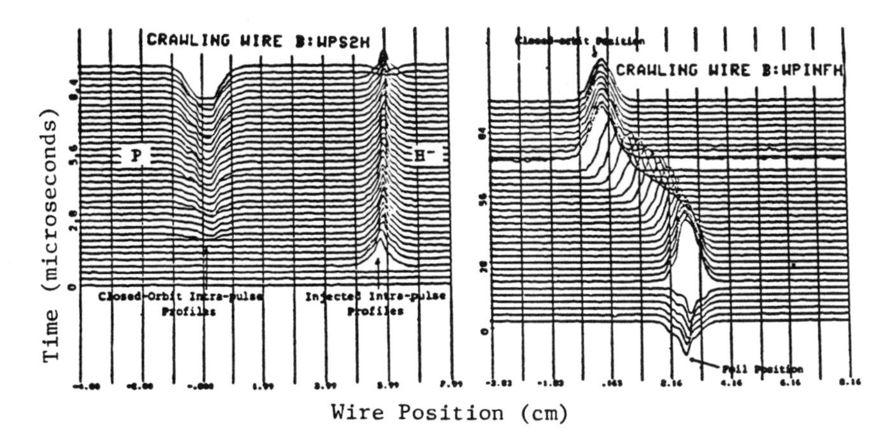

Figure 1. Turn-by-turn profiles generated in the Fermilab Booster at the injection septum (left) and at the stripping foil (right). At the injection septum the profiles were inverted for data analysis purposes. At the stripping foil the negative profiles are the incoming H^- beam.

Because of large magnetic fields in the vicinity of the ORBUMPs, a wire positioned about 40 m downstream of the injection point had to be used to obtain the profile data used in the final analysis. This is a vertically mounted tungsten wire, 5 mils thick, located in a 6 m drift section away from fringe fields. A stepping motor capable of stepping the wire horizontally in submillimeter increments is used to control the wire position. For the measurements, the wire was stepped in 1 mm steps across the 10 cm Booster beam aperture. Since the rotational frequency of the beam in the Booster is 2.8*μsec* at the injected energy of 200 MeV, this is the frequency at which the wire was read out in order to obtain turn-by-turn injection profiles. The raw wire data is then sent through a 1 MHz filter and amplified before being input to a fast digitizing Gould scope. The Gould scope is read out through a GPIB interface and into Fermilab's main accelerator controls system by frontend PDPs. The PDPs then transfer the data into a filesharing utility on a VAX. Because of the limited resources on the PDP frontends, only 36 time ticks of data can be conveniently stored at one wire position, so only 36 profiles are generated per data run. For long term data storage and ease of analysis the data are compressed into binary ZEBRA[1] data structures.

Numerous sets of wire profile measurements were taken for 1- through 6-turn injection in 1-turn steps. These corresponded to circulating intensities of .5E11 protons for 1-turn injection up to 2E12 protons for 6 turns. The length of the injected H^- beam pulse ranged from 2.8 to 16.8 microseconds. One set of measurements was also performed on 7-turn injection.

III. THE DATA ANALYSIS

The stored ZEBRA files were read, analyzed, and histogrammed using a software physics analysis package, PAW[2]. Characteristic profiles from low-intensity and high-intensity data are illustrated in Figure 2. The amplitudes plotted depend on the scope settings and a knowledge of these must be included when comparing intensities from different runs. The raw profiles shown were fit with a Gaussian which overlays the data in Figure 2. The Gaussian proved to be an excellent description of the data, generally giving a χ^2 of less than 1 for the fits.

From each of the Gaussian fits, a central peak position, an overall normalization, a one-sigma width, and a constant background were found. The beam position and width of the Linac pulses remained stable throughout the measurements. The average beam position varied less than .5 mm and the one-sigma

width of the injected pulse was stable to 5% between different profile runs. (Larger variations were observed, of course, for different injection tunes.) Since beam intensity is proportional to the total area under a measured profile, the overall normalization multiplied by the standard deviation was used to represent the relative beam intensity of each profile (ignoring the constant factor of 4π in the integral of the area).

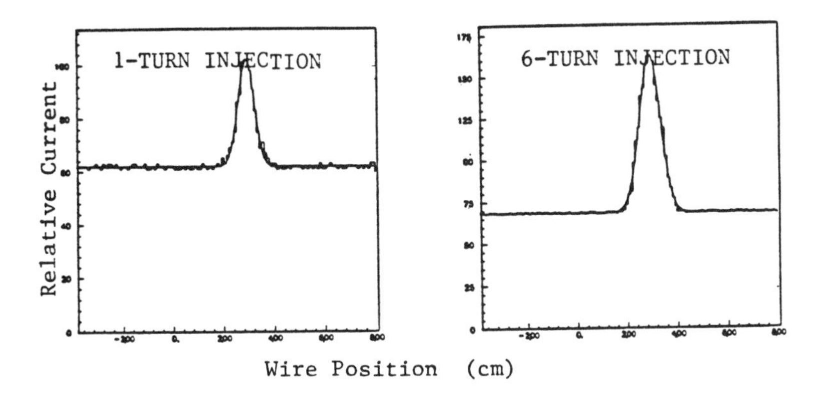

Figure 2. Characteristic low-intensity and high-intensity profiles overlaid with Gaussian fits to the data.

IV. RESULTS

Once all the data were fit with Gaussians, injection efficiency could be studied as a function of specific beam parameters, the peak position and the one-sigma width of the injected and circulating beam, (referred to as the dipole and quadrupole components of the beam, respectively). The 1-turn data provided a reference trajectory and an unambiguous beam profile on a turn-by-turn basis.

In this work the injection efficiency of a turn is defined to be the beam profile area of that turn minus the profile area measured on the previous turn. The remainder is then normalized to the area of the first injected turn. Since the Linac beam pulse is stable to a few percent in intensity throughout the part injected into the Booster, the amount of beam actually injected can be measured if the wire is close enough to the injection point. However, by subtracting off the intensity measured on the previous turn, the remainder is artificially low due to machine losses experienced by an injected turn in its initial trek around the Booster. (The majority of beam loss due to aperture restrictions or lattice mismatches occurs on the first turn after injection. Such losses can be resolved using the single-turn profile data.) This initial first-turn machine loss drops out in a comparison between subsequent injected turns. That is, the injection efficiency is a relevant quantity when comparing injected turn #2 and higher on multiturn injection. As calculated, the injection efficiency becomes a measure of how well the H^- beam is injected into the circulating proton beam.

With this in mind, it is most informative to view the 5-, 6-, and 7-turn injection data. (The results for the higher-turn injection data were consistent with the lower, 2-, 3- and 4-turn injection profiles.) A comparison of the injected efficiency with dipole motion revealed no significant change in the beam position between turns and correspondingly no dependence on the beam position at injection. Figure 3 shows the injection efficiency versus injection turn number for 5-, 6-, and 7-turn injection along with the peak position of the beam versus turn number. (The first profile on the 7-turn data was not measured so data is normalized to the second turn.) While the injected intensity shows clear oscillating structure, the dipole component is zero after an initial shift which occurs in the first couple of turns. This initial shift in beam position is most likely due to various pulsed and switching power supplies in the injection line which slice the beam destined for Booster out of the Linac pulse. The vertical scale is in tenths of millimeters so the shift is almost a centimeter for 6- and 7-turn injection.

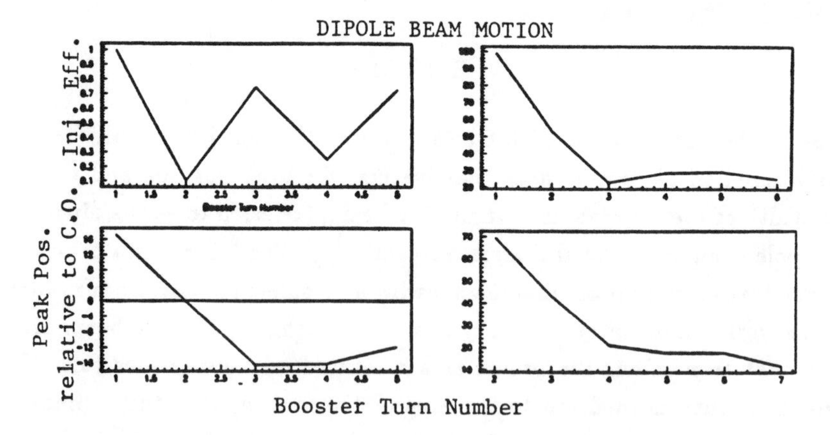

Figure 3. Dipole motion at injection for 5-, 6-, and 7-turn injection.

Quadrupole motion, however, was observed on consecutive injected turns to be in synch with the injection pattern. In Figure 4, the one-sigma width of the same beam profiles used for the dipole analysis is also plotted against the injected turn number for 5-, 6-, and 7-turn injection. Variations of about 20%

in the beam profile width were observed on alternate turns in the Booster which affected the injection efficiency. Narrow widths were observed on even-numbered injected turns and larger widths were measured on odd-numbered injected turns. Because of this consistancy, the injection trend is observable even in the closed-orbit circulating beam well after injection (Figure 5). Conservative estimates based on Figure 5 indicate this breathing caused around a 20% drop in the injection efficiency of a turn when injected atop the narrower distributions. The increase in injected beam between 3 to 4 turns and 5 to 6 turns is approximately 20% less than the increase observed for 2- to 3- and 4- to 5- turn injection.

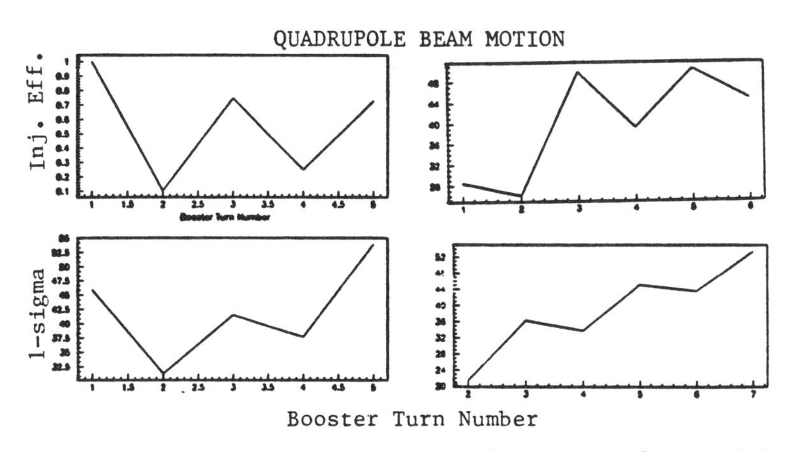

Figure 4. Quadrupole motion at injection for 5-, 6-, and 7-turn injection.

The drop in injection efficiency as a function of odd *versus* even turn number was based on profile data taken well away from injection because similar, but presumably weaker, oscillations could be manufactured from nonlinearites in the wire currents as a function of beam attributes. In our case, a narrow beam profile can result in too small an amplitude overall in the measured profile due to nonlinearities in the wire response. Wider profiles on the average produced larger areas and higher calculated intensities than narrower profiles of equal intensity. For quantative comparisons between runs, data were used where beam attributes were similar, which occurs about 12 turns after completion of injection. At this time the breathing mode ceases and the profile widths settle at a sigma of 4.8 +/- .2 mm, independent of the number of turns injected up to 6. Therefore, more reliable values for the injection efficiency differences were extracted from late, stable-width data as displayed in Figure 5. Only on seven-turn injection did there appear to be a 10% increase in the width of the closed-orbit profile.

This may be indicative of space-charge beam blowup.

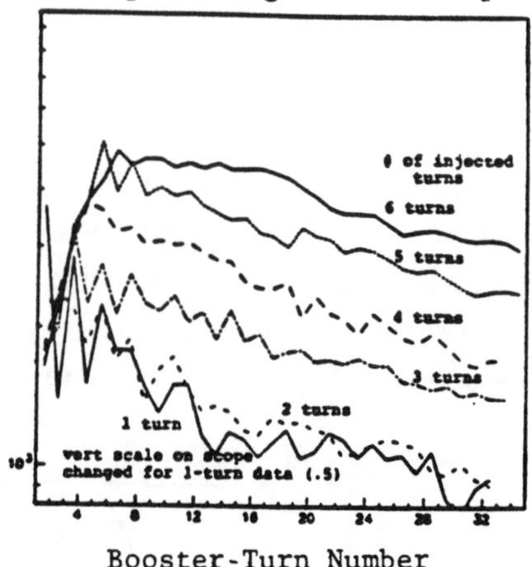

Figure 5. Relative beam intensity as a function of turn number in the Booster for 1- thru 6-turn injection.

The 1-turn data were analyzed as a check on the multiturn analysis. Beam intensity oscillations (Figure 6) were also observed consistently in the 1-turn data as a function of turn number. Since it was not possible for the beam intensity to increase when only one turn is injected, the raw profile data (Figure 7) was examined for an explanation. The quadrupole breathing mode is most apparent in the raw 1-turn data and the Gaussian fits clearly do not represent the profile area accurately. Peak positions and one-sigma widths remain a reasonable description of beam behavior.

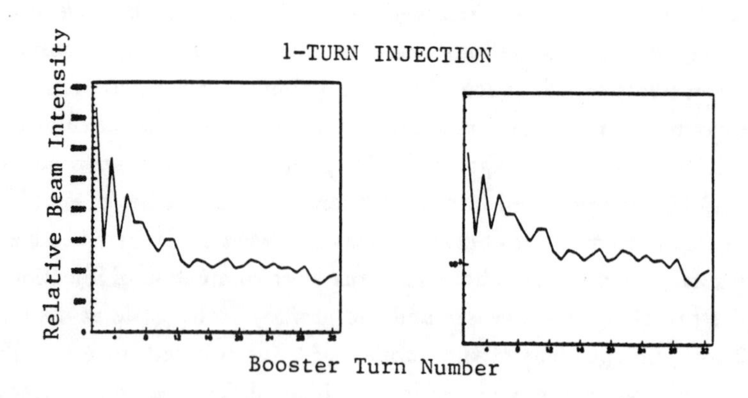

Figure 6. Intensity oscillations in 1-turn injection plotted on both a linear and log scale.

Starting in the second turn, structure appears, and at first it was attributed to the lower statistics of 1-turn data and possibly the systematics of the wire, filter, and amplifier used. Another 1-turn data set (Figure 7) was also examined, except in this measurement the injection position and angle had not been properly tuned. It became clear from the mistuned data set that an aperture was being hit downstream of the wire position because the first wire profile is clean. Since the first turn profile measurement is upstream of the aperture, its area represents the full injected intensity of the H^- beam. The second turn shows an unmistakable gap in the profile where the beam hit something and a corresponding drop in the intensity. Subsequent turns of the beam around the Booster just illustrate a betatron oscillation, this time of a hole in the beam rather than the beam itself. An aperture restriction provides a clean incoherent betatron kick to the beam which does not couple the transverse directions. The end result of an incoherent kick is a quadrupole breathing mode in the beam.

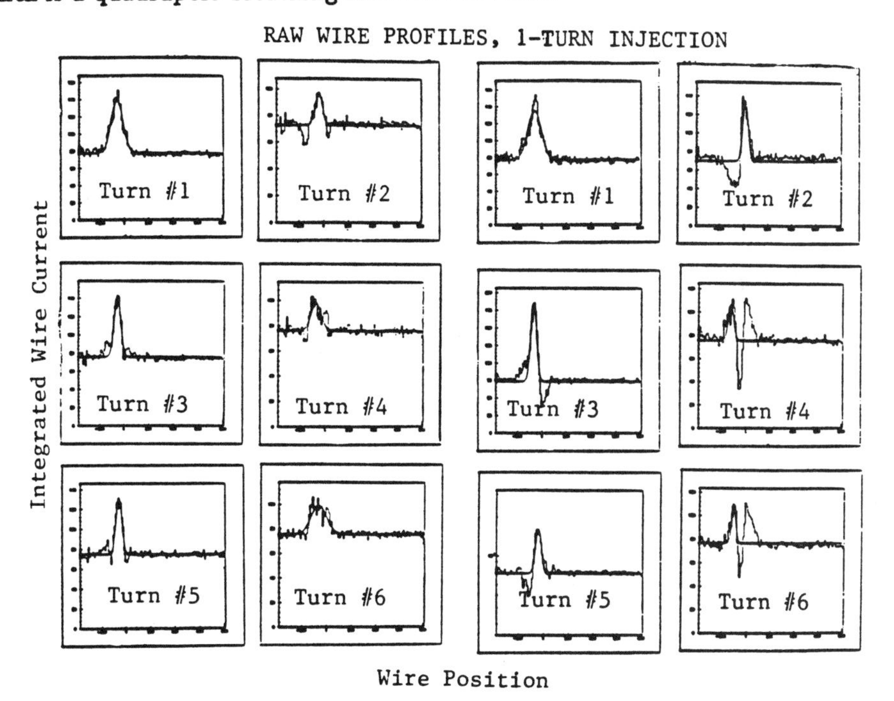

Figure 7. Raw wire profiles taken before (right) and after (left) position and angle of the beam were tuned at injection.

Depending on the strength of the kick, the breathing mode damps about 12 turns after injection is finished, or about 6 turns after the ORBUMP magnets

turn off. Position and angle tuning through the aperture at injection made a dramatic improvement in the intensity, dipole, and quadrupole oscillations of the beam although from Figure 8 it can be seen scraping was not completely eliminated with the standard injection tuning procedure.

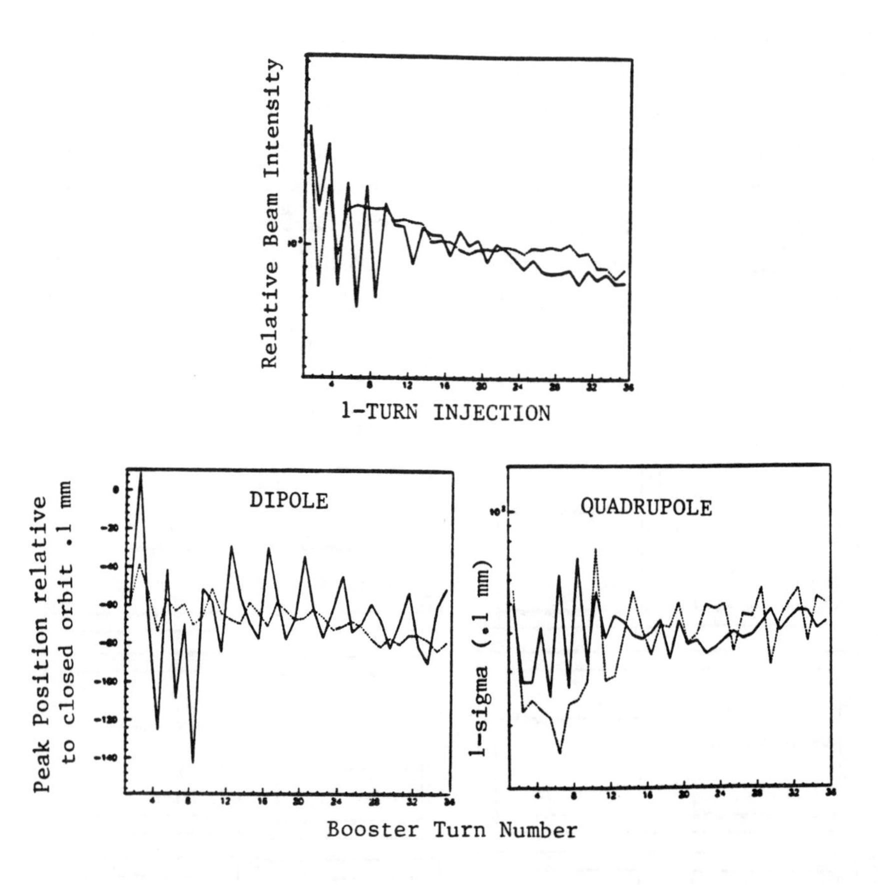

Figure 8. Beam intensity, dipole motion, and quadrupole motion before and after injection position/angle tuning. Solid lines represent data with a significant position/angle area and a significant aperture restriction. Dashed lines are the same profiles taken after tuning through the aperture was performed.

The exact location of the aperture restriction was not known. Three known apertures exist in the Booster Synchrotron, two on extraction septa and one on the injection septum. At a later date position bumps used for aperture scans were placed at known locations around the Booster ring and wire data taken for the various bumps. The aperture restriction responsible for the quadrupole

breathing mode in the Booster was discovered to be the injection septum. (Injected beam does not encounter the injection septum until after it has almost completed its first loop around the Booster.) This beam loss in the initial Booster turn has been corroborated by data obtained from the injection toroid, which is the standard diagnostic used to monitor Booster intensity at injection. Depending on the injection tune, a 10-20% beam loss was observed on the toroid for the first turn of 1-turn injection. This is consistent with the wire profile analysis. Multiturn injection data sets have not yet been taken using the injection toroid.

V. CONCLUSION

A quadrupole breathing mode in the Booster Synchrotron beam can cause the incoming H^- beam to be mismatched to the circulating proton beam at injection. When a mismatch in the profile width occurs it was observed to lower injection efficiency by as much as 20%. The breathing mode resulted from an incoherent kick or hard aperture in the beam caused by a restriction rather than lattice mismatches or coupling. A nondestructive diagnostic which records both dipole and quadrupole beam modes is needed to monitor the reproducibility and integrity of the beam for proper Booster operation. The wire data show that a beam position measurement alone does not ensure beam quality. Currently the possibility of using BPMs for both a position and a relative width measurement is being investigated.

VI. REFERENCES

1. R. Brun, O. Couet, C. Vandoni, P. Zanarini, "*Physics Analysis Workstation,"CERN Computer Center Program Library, Long Writeup, Q121, Oct. 1989.*
2. R. Brun, J. Zoll, "*ZEBRA-Data Structure Management System,"CERN Computer Center Program Library, Q100, 1989.*

Tune and Chromaticity Measurements in LEP

H. Schmickler

European Organization for Nuclear Research, CERN Division SL-OP,
CH-1211 Geneve 23, Switzerland

Abstract

The LEP Q-Meter allows the excitation and observation of transverse beam oscillations. The instrument is used for continous tune measurements but also for single shot precision measurements of spectra which contain all beam resonances. As the repetition rate of signals at LEP is low, the Q-Meter has been conceived around two fast processors treating the beam data online at the pace of the LEP revolution frequency. The presentation will cover the following subjects: Measurement of beam spectra using the Fourier Transform of the beam motion, precision tune measurements using swept frequency beam excitation and harmonic analysis of the beam motion, continous tune monitoring by a phase locked loop beam excitation, closed loop control of the machine tunes by acting on the main quadrupole strings of LEP and chromaticity measurements using a 3 sec long RF frequency modulation and continous tune monitoring.

1 Instrument description

The LEP Q-Meter measures the fractional part of the betatron tunes by observing coherent transverse oscillations in the horizontal and vertical planes with a single dedicated beam position monitor. The oscillations are excited by small kicker magnets, called beam shakers, which act selectively on one bunch. Fig. 1 shows the main sub-systems of the instrument for one plane : The beam position measurement (pick-up electrodes, preamplifiers and A/D conversion electronics), the signal processing part (DSP) and the beam shaker with its pulsed power supply.

The beam position measurement uses the standard electronics of the wide band beam orbit measurement system [1]. The beam position is calculated from the four button signals that were digitized individualy. By an automatic adjustment of the gains of the preamplifiers as a function of the beam current the full range of the 16-bit A/D converters is exploited and coherent beam oscillations down to 1 μm can be observed.

In 1988, the year of the conception of the LEP Q-Meter, the fastest microprocessors available on the market were able to run at execution speeds of

5 Mips. Given the revolution time of LEP with 89 μsec such a microprocessor could execute about 400 instructions per machine turn. That is why the LEP Q-Meter has been conceived as a pure digital solution around 2 identical microprocessors (Motorola 68020) treating the information of the horizontal and vertical beam motions independently. Apart from noise free data treatment this offers a great flexibility in the applications of the instrument as any different operation mode can be achieved by pure software commutation.

The beam shakers are single turn ferrite magnets built around a ceramic vacuum chamber with a thin internal metallisation. Their pulsed poer supplies are located in the support frame of the shakers. Because of the moderate requirements of driving voltage and commutation speed, fast MOSFET power transistors are used in a current regulated amplifier circuit. [2]

More details on the conception of the instrument can be found in the design report [3]

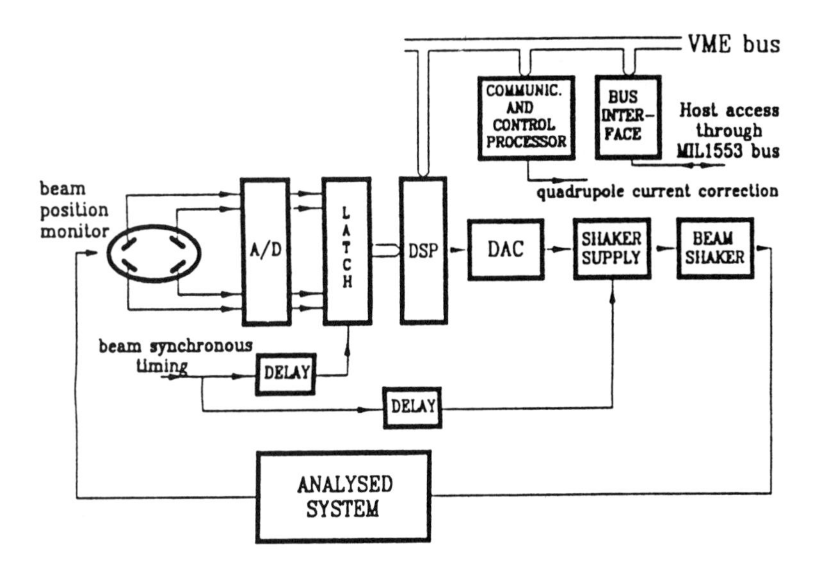

Figure 1: Hardware layout of the LEP Q-Meter

2 Tune Measurements

The LEP Q-Meter can be used in 3 operating modes applying different beam excitations and different data treatment of the observed beam motion. Depending on the available measurement time and the desired accuracy one of the modes is choosen. Table 1 shows some characteristics of the operation modes and their principal application.

In FFT-Mode (Fast Fourier Transform) the beam is excited with white noise kicks and the spectral distribution of the induced beam motion is obtained by a Fourier Transform. This fast measurement mode is frequently used in machine operation to control and adjust the tunes.

In SWF-Mode (Swept Frequency) the beam is excited with a sin wave where the excitation frequency is slowly increased or decreased through the tune range of interest. Due to the nature of the excitation and the larger beam motions this method is more precise than the FFT Analysis, but needs more measurement time and can not be applied in all machine operation modes. Therefor it is mainly used during MD shifts where the high resolution is needed.

In PLL-Mode (Phase Locked Loop) the beam again is excited with a sin wave, but by a phase lock the excitation frequency is forced to follow the machine tune. This method provides a continous tune measurement which is of particular interest during critical operation steps like energy ramping or beta squeezing. Based on these continous tune measurement a closed loop tune regulation can be switched on. This loop acts on the power converters of the main ring quadrupoles and aims at maintaining the measured tunes constant at given reference values.

mode	meas. time [sec]	accuracy δq	application
FFT	1 ... 10	$\simeq 10^{-3}$	machine operation
SWF	10 ...100	$\simeq 10^{-4}$	machine development
PLL	continous	$\simeq 10^{-4}$	operation, MD, Q-Control loop

Table 1: Typical characteristics of the Q-Meter operating modes

2.1 FFT Measurements

The desired frequency resolution δqof the measurement is programmed by selecting the number of position measurements N_s of the input to the Fourier Transform, since $\delta q = 1/N_s$. The statistical measuring errors can not be reduced by increasing the number of of position measurements, but by taking the average of successively measured spectra. Fig. 2 shows an amplitude spectrum of the beam oscillations in the horizontal and vertical planes beased on data from 4096 beam revolutions and an averaging count of 8. The data has been taken during a physics coast, so apart from the machine tunes beam-beam modes are visible in the spectra. Beam oscillations as small as 3 μm can be measured with a signal to noise ratio better than 10 dB.

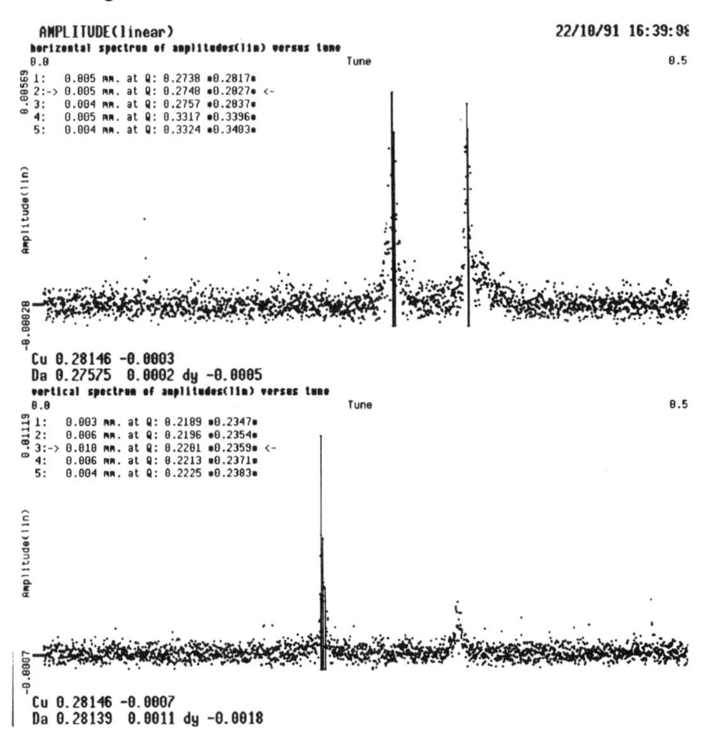

Figure 2: amplitude spectra of transverse beam motion measured in FFT mode

2.2 SWF Measurements

This method is commonly used in frequency response analysers. The beam excitation is a sinwave of variable frequency. The frequency is swept according to a staircase function with variable frequency increment. The beam response is analysed by standard techniques (Harmonic Analysis) over a large number of signal periods. The harmonic excitation may always produce beam oscillations of sufficient amplitude. The signal to noise ratio is very high und hence we obtain measurement resolutions of δq= 10^{-4}. This error is of the order of the short term stability of LEP. Fig. 3 shows an accurately measured beam transfer function. The frequency sweep was programmed to zoom a small domain around the horizontal tune.

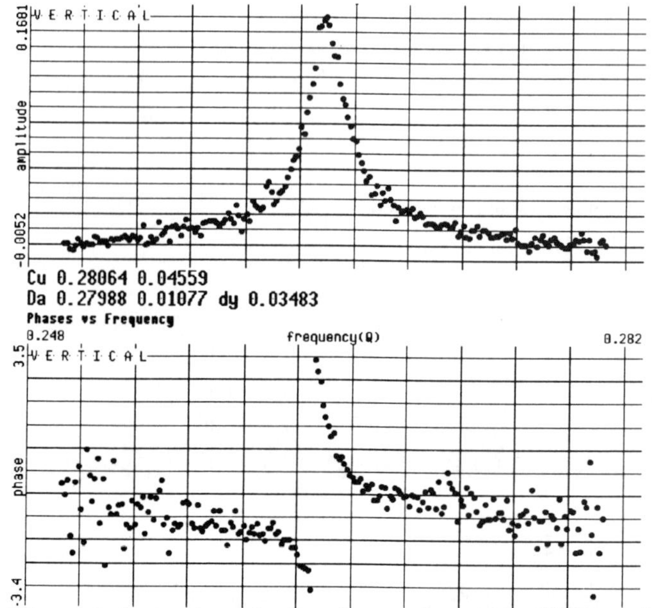

Figure 3: beam transfer function measured in SWF mode

2.3 PLL Measurements

A block diagram of this method is shown in fig. 4. The beam oscillations are tracked by a phase locked loop consisting of its classical building blocks: The phase detector, the loop filter and the voltage controlled oscillator or its digital equivalent, a phase controlled oscillator (PCO). The PCO output is used for harmonic excitation of the beam through the beam shaker.

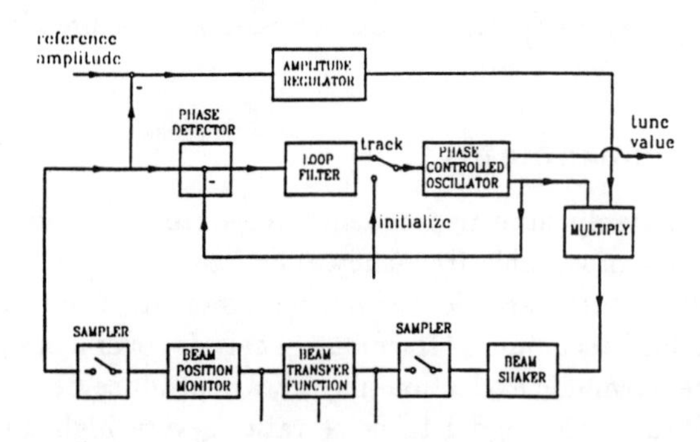

Figure 4: signal flow diagram for PLL mode

This coupling between the PCO and the PLL input variable (the transverse bunch position phase) constitutes a very particular loop configuration. PLL's are normally used to track external signals, not to react back on them. With a correctly chosen phase shiftbetween the PCO output and the beam shaker,

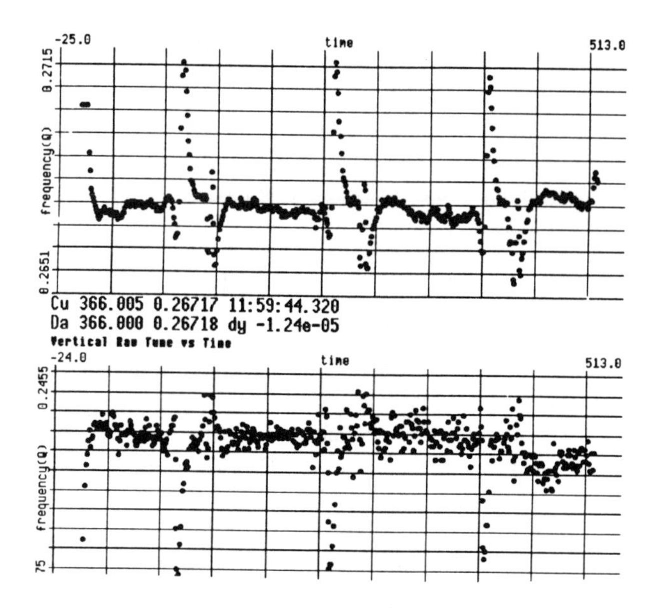

Figure 5: tune modulation of LEP by SPS magnetic cycle measured in PLL mode

The PLL will capture and track a resonance. The amplitude regulation, shown in fig. 4 is necessary to avoid an over excitation of the beam when approaching the betatron resonance.

By its nature the PLL comprises a continous tune measurement. At fixed time intervals tune values may be stored in a buffer (a so called tune history) and can then be used for further analysis. As presently set the PLL is able to follow tune changes with a bandwidth of 40 Hz. The measurement accuracy depends very much on the beam conditions, i.e. the presence of adjacent beam resonances, but in general we find a resolution of better than δq= 10^{-3}.

PLL measurements are exploited for many machine studies and diagnostics. Whenever tune variations due to changes in machine settings are expected a tune history allows to follow their evolution in time. As a measurement example fig. 5 shows the LEP tunes during a physics coast. Normally one would expect constant values, but due to a magnetic coupling between LEP and the SPS performing a 14.4 sec acceleration cycle, the tunes are modulated at this periodicity. More details on this effect can be found in [4]

2.4 Dynamic Chromaticty Measurements

A Chromaticity Measurement in LEP usually takes several minutes of measurement time. The tunes are measured, then the Rf frequency is shifted by Δf, the tunes are remeasured and chromaticity is calculated from the observed tune difference Δq using the approximate equation

$$Q' = 143000 \cdot \frac{\Delta q}{\Delta f\ [Hz]}. \quad (1)$$

In order to measure the chromaticity during critical machine phases like energy ramping a faster method is proposed that allows to measure chromaticity dynamically at about 3 second intervals. The Rf frequency is continously ramped up and down in a sawtooth like waveform and the resulting tune changes are tracked with the Q-Meter in PLL mode. Fig. 6 shows the waveshape of the fast Rf frequency modulation with a frequency shift of ± 40 Hz and the corresponding tune variation in 40 msec sampling intervals. The observed tune changes are only about 0.003, but the noise level of the measurement is of the order of 10^{-4}. The deduced chromaticities are $Q_H' = 5.5$ and $Q_V' = 10.5$

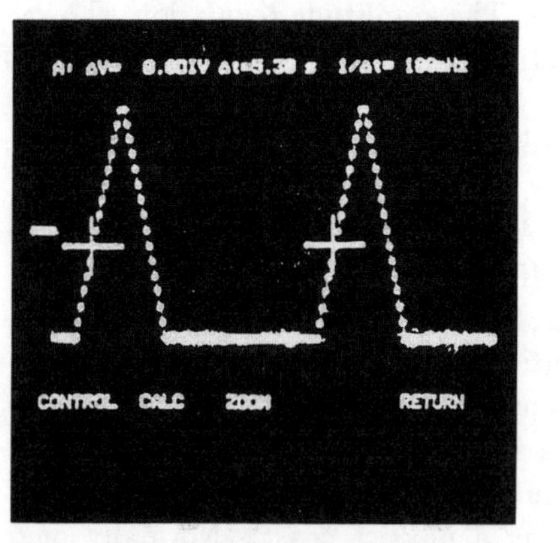

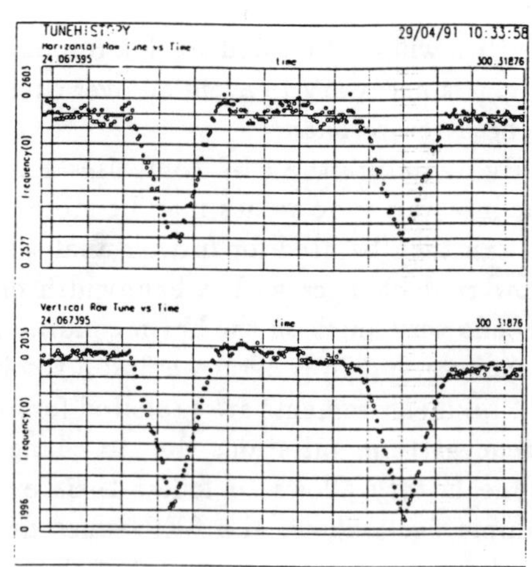

Figure 6: Fast Rf frequency modulation of ± 40 Hz and observed tune variations at 40 msec sampling intervals.

2.5 Tune Control Loop

A continous tune measurement in PLL mode may be used for a closed loop control of the tunes by acting on the currents of the main quadrupoles. This feedback is of particular interest during acceleration to avoid beam looses due to tune variations. Fig. 7 shows the signal flow of this regualtion. The difference between the measured tunes is processed by two digital regulators (PI-Type). A correction applied to one of the power converters causes always a change of both the horizontal and the vertical tune. This coupling between process variables is compensated by mixing both regulator outputs with coefficients computed from the ratio of the beta values of the main quadrupoles. The Tune Control Loop has sucessfully been used to keep the tunes constant within a range of $\Delta q = 0.002$ during acceleration. As illustration fig. 8 shows the machine tune during a period of accelertion and the control signal to keep the tune constant.

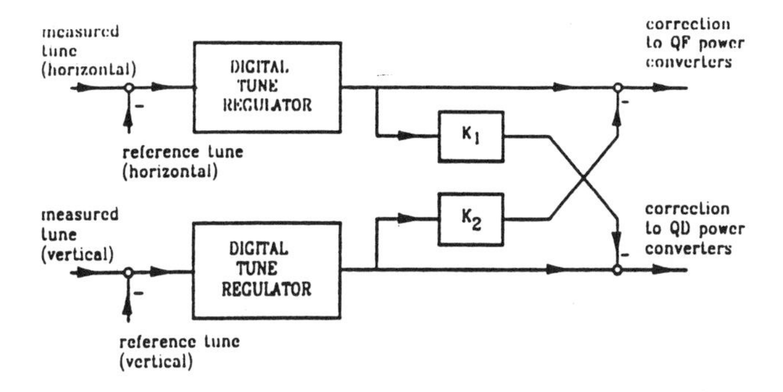

Figure 7: Closed loop tune regulation

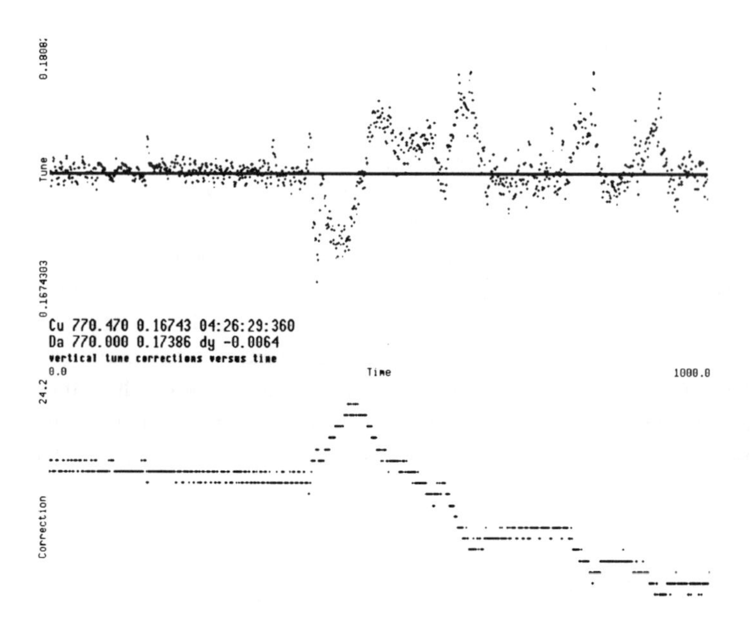

Figure 8: trace of vertical tune during LEP acceleration with q control switched on. The second curve shows the correction signal applied in order to keep the tune close to the reference value (horizontal line)

References

[1] J. Borer et al.
The LEP Beam Orbit Measurement System,
Proceedings of the IEEE Particle Conference, March 16-19, 1987.

[2] M. Desroziers
Amplificateur de puissance pour shaker de mesure de Q,
Internal Note LEP/BI/MP/Note 89-5, CERN, Geneva.

[3] K. Lohmann, M. Placidi and H. Schmickler, Q-Monitoring in LEP
Internal Note LEP/BI/MP/Note 88-45, CERN, Geneva.

[4] O. Berrig et al., Magnetic Coupling between SPS and LEP,
LEP Performance Note 54, CERN, Geneva.

TRANSVERSE FEEDBACK SYSTEM IN THE CERN SPS

L. Vos
CERN
CH-1211 Geneva 23

INTRODUCTION

The transverse feedback system in the CERN SPS has many different tasks in an environment that is in constant evolution. It has thoroughly been described in the past [1,2]. In order to keep some clarity in this description we will proceed as follows. To set the scene we list first some of the relevant machine parameters together with the properties of the beams that the feedback will be faced with. Then we present the operational functions of the system. From the main functions, the machine properties and beam properties we will derive the specifications of an ideal system. The performance of the existing SPS system will then be compared with the ideal one. We will point out some important shortcomings. Finally we will turn our attention to a number of special features.

SUMMARY OF RELEVANT MACHINE AND BEAM PARAMETERS OF THE SPS

The following Table I contains the parameters relative to the machine and to the beams that play a role in the transverse feedback system.

Table I Machine and beam parameters in the SPS

Parameter				
Radius	R	1100 m		
Revolution time	T	23 μs		
Revolution frequency	f_{rev}	43 kHz		
Harmonic number		4620		
RF frequency		200 MHz		
Max of β	β	110 m		
Phase advance/cell	$\Phi/2\pi$	0.25		
Half aperture H	a_H	78 mm theoretical, ~40 mm in practice		
Half aperture V	a_V	26 mm theoretical, ~18mm in practice		
Machine mode		LEP injector	fixed target	collider
Beam momentum inj		3.5	14	26 GeV/c
Beam momentum op		20	450	315 GeV/c
Beam particles		e+/e-	protons	p-pbar
Transverse tune	Q	26.6	26.6	26.7/27.7
Number of bunches		4	4620	6-6
Intensity / bunch		$5\ 10^{10}$	-	$1.5\ 10^{11}$
Total intensity	n	-	$3.5\ 10^{13}$	-

It may be noted that one very important machine mode is absent from Table I, i.e. the fixed target mode with heavy ions. The reason is that the beam intensities are so weak (in the order of a few 10^9 elementary charges distributed over one machine turn!) that the system resolution becomes insufficient to be of practical interest.

TASKS OF THE SPS TRANSVERSE FEEDBACK SYSTEM

Historically the first task of the feedback system was to stabilise the very high intensity proton beams for fixed target operation against the resistive wall coupled bunch instability. A second task derives naturally from the first one, that is the damping of injection oscillations with the main objective of limiting the beam losses. This task became very important with the advent of the p-pbar collider. The low number of bunches and the low average intensity make the resistive wall instability harmless in this case. However, the conservation of the transverse emittance is very important in the collider performance (luminosity) and this made the issue of the damping of injection oscillations of prime importance. The third task of the feedback system takes its origin in the increasing complexity of the tuning process of the SPS. The speed and power of the system made it possible to kick the beam at very short time intervals so that several seconds of machine cycle can be measured in one single measurement sweep. The limit to the measurement is the capability of the control system to absorb and analyse the data. We will not go into further details on this aspect of the feedback but it is a very important one [3].

SPECIFICATION OF AN IDEAL SPS TRANSVERSE FEEDBACK

The input of a transverse feedback system is the measurement of the transverse position of a longitudinal fraction of the beam. This position information is treated in a signal processing chain and ends up in a beam deflector. This beam deflector should transmit the information to the same fraction of the beam that was measured. Since the deflection can only act on the angle of the transverse motion it follows that the position monitor and the deflector are to be separated by an uneven number of quarter β-oscillations. To ensure that the information is transmitted to the correct beam portion it is further necessary that the electronic phases or delays of the electric signal are well matched to the speed of the particles in the machine. It is always possible for an existing feedback chain with some built-in flexibility to be tuned such that the requirements on time synchronism and β-phases are satisfied. We will come back on that subject later. However, a set of basic properties have to be available to guarantee the functionality of the system. These properties are bandwidth, gain and deflection strength. Their specification is the subject of the following paragraphs.

Bandwidth requirements derived from stability aspect

The feedback has to fight the resisitive wall instability. This instability belongs to the family of the coupled rigid bunch instabilities. It may be triggered in many different ways or modes. The lowest frequency mode is the one where all bunches move in the same fashion. The highest frequency mode is the one where two consecutive bunches move in opposition to each other. The frequency of the highest mode is simply half the bunch repetition frequency. This upper frequency limit is half the RF frequency (100MHz) when all the buckets are filled, as is the case for the fixed target beam in the SPS.

It is slightly more complicated to determine the gain of the feedback. We consider its two basic functions: feedback against resistive wall instability and damping of high density *single* bunch injection oscillations.

Gain requirements derived from stability aspect

The action of the resistive wall effect can be calculated as a transverse tune shift:

$$\Delta Q_{\perp} = \frac{neZ_{\perp}}{8\pi^2 Q(E/ec)}$$

where n is the number of particles and $Z_{\perp}$ is the transverse impedance. The opposing action of a feedback system can be associated with an equivalent transverse impedance, but with the opposite sign. We proceed as follows.

The definition of transverse impedance is:

$$Z_{\perp} = j\int \frac{(\bar{E} + \bar{B} \times \bar{c})}{I < x >} ds$$

The action of the deflecting field *E* over a length *ds* can also be expressed in terms of kick strength (units rad) and particle rigidity ($B\rho$):

$$Z_{\perp} = j\frac{kB\rho c}{I<x>}$$

where k is the kick delivered by the system when a position x is detected. The total beam intensity is I. The kick <k> and the position <x> are taken at an average value of β. We now define the gain G of the system :

$$G = \frac{<k><\beta>}{<x>} = \frac{k}{x}\sqrt{\beta_K \beta_{PU}}$$

which this leads to a tune shift induced by the feedback:

$$\Delta Q_{\perp} = \frac{Gc}{4\pi Q\Omega <\beta>} = \frac{G}{4\pi}$$

and this can also be written in terms of an inverse damping time :

$$\tau^{-1} = G\frac{\Omega}{4\pi} = \frac{G}{2T}$$

This formula yields the necessary gain $G = 2\frac{T}{\tau}$ to damp an instability which produces a growing oscillation at a rate of τ^{-1}. The required minimum gain G is a function of the intensity since the damping time depends on the total number of particles n in the beam. It is also a function of the frequency since the resistive wall effect decreases as the root of the frequency (skin effect). For the SPS fixed target conditions we find for 3.5 10^{13} protons G>0.2 in the vertical plane. The horizontal effect is half the vertical one which yields G>0.1 .

Gain requirements derived from injection aspect

To conserve the emittance of the injected beam the feedback damping rate should be larger then the inverse of the coherence time of transverse oscillations. Indeed the feedback can only act on coherent signals. These signals will die away in a time which will be shorter when the spread of b-frequencies in the beam will be larger (coherence time). The most difficult situation arises at 26 GeV/c. The coherence time of the dense single bunches in the SPS is rather short. The dominating effect is the space charge effect. For typical collider bunches we find $\Delta Q_{scH} = 0.012$, $\Delta Q_{scV} = 0.032$, and the coherence time follows : $T_{coh} = T/4\Delta Q$. Putting in the numbers we find a gain G = 0.25 in the vertical plane and G = 0.1 in the horizontal plane. It is of course a coincidence that both gain requirements (instability and injection) lead to nearly identical specifications!

Deflection strength and system acceptance

The acceptance of the electronic processing chain should be well matched to the transverse machine aperture. The maximum amplitude of the coherent signals produced by the position monitor are determined by the maximum possible amplitude of coherent β- oscillations in the machine. These are limited to the half aperture reduced by the beam size (radius). The largest acceptable injection errors in proton machines occur at the highest injection energy where the beam size is minimal. Putting in the numbers for the SPS we arrive at a betatron acceptance of 15 mm (V) and 30 mm (H). This acceptance is shared between closed orbit and coherent β-oscillations. The closed orbit is determined by the state of the machine (set-up, well corrected....) and the operational state (orbit bumps, separator bumps). We lack a clear criterion for the fraction of the aperture to be attributed to β-oscillations which will have to be digested by the feedback. Tentatively we take a fraction of 1/3. From the formula which relates the measured displacement x to the deflection k ($k = Gx/\sqrt{\beta_K \beta_{PU}}$), we find a maximum kick of 20 μrad in V and 13 μrad in H at 26 GeV/c .

We can now summarise the basic specifications for an ideal transverse feedback system : bandwidth of 100 MHz, gains of 0.25 (V) and 0.1(H) and a deflection strength of 20 μrad(V) and 13 μrad(H) at 26 GeV/c .

DESCRIPTION AND PERFORMANCE OF SPS TRANSVERSE FEEDBACK

The SPS feedback is located in long straight section 2. It consists of four independant channels, two in the vertical plane and two in the horizontal plane. The kickers are of the electrostatic type. They are installed one right next to the other. Their relevant parameters are listed in Table II. A set of eight regularly spaced electrostatic pick-up monitors (four in each plane) surrounds the deflectors. It is important to note that both pick-up and deflector are non-directional devices. Hence no polarity change is required between positive particles in one direction and negative particles in the opposite direction. This simplification is very much appreciated in the complex environment of the SPS! By convention we say that the protons are travelling in the positive direction. It should be mentioned that the polarity of the magnetic field is never changed in the SPS.

Table II Deflector and monitor parameters

Deflector/monitor	β	gap	length
H1/H2 V1/V2	76m 45m	142mm 38mm	2.396m 1.536m
PU monitor 4H,4V at π/2	100m		

The layout of one chain is shown schematically in Figure I.

Figure I Typical feedback layout for particles travelling in positive direction

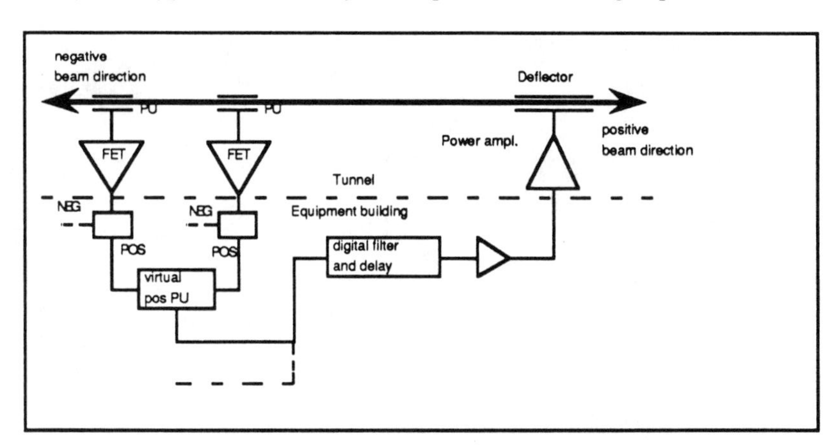

Bandwidth of feedback system

The bandwidth of the system is 10 MHz. The most important limitations in this respect are the electrostatic deflectors. To compensate the missing bandwidth octupoles are used to introduce the necessary non-linearities so that the high intensity fixed target proton beam is stabilised for frequencies higher then 10 MHz[4]. The high impedance low noise FET amplifiers are specially matched to the electrostatic monitors. The usefull bandwidth extends from a few kHz to about 10 MHz. An upgraded version of the amplifiers has a better noise characteristic and a larger bandwidth.

Gain

The gain is as expected, one might say per definition. Indeed, the system is intended to fight the resistive wall instability. This can only be successfull when the minimum gain requirements are met. It may be worthwhile to mention one special feature at this point. The position information as it is taken from the monitor is not normalised for intensity. In other words, the signal is proportional to the product of position and intensity. We have seen that the required gain is proportional to the beam intensity. When this gain is adjusted correctly for a given intensity it is automatically correct for all other intensities. Care has to be taken that the system does not saturate for the highest beam intensities.

Deflection strength

The deflection strength at 26 GeV/c of the kickers is 4.6 μrad in the vertical plane and 2 μrad in the horizontal plane. This is well below the theoretical requirements that were found in the previous chapter. It may be interesting to mention that this *weak* deflection power is produced by 40 kW amplifiers which deliver 3 kV between the kicker plates! Figure II gives a physical idea of the dimensions of the high power part of the system. The lack of kicking strength is partially overcome by using two chains simultaneously in each plane to damp the injection oscillations of the single bunch 26 GeV/c collider beams. The effect of this limitation is that the feedback operates in saturation mode (bang-bang) when the vertical amplitude exceeds 2.5 mm and the horizontal amplitude 3 mm. The SPS feedback can only handle efficiently coherent b-oscillations which stay within 10 to 15 % of the transverse aperture. The operational experience is that this acceptance is too low so that frequent static tuning is necessary to ensure that the injection oscillations remain within the capabilities of the feedback system.

Figure II Deflector tanks and power amplifiers in SPS tunnel

SPECIAL FEATURES OF THE FEEDBACK SYSTEM

As an introduction to this chapter we address the question of the time and β-phase synchronism [5]. To do this we examine a feedback system where the deflector is located downstream of the position monitor. The β-phase difference between the two is $\Delta\mu$. The amplitude of the uncorrected β-motion is A. The feedback calculates a correcting kick with amplitude K. If we take this kickvector as a reference then it can be shown that the amplitude A_c of the corrected β-motion is the result of the following vector sum :

$$A_c = K + A^{j(\Delta\mu + \pi/2)}$$

It is clear that the correcting kick is most efficient for $\Delta\mu=\pi/2$. Next we assume that the processing chain of the feedback has a time delay error dt with respect to the flight time of the beam . The β-motion and the position signal can be decomposed in two waves, the so called fast and slow waves. A time delay error in the signal will increase the phase of one wave and will decrease the phase of the other. This kind of errors will affect the feedback efficiency in the same way as β-phase errors do. The phase error for a constant time error increases linearly with frequency. Hence the effect becomes worse and worse with increasing frequency and should be carefully controlled to avoid anti-damping of high mode numbers.

Virtual position monitor

The SPS feedback is faced with different machine and beam conditions. For a given set of monitors and deflectors it is unlikely that the requirements on the β-phase can be met in all cases. To solve this problem advantage was taken from the fact that the phase advance per cell in the SPS is π/2. A virtual position monitor can be configured by combining the signals of two consecutive monitors. By varying their relative contribution arbitrary b-phases can be produced. The combination of the signals takes place at that point in the processing chain where the time synchronism is correctly established. The way this is done will be the subject of the next paragraph.

Signal delays and time of flight

The deflectors and the position monitors are located close together for practical reasons. This configuration makes it impossible for the system to sample the position of a portion of the beam and act immediately on the same portion when it passes the deflector. The electrical length of the signal path is much longer then the distance that separates the monitor from the kicker. In some configurations we even find the kicker upstream of the monitor! Since the correcting kick cannot be applied in the same turn it is delayed by one turn. This delay does not harm since the instability risetime is 10 turns and more.

The very existence of this delay is the source of the next problem. Indeed, while the proton momentum changes from 14 GeV/c to 450 GeV/c, the particle velocity increases ($v = \beta c \quad, \beta = \sqrt{1-\gamma^{-2}} \quad \gamma = E/E_0$). To keep up with the beam the signal delay should change by 51 nsec during an acceleration cycle. Given the bandwidth of the SPS feedback we require the synchronism to be better then a few ns so that the higher modes (near 10 MHz) remain correctly damped (see higher). Hence the change in particle velocity during the acceleration cycle has to be compensated. This is achieved with a digital delay. The clock frequency (33 MHz) of this delay is synchronous with the machine revolution frequency derived from the rf accelerating frequency (200Mhz). This feature ensures sufficiently accurate time tracking during the acceleration. The use of a digital delay enhances the flexibility of the system considerably (see also below under *Rejection of common mode*). However it has an important drawback. The digitalisation (we use an 8 bit system) introduces discrete (single bit) noise levels which are larger then the analog noise in the system. For this reason it was impossible to leave the feedback active during p-pbar coasts since the continuous perturbation by the feedback blew up the beam causing high background rates and low luminosity lifetime. It was fortunate that this particular kind of beams do not require the feedback for stabilisation since their average intensity is small enough to stay below the Landau damping threshold.

One of the complications in the SPS is the existence of beams in the two directions. They may even be simultaneous($p-\bar{p}$). To facilitate the operation of the complex the signals of the monitors were split into two branches one for the positive beam direction and one for the negative beam direction. For example for a monitor located upstream of the kicker (positive beam direction) it is necessary to add an additional delay in the positive branch with respect to the negative one. This delay is known very accurately and is equal to the time that the beam takes to cover twice the distance from monitor to deflector. For a given optics both the positive branch and negative are permanently connected to the rest of the signal processing chain except for the fixed target proton operation. Indeed with the two *monitors* (positive and negative) active at the same time every bunch passage produces two distinct correcting kicks at two different times. For the scarcely filled machine ($p-\bar{p} \quad e^+ e^-$) the extra kick falls in between bunches and is harmless. It is clear that this situation cannot be tolerated for a full machine!

Rejection of common mode

The feedback system is supposed to act on coherent β-oscillations. The position monitor provides a signal which is proportional to the beam position which contains the β-oscillation together with the closed orbit component. The closed orbit information is processed together with the b-oscillation if no special measures are taken. The closed orbit component (common mode) would produce a constant kick at the deflector,

Figure III Principle of notch filter

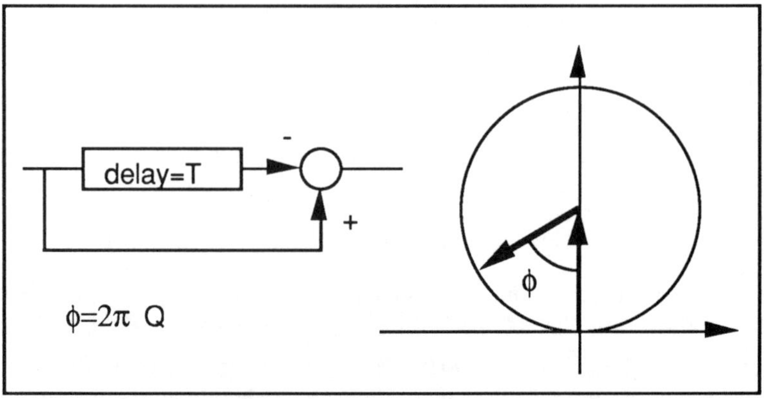

delay=T

-

+

ϕ

$\phi=2\pi$ Q

causing an additional orbit deformation and worse, it could force the system in a permanent saturation state. It is therefore imperative to eliminate this information from the signal. The common mode can be rejected in a very efficient way with a notch filter which is incorporated in the digital delay mentioned earlier. The principle of the notch filter is rather simple. The output is the difference between the position information of two consecutive turns. The principle is shown in Figure III.

Figure IV Damping of injection oscillations

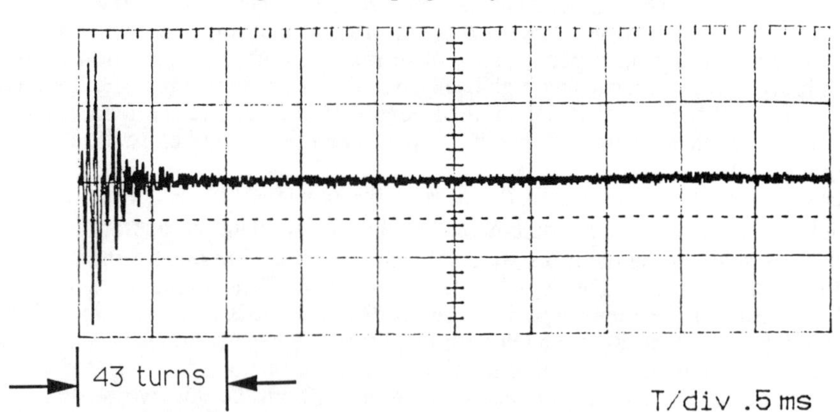

The closed orbit is common to both turns and cancels out in this operation. However the apparent β-phase between PU and kicker changes and this effect should be accounted for in the β-phase balance. This phase change can be computed as follows. The transfer function of the filter in betatron phase is:

$$F = \left(1 - e^{-j2\pi Q}\right) = 2 \sin \pi Q e^{-j\pi\left(Q-\frac{1}{2}\right)}$$

The apparent β-phase change introduced by the notch filter is:

$$\frac{\Delta\phi}{2\pi} = \frac{Q}{2} - 0.25$$

The correction of this effect in the SPS happens to be relatively small for the normal fixed target and lepton optics where the tune is just below the third integer resonance (Q=26.6) so that $\Delta\phi/2\pi \sim 0.05$. The tune for the collider optics is just above the third integer resonance so that $\Delta\phi/2\pi \sim 0.15$ which is not negligable.

In Figure IV we show an example of damped injection oscillations. The measurement is taken after the notch filter.

ACKNOWLEDGMENTS

The SPS transverse feedback system is now in operation for more then 10 years. It has constantly been modified and upgraded to cope with ever changing conditions and requirements. This process is still going on today. Many people have been associated with it now and in the past. Special credit should be given to R. Bossart who designed the original feedback and to R. Louwerse, J. Mourier and J. Pointet who keep the equipment in good shape and implement improvements and modifications in the hardware.

REFERENCES

1. R. Bossart, J-P. Moens, H. Rossi, The new damper of the transverse instabilities of the SPS at high intensities, SPS/ABM/RB/Report 81-1, (1981).
2. R. Bossart, R. Louwerse, J. Mourier, L. Vos, Operation of the transverse feedback system at the CERN-SPS, IEEE Particle Accelerator Conference, 763 (1987).
3. R. Bossart, et al., Tune measurement and control at the CERN-SPS, IEEE Trans. on Nucl. Sci. Vol. NS-26, N 5 (1985).
4. L. Vos, Transverse stability of the high intensity SPS proton beam, European Particle Accelerator Conference, 896 (1988).
5. L. Vos, Transverse stability considerations for the SPS beam in fixed target operation, CERN SPS/86-3 (DI-MST) (1986).

BEAM INSTRUMENTATION IN THE AGS BOOSTER*

R. L. Witkover
AGS Department, Brookhaven National Laboratory
Upton, NY, 11973

Abstract

The AGS Booster was designed to accelerate low intensity (2×10^{10}) polarized protons, high intensity (1.5×10^{13}) protons and heavy ions through Au^{+33}. Coping with this wide range of beams, the 3×10^{-11} Torr vacuum and the radiation environment presented challenges for the beam monitors. Some of the more interesting instrumentation design and performance during the recent Booster proton commissioning will be described.

Description of the Booster

The AGS Booster will: (1) Increase AGS proton intensity by injecting 4 Booster pulses per AGS cycle at 1.5 GeV versus 200 MeV. (2) Raise AGS polarized proton intensity by accumulating 20 Linac pulses each AGS cycle. (3) Accelerate ions through Au^{+33} to AGS injection energy for further acceleration and transfer to RHIC. Figure 1 shows the Booster layout. The 201.8 m circumference has 48 half-cells with 6 missing dipoles. The flexible magnet supply typically follows one of three cycles during an AGS cycle: 4 pulses at 7.5 Hz for protons, a 2.667 sec injection porch with 0.133 sec up-down ramp for polarized proton accumulation and a 1.4 - 2 sec ramp for heavy ions. These can be interleaved freely and special study cycles added within these. The RF duration ranges from 60 msec for protons (2.5 - 4.1 MHz) to 620 msec for Au^{+33} (0.852 - 3.06 MHz). Protons are accelerated on the 3rd harmonic but ions will begin on the 12th and switch to the 3rd. Protons from the Linac are chopped at the Booster RF frequency with the width varied during injection to optimally paint the bucket. There will be four bucket-to-bucket transfers to the AGS for high intensity protons, but be only 1 for heavy ions or polarized protons.

Booster beam intensity is measured using a fast current transformer for injection and capture, with a lower frequency transformer for the remainder of the cycle. Bunch longitudinal density is observed using wall

---

*Work performed under the auspices of the U.S. Department of Energy.

current monitors. Beam loss is monitored in each half-cell by a 4-m long coaxial ion chamber described in the last Instrumentation Workshop.[1] An IPM (ionization profile monitor) is used to measure beam profile. It uses two layers of micro-channel plates[2] to increase the signal in the 10^{-11} Torr vacuum. It is described in detail in another paper at this Workshop.

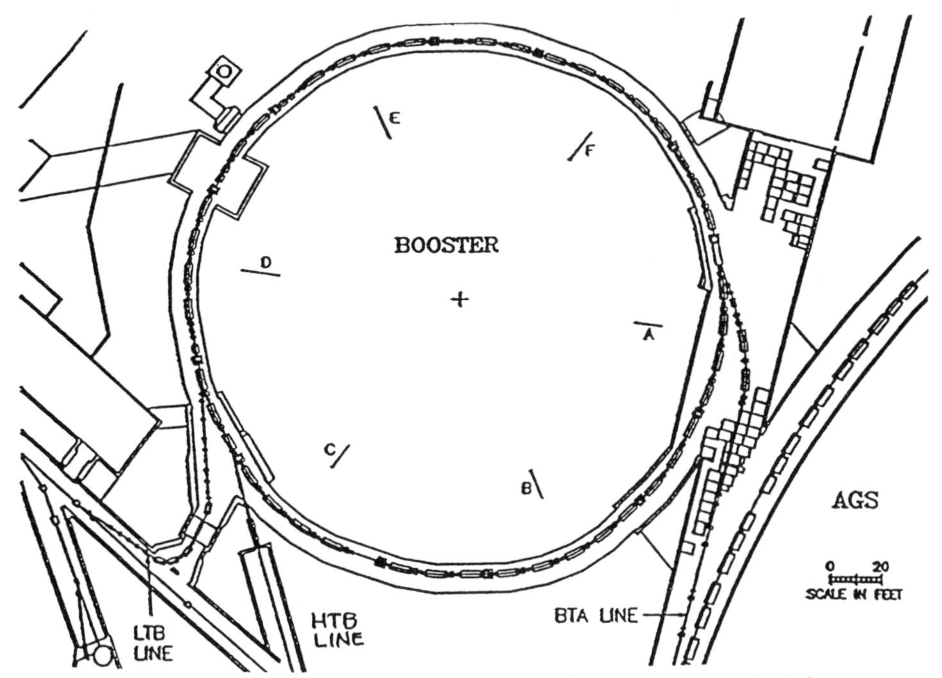

Fig. 1. Layout of the Booster and its transport lines.

High intensity and polarized H^- beams are injected via the LTB line, a branch off the line to the AGS. HARP multiwire profile monitors[3] located upstream of the 126° bend and just before the Booster are used to measure emittance. Beam current is measured at the ends of the LTB line by conventional beam transformers with 10 and 100 μA/V and 10 mA/V ranges, 70 nsec risetime (low gain) and 1% /msec droop. Eight 4-m long coaxial ion chambers measure radiation and stop the Linac beam for high losses. Stripline couplers[4] with wideband electronics measure beam position at 7 locations.

Ion beams from H^+ to Au^{+33} are transported from the Tandem Van de Graaff to Booster in the HTB line, an extension of HITL.[5] Eight HARPs and 3 beam current transformers of the type used in LTB and HITL monitor the beam. Faraday Cups at each HARP location measure current from 1 μA to several hundred μA in the pulsed beam down to the nano-Amp DC beam used for tuning HTB. This line now being commissioned.

Beam is extracted to the AGS in the BTA line. HARPs provide profile data at 4 locations. Total charge in the 3 bunches is measured at 3 locations in the BTA line by wideband current transformers with fast integrator electronics. Position is monitored at 6 locations using the same PUEs as in the Booster Ring. Eight fixed location 7-m long coaxial loss monitors measure beam loss aided by 8 relocatable detectors.

The Design Problems

Most of the instrumentation was designed to operate from $<2 \times 10^{10}$ to $>1.5 \times 10^{13}$ charges per cycle with 2 decades of resolution. Some electronics had to be in the tunnel to provide the bandwidth and low noise required, exposing them to radiation from high intensity protons which can be inter-leaved with low intensity beams. For 4800 hours of proton operation (an over estimate) the typical dose at floor level was estimated[6] to be 5.4 kRad and 3.4×10^{12} n/cm^2. The fiber optic links used for the Ring PUE signals were tested in the AGS to over 50 kRad, similar to 5 years in the Booster. At hot locations, such as the internal beam dump, shielding is required. High dose radiation monitors (TLDs) were placed with the electronics to provide data for failure analysis. Higher radiation at beam height precludes materials such as PTFE for cables or connectors. Since PTFE is used in most high temperature cables, bakeout of the Ring vacuum system at 300°C presented problems. The 3×10^{-11} Torr Ring vacuum requirement limited detector materials to ceramic and stainless steel. The PUEs had to retain tight tolerances even after vacuum firing at 950°C. Some of the solutions to these challenges will be described.

The Loss Monitor System

An array of 64 fixed-position and 16 relocatable coaxial ion chambers measure beam loss. The detectors are made from 7/8-inch diameter air dielectric coaxial cable (Andrew type HJ5-50), closed at the ends and pressurized with Argon to 10 PSIG. A 200 V DC bias is applied to the center conductor of each detector by its own floating DC-to-DC converter, allowing a single signal coax for both bias and signal and response down to DC. The output from each detector is integrated by a low leakage circuit (<10 pA), digitized and put into dual ported memory at a 140 kilo-word rate by a scanning ADC. The system is fully described in Reference 1.

The loss monitors were an important tool in the Booster commissioning. Typical current in the LTB line was 25 mA, but losses from a 1.5 mA, 25 μsec beam were easily seen. Most of the Ring commissioning used only a single injected turn (1.2 μsec) for a nominal intensity of 2 x 10^{10} protons, the lower design limit. Offset and noise in the high gain state was typically 10 counts of a maximum of 2047. Figure 2 shows the integrator output for an interesting loss in LTB. The rapid (1-2 μsec) rise at the edges of the pulse were explained when the LTB BPMs showed large position excursions at the same time (Figure 3). These are believed to be beam swept by the chopper which survived the Linac and HEBT. The pulse-to-pulse losses were displayed as a bar graph in which any 25 selected units or groups can be displayed.

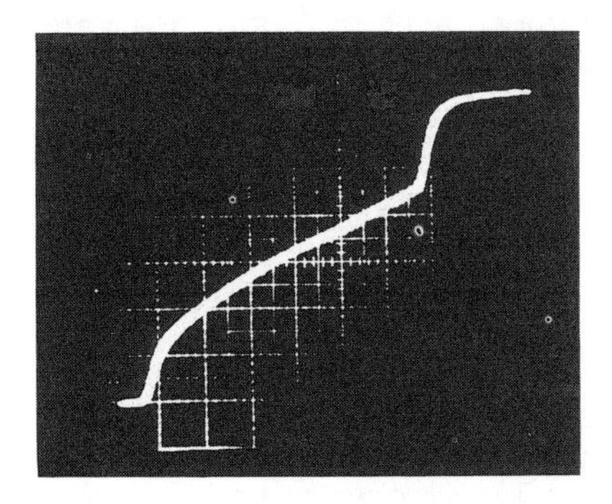

Fig. 2. Integrated LTB Beam Loss Monitor Output. 200 mV/Div., 5 μsec/Div.

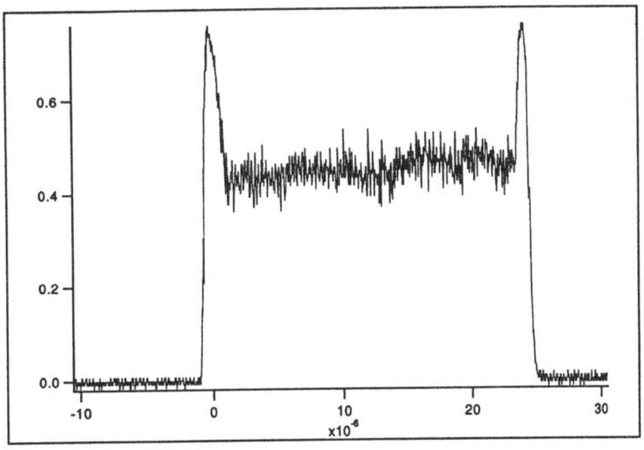

Fig. 3. Corresponding LTB Beam Position Monitor Output. Y-axis in mm, X-axis in μsec.

Ring Beam Current Transformers

Beam current in the Ring can be less than 100 μA at injection to 2.9 A at full intensity. To meet the time response requirement of < 1 μsec to observe stacking, to almost 3 sec for accumulating polarized beam, an injection beam monitor (BIBM) and a separate circulating beam monitor (BCBM) were installed, each covering the full intensity range.

The BIBM is a conventional beam current toroid[7] with a sensitivity of 0.1 V/A into 50 Ohms and bandwidth from 2 Hz to 8 MHz, enclosed in a 170 mm ID aluminum case. The electronics, located at floor level, are similar to those in LTB and HTB, but with gains of 1 mA/V, 10 mA/V and 1 A/V. The rise time is 70 nsec for the high and 1 μsec for the low intensity range.

The BCBM is a current transformer (DCCT) employing magnetic modulation at 6.928 kHz with synchronous 2nd harmonic detection[8] to provide DC to 15 kHz bandwidth. This commercial unit[9] uses specially annealed metal glass tape cores. The BCBM is 170 mm ID and 62 mm long including magnetic shielding. The Front End Electronics (FEE) must be within 3 m of the detector but the Back End Electronics (BEE) can be up to 300 m away. The standard ranges of 1 A/V and 10 mA/V produced only 1 mV for a 10 μA beam, so a relay switched gain of 10 amplifier was added after the FEE. DCCTs claim 1 μA resolution when viewed with a 1 sec integration window during which modulation, Barkhausen and environmental noise average out. This is applicable for storage ring beams but not for pulsed beams. Two of 3 samples showed modulation noise equivalent to ± 150 μA, consisting of spikes and components at the fundamental and odd harmonics. A filter with -40 dB notches at 6.928 and 20.782 kHz reduced this to ± 10 μA of 5th harmonic. This was removed by a 4th order 21 kHz Butterworth low pass leaving about 10 μA pp random noise. The third unit showed much less initial noise and required only the low pass filter.

The two transformers are housed in a 1.27 cm thick 1006 steel magnetic shield mounted on vibration isolators. A water cooled copper sheet keeps the detector temperature under 80°C during 300°C bakeout. A ceramic vacuum pipe section forces the wall current to a lower impedance path outside the transformers. The FEE and BIBM circuits are located near the floor in a steel electrical box with steel conduit shielding the cables to the detectors. An injection kicker, which is pulsed to 500 A, located 30 cm away, produced about 100 μA pickup on the BCBM and somewhat more on the BIBM. Additional shielding will be provided. No RF pickup was observed up to 3.5×10^{12} but similar units in the AGS showed some above 5×10^{12}. Figure 4 shows the BCBM output for a single turn of injected beam with 1.65×10^{10} protons accelerated to 1.2 GeV then decelerated down to 200 MeV. The lower trace is the same signal "normalized" to remove the velocity dependence.

LTB Beam Position Monitors

The 7 dual plane stripline BPMs the LTB line were designed for use with the high intensity H^- beam, but may be extended for polarized beam by redesigning and

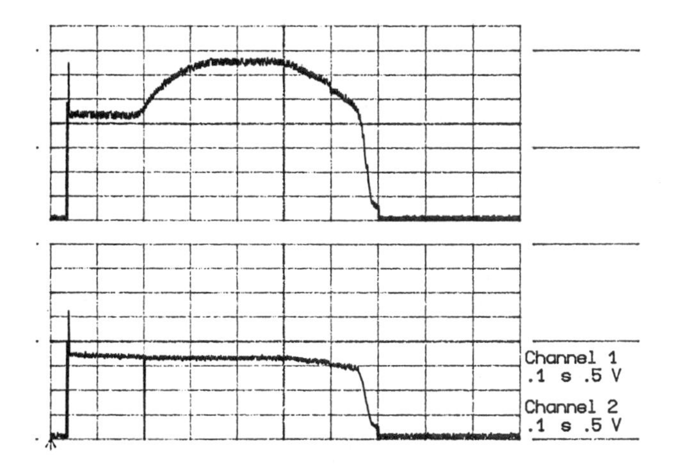

Fig. 4. Channel 1: raw BCBM output during Booster acceleration and deceleration. 0.5 mA/Div., 0.1 sec/Div.; Channel 2: same normalized to give charge. $5x10^9$/Div., 0.1 sec/Div.

resonating the striplines[10] and adding input preamplifiers. The H and V planes were displaced longitudinally to reduce coupling, which can be important if the strips are resonated. To save space the detectors were designed for twice the RF frequency, 402.5 MHz, cutting the length almost in half with a loss of only 5% in amplitude. This also moved the frequency away from the Linac RF, reducing the background noise.

The signals are AM/PM processed[11] directly at 402.5 MHz without down conversion.[12] This increased the bandwidth but made the limiter stage which remove the amplitude dependence of the phase modulated signal very difficult. By selecting the Plessey 532C limiters in pairs matched better than 0.18^o, the 4 stage limiter provides ± 0.5 mm precision over a 26 dB dynamic range as measured with beam. The position bandwidth was found to be 40 MHz in bench tests. During early Booster commissioning the BPMs appeared to give reasonable results with the beam nearly centered in the LTB line. <u>Figure 3</u> shows the horizontal position after the big bend in LTB with the "horns" described above. Later in the commissioning, when the beam was run 20-30 mm off center to optimize Booster injection, the computed position appeared to saturate. Position calculated from scope traces did not show the apparent saturation. Later tests indicated that while small errors in calibration did exist, the primary cause was probable mis-timing of the data acquisition.

Booster Ring Beam Position Monitors

Ring beam position is measured using split plate, capacitive pick up electrodes (PUEs)[13] to provide low frequency response for the heavy ion bunches. Measurements have shown the linearity to be <± 0.1 mm over ± 30 mm of the 75 mm radius.[14] A 20-cm long single plane PUE is located at the quadrupole (beta-max) in almost every half-cell and at six places in the BTA line to the AGS. The single turn trajectory or the average orbit can be acquired by varying the number of bunches integrated. The electronics are separated from the PUE by 10 feet of cable to maximize the voltage, minimize the noise and extend the low frequency response. All timing, commands, digital data and analog signals are carried over fiber optic links to prevent ground loops. Data can be read from all PUEs every 10 msec.

To meet the vacuum specification the complete assembly had to be vacuum fired at 950°C but maintain the mechanical stability to within ±0.1 mm of the original, as confirmed by measurements. The resulting double gimbal support (Figure 5) also provided a means of

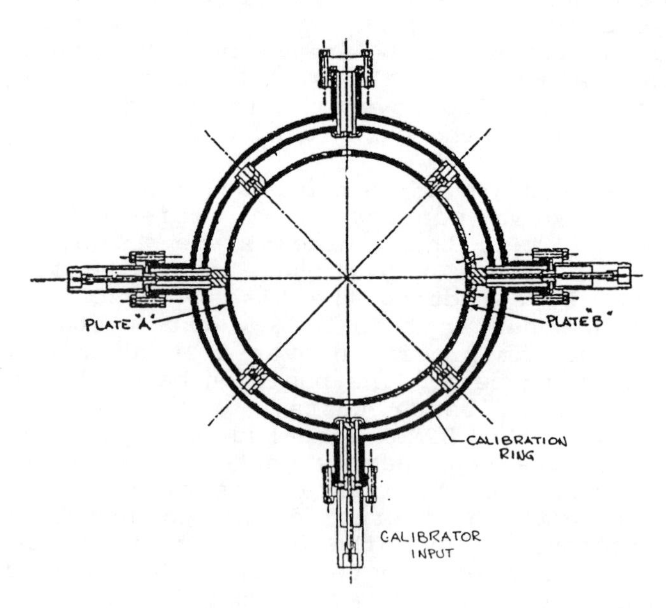

Fig. 5. Booster PUE showing double Gimbal support.

coupling calibration signals to the electrodes. The 316N stainless steel cylindrical electrodes are locked together at two end points with ceramic posts. Four radial ceramic posts fastened to the plates pass through close-fit holes in the calibration ring surrounding the electrodes. Four additional holes, closely fitted to ceramic sleeves on the vacuum feedthroughs, support the calibration ring from the vacuum shell. The electrodes are free to expand and contract axially while sliding radially along the ceramic posts.

Cables to withstand the 300°C bakeout are usually made of PTFE, which would soon fail in the expected radiation. A thermal isolator was designed which dropped the temperature to 60°C allowing standard polyethylene RG-62 cable to be used. The isolator consisted of a 15 cm stainless steel type-N adapter with a stainless inner conductor supported by ceramic disks. For additional RF shielding the cable was run in copper tubing.

The electronics cover 3 decades of intensity with resolution of ± 0.1 mm. They were designed for bunch lengths from 3750 to 50 nsec as the RF went from 0.21 - 4.1 MHz using 3rd harmonic acceleration. Now heavy ions will start on the 12th harmonic (0.84 MHz RF) and later switch to the 3rd. A block diagram of the electronics is shown in Figure 6. Beam synchronized timing is brought in on a fiber-optic link. Division of the difference by the sum is done in the high level application code.

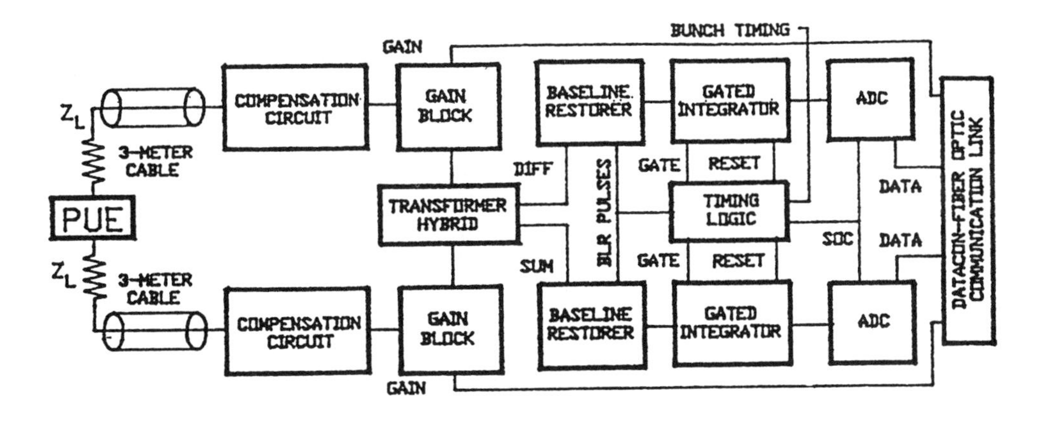

Fig. 6. Block diagram of PUE local electronics.

The Front End Board (FEB) provides gain and takes the sum and difference of the bunch signals. The Compensation Circuit provides a match at high frequency but high impedance at low frequency, flattening the response to ± 1.5 dB from 10 kHz to 30 MHz. An AD9610 low noise amplifier provides gains of 0.1, 1 and 10. A 4-port trifilar transmission line transformer with more than 60 dB common mode performs the sum and difference. Overall, the FEB provides a difference signal of -50 dB with respect to the sum from 50 kHz to 20 MHz, with most reaching -55 dB. Either the amplified single plate or sum and difference bunch signals can be sent on analog fiber optic links[15] from the Booster tunnel. These analog links have a bandwidth of 6 Hz to 35 MHz and linearity better than 3% over a 26 dB dynamic range.

The Acquisition Board performs the baseline restoration, integration and digitization of the bunch data. The BLR sets the zero of the AC-coupled signals by using a balanced diode bridge (HP 5082-2813) to ground the line between bunches. The Gated Integrator accepts the desired number of bunches (1 - 240), integrating and holding the data until read (50 μsec) by an on-board 16-bit ADC. The digital data goes directly to the controls interface in each crate.

The 60 nsec BLR pulses must fall between bunches within a 20 nsec error over the RF range. The equally spaced PUEs require BLR pulses at (48/harmonic) phases, which must ripple around the Ring to measure the single turn trajectory. BLR triggers and integrator gate are generated by the Bunch Timing Sequencer. The Low Level RF (LLRF) system generates a signal, phase-locked to the beam, at 48 times the revolution frequency using a direct digital synthesizer (Stanford Telecommunications Inc., Santa Clara CA, Model STEL 9273). These pulses clock a 48-bit ECL shift register to make the evenly phased triggers. The gate time is encoded by counting the RF frequency for the desired number of bunches, starting at the desired time. These are the data clocked through the shift register. This allows each PUE to get all its timing information on a single fiber optic link, avoiding high power line-compensated drivers while providing ground isolation.

Hewlett-packard HFBE-1404 GaAlAs transmitters drove the signals down 200 m of 200 μm Ensign Bickford HCR step index silica cable to HFBR-2404 GaAlAs photo-diode receivers. Tests in the AGS[16] showed < 2 dB loss for the receiver and < 1 dB loss for the transmitter for 28 kRad(Si). Attenuation for cables exposed to this flux was under 0.1 dB/meter.

All of the electronics crates were installed in the tunnel for commissioning, but only 34 of 46 were able to provide data through the computer. Phase locking the Timing Sequencer to the beam through the LLRF was more difficult than expected and in the beam studies time available, was achieved only for a single frequency. A beam bunch simulator has since been built and is now being used to synchronized the Sequencer over the full frequency range. Another problem first observed during commissioning was noise from the BLR pulses coupling to the FEB, which, at the low intensity (2×10^{10}) used for most of the studies, made the orbit data meaningless. As the intensity was raised, the sum signal began to behave reasonably but no orbit could be seen until the intensity was over 10^{11}. Following the beam runs the major source of the noise was found and steps are being taken to eliminate it. The amplified bunch signals sent on the analog fiber optic lines were quite clean even at 2×10^{10} (if the BLR pulses were turned off) and were heavily used during the commissioning. Figure 7 shows a typical signal with some BLR noise.

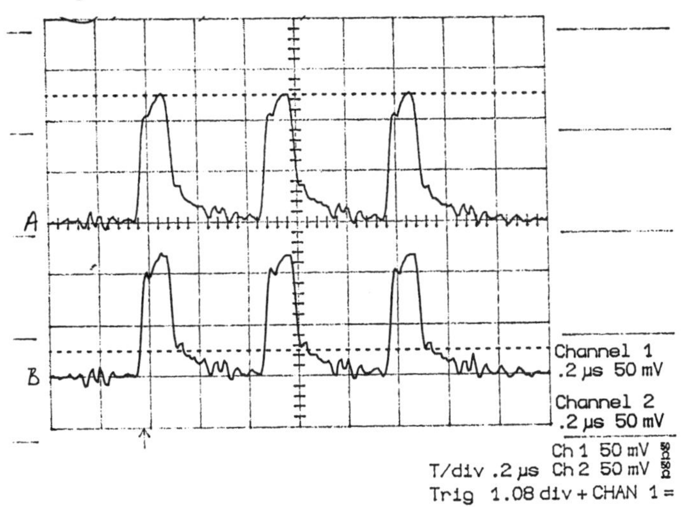

Fig. 7. Bunch signals from "A" and "B" plates of horizontal PUE at C6 (H^- injection). Single turn injected (3 bunches). Total charge = 2×10^{10}.

The Ring Profile Monitor

Flying wires were rejected as the Booster profile monitor because partially stripped heavy ions would be further stripped and lost. A residual gas ionization profile monitor (IPM) would work with heavy ions but even with enhancement by micro channel plates (MCPs) there are too few interactions in 1 msec by $2x10^{10}$ polarized protons at $3x10^{-11}$ Torr to be meaningful.[17] Since the Booster will accumulate 20 such pulses it would be useful through most of the cycle. An IPM has been built for the Booster.[18]

IPMs which used MCPs did not have to operate at the high vacuum or over the intensity range of the Booster. Krider[19] used a 2 layer MCP with a 1 msec integration time to observe 1 μA anti-protons in a $5\text{-}10x10^{-8}$ Torr vacuum. This produced a number of ions similar to that of the accumulated polarized beam. MCP degradation with increasing charge output was observed by Krider and by Kawakubo, et al.[20] In the AGS Booster an UV light will be used to track the calibration of the MCPs. To extend the life of the MCPs the controls adjust the bias voltage to limit anode current. The bias voltage will be rapidly gated on only during measurements.

Because of the high vacuum, the resistive divider normally used to produce a uniform electric field was unsuitable so specially contoured stainless steel electrodes were designed.[21] A POISSON equipotential plot is shown in Figure 8. A third set of electrodes mounted at

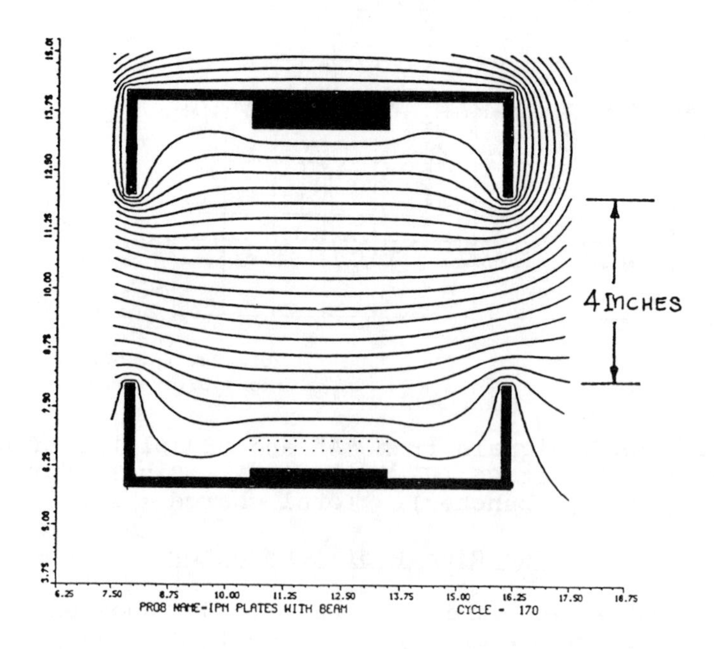

Fig. 8. POISSON plot of IPM equipotentials.

45°, is used to compensate the effect of the other two on the beam. The electrodes are 2 inches deep and 8 inches square with a 4 inch gap. The MCP array sits in an opening in the grounded electrode. Each detector plane consists of a 2 layer MCP capable of 10^7 gain, with an integral 64 channel anode array. The collectors are 1.1 mm wide on a 1.47 mm pitch, 75 mm long.[22] Further details of the IPM are presented in another paper at this conference.[23] Everything but the detector was vacuum fired to 950^oC. The entire IPM was baked to 300^oC.

The VME-based electronics for the IPM scans both 64 channel planes into memory in less than 1 msec. Each anode is connected to a low leakage (< 10 pA) gated integrator[24] followed by a Sample and Hold (Datel DVME-645) allowing the integrator to be reset and started again while the data is being processed. A microprocessor controlled 12-bit ADC VME board (Datel DVME-601) digitizes the voltages and puts the data into dual port memory at a 140 kHz rate. A bus translator (Bit-3 Corp.) maps the dual port RAM into the address space of the remotely located control computer. During commissioning the IPM was controlled by a "286 AT-Clone" but it will be connected to the AGS Apollo based control system.

The IPM MCP bias was on for about 30 minutes during Booster commissioning, during which beam profiles were obtained. Beam time was much less since the duty factor was under 10%. No obvious degradation was observed although the plates have not been re-calibrated. Figure 9 shows the raw data for a pair of profiles taken 64.5 msec into the injection flat-top. The proton intensity was $8x10^{10}$ at a vacuum of $2x10^{-10}$ Torr, 4 times the minimum intensity and over 6 times the minimum pressure to be expected. The signals are very clean except for some repeatable noise following the horizontal profile, which could have been subtracted from the baseline. The peak counts were somewhat over half scale but the MCP could have provided another 50-100 times more signal. Up to 16 scans were made per Booster cycle. While there were no failures during commissioning, the IPM was not completely operational, using temporary timing and roughly calibrated voltage readbacks. Being off the main control network prevented access to the database so there was no on-line momentum, vacuum or intensity information. Testing time was limited because the bias voltage pulser was not installed.

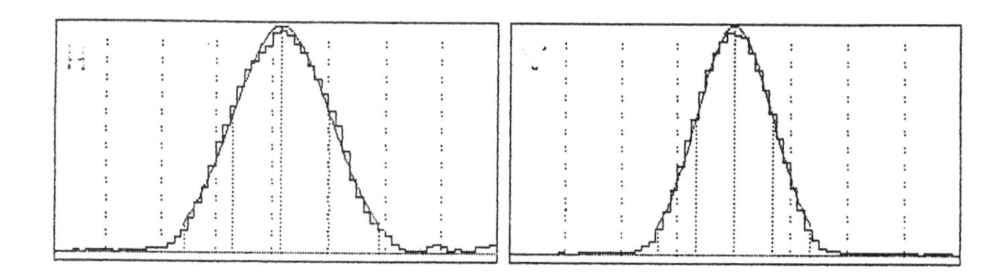

Fig. 9. IPM generated profiles for $8x10^{10}$ protons, at 200 MeV, $2x10^{-10}$ Torr, 1 msec integration. Max. horizontal signal = 5.02 V, max. vertical signal = 6.39 V.

Acknowledgments

While the entire AGS Instrumentation Group contributed to the construction of the Booster, certain individuals were responsible for the specific work reported here. R. Witkover, E. Zitvogel and S. Naase worked on the Ring beam current transformers, The LTB BPMs were designed by R. Bossart of CERN and J. Brodowski of BNL. The circuits were by T. Shea, V. LoDestro and C. Degen. The Ring PUE was designed by W. Van Zwienen, and tested by R. Thomas. The PUE electronics were designed by D. Ciardullo and G. Smith of BNL with help from E. Schulte of CERN. E. Beadle did the fiber optic work. A. Stillman was responsible for the IPM, with R. Thern working on the AT computer interface.

REFERENCES

1 E. R. Beadle and G. W. Bennett, "The AGS Booster Radiation Loss Monitor System", Proceedings of the Second Annual Workshop on Accelerator Instrumentation, Batavia, IL, p 35-47. AIP Conf. Proc. 229, American Institute of Physics, NY, 1990, p 35-47.

2 A. N. Stillman, R. E. Thern, R. L. Witkover and W.H. Van Zwienen, "Design of the AGS Booster Ionization Profile Monitor", Proceedings of the 1991 Particle Accel. Conf., San Fransisco, CA.

3 Model DL-150BR/DG-01.5BR, Manufactured by both: NTG, Im Steinigen-Graben 12-14, D-6460 Gelnhausen-Hailer, W. Germany and PET Ing.-Buro, D-6100 Darmstadt, W. Germany

4 T. J. Shea, C. M. Degen, D. M. Gassner and V. LoDestro, "A Beam Position Monitoring System for Brookhaven's Linac to Booster line", Proceedings of the Second Annual Workshop on Accelerator Instrumentation, Batavia, IL, 1990, p. 273-9, AIP Conf. Proc. 229, American Institute of Physics, NY, NY, 1991.

5 R. L. Witkover, et al., "Beam Instrumentation for the BNL Heavy Ion Transfer Line", IEEE Cat. No. 87CH2387-9, Wash., DC, 567 (1987).

6 A. Stevens, private communication.

7 Manufactured by Ion Physics Corporation, 323 Andover Street, Wilmington, MA, 01887.

8 K. B. Unser, "A Toroidal DC Current Transformer with High Resolution", IEEE Trans. Nucl. Sci., NS-28, No. 3, 2344 (1981).

9 Manufactured by BERGOZ, Crozet, 01170 GEX, France.

10 R. Bossart, "Resonant Beam Position Monitor for Low Beam Intensity", CERN SPS/88-4 (ABM), December 1987.

11 S. P. Jachim, R. C. Webber, R. E. Shafer, "RF Beam Position Measurement Module for the Fermilab Energy Doubler", IEEE Trans. Nucl. Sci., NS-28, No. 3, 2323 (1981).

12 F. D. Wells and S. J. Jachim, "A Technique for Improving the Accuracy and Dynamic Range of Beam Position Detection Equipment", IEEE Cat. No. 89CH2669-0, Chicago, IL, 1595 (1989).

13 D. J. Ciardullo et al, "The AGS Booster Beam Position Monitor System", Proceedings 1991 Particle Accel.

Conf., San Fransisco, CA., to be published.

[14] R. Thomas, et al., "Design and Tesing of the AGS Booster BPM Detector", Proceedings of the 1991 Particle Accel. Conf., San Fransisco, CA, to be published.

[15] Meret Inc., 1815 24th Street, Santa Monica, CA, 90404. Model No. MDL288TV.

[16] E. Beadle, private communication.

[17] R. L. Witkover, private communication.

[18] A. N. Stillman, R. E. Thern, R. L. Witkover and W. H. Van Zwienen, "Design of the AGS Booster Ionization Profile Monitor", Proc. 1991 Part. Accel. Conf., San Fransisco, CA, to be published.

[19] J. Krider, "Residual Gas Beam Profile Monitor", Nucl. Instrum. Meth., A278 (3), 660 (1989).

[20] T. Kawakubo, T. Ishida, E. Kadokuro, Y. Ajima, T. Adachi, "Fast Data Acquisition System of a Non-Destructive Profile Monitor for a Synchrotron Beam by Using Micro Channel Plate with Multi-Anodes", to be published in Nucl. Instrum. Meth.

[21] E. Leal-Quiros and M. A. Prelas, "New Tilted Poles Wein Filter with Enhanced Performance", Rev. Sci. Instrum., 60, No. 3, 350 (1989).

[22] Galileo Mosel 3810 with 64 element anode. Manufactured by Galileo Electro-Optics Corp., Galileo Park, PO Box 550, Sturbridge, MA 01566.

[23] A. N. Stillman, R. E. Thern, R. L. Witkover and W. H. Van Zwienen, "Design and Operation of the AGS Booster Ionization Profile Monitor", these proceedings.

[24] Manufactured by Advanced Technology Laboratories, Inc., 1111 Street Road, Southampton, PA 18966.

PRECISION BEAM ENERGY MEASUREMENT AT CEBAF USING SYNCHROTRON RADIATION DETECTORS

B. Bevins
CEBAF, 12000 Jefferson Ave., Newport News, VA 23606

ABSTRACT

Continuous measurement of the mean energy of the extracted electron beam is required with a precision of 10^{-4}. It is proposed to achieve this precision by measuring the beam bend angle through a dipole magnet. The angle will be measured by magnetically kicking the beam in a direction perpendicular to the plane of the main bend just before and just after the beam passes through the dipole. The narrow beams of synchrotron radiation (SR) emitted in these kicks will precisely indicate the angle through which the beam has been bent. With sufficient care in the design and construction of the SR detectors and in the measurement of the magnetic field integral, it will be possible to measure relative energy deviations at the 10^{-4} level and the absolute energy at nearly this precision.

INTRODUCTION

The need for precise beam energy measurement encompasses questions of both the absolute energy and relative energy deviations. While an accurate absolute measurement is also desirable, it is the relative measurement that is required with the highest precision. In particular, it must be possible to reproduce a previously obtained energy setting to a very high accuracy. Factors that will affect the precision of both measurements are discussed along with the design considerations that result.

There are two distinct components of this measurement, the determination of the bend angle and the determination of the magnetic field integral of the measuring dipole. The field integral measurement will be accomplished by comparing several high precision techniques, including integration of nuclear magnetic resonance (NMR) probe data and measurement of the EMF induced in a long coil rotating within the field and a wire translating transversely through the field. Field integral measurements to precisions better than 10^{-4} have been achieved at the Stanford Linear Collider (SLC) using such methods.[1]

The angle is to be determined by detecting the synchrotron radiation emitted as the beam passes through kicker magnets placed just before and just after the measuring dipole and oriented perpendicular to it. The SR beams will intersect two phosphor targets along each arm placed several meters downline of the magnets and produce a glow that will be observed by video cameras mounted out of the path of the radiation. While the apparatus may be oriented at any angle around the longitudinal beamline axis, it seems most convenient to orient the large dipoles so as to produce horizontal deflections and the kickers so as to produce vertical deflections. See Figure 1.

MEASUREMENT UNCERTAINTY ANALYSIS

The fundamental governing equation is:

$$B\rho = \frac{p}{e},$$

where B is the magnetic field intensity, ρ is the radius of curvature of the path of the electron, p is the electron momentum, and e is the electron charge.[2]

A more useful form for this analysis is:

$$p = \frac{e}{\Theta} \int B dl,$$

where $\int B dl$ is the magnetic field integral over the path of the electron and Θ is the angle through which the electron is deflected. This leads to the uncertainty relation:

$$\frac{\Delta p}{p} = \sqrt{\left(\frac{\Delta \int B dl}{\int B dl}\right)^2 + \left(\frac{\Delta \Theta}{\Theta}\right)^2}$$

for the momentum.

Θ is determined from the detector parameters as follows:

$$\Theta = \sin^{-1}\left(\frac{D}{S}\right),$$

where

$$D = d_2 \sin \alpha_2 - d_1 \sin \alpha_1,$$

and the other parameters are defined as in Figure 2. α_1 and α_2 will nominally be right angles. They are included in the formula in order to allow the uncertainty analysis to take angular misalignment of the targets into account. They are shown exaggerated in the figures.

The total uncertainty in the angular measurement is given by:

$$\Delta \Theta = \sqrt{\frac{\left(\frac{\Delta D}{S}\right)^2 + \left(\frac{D \Delta S}{S^2}\right)^2}{1 - \left(\frac{D}{S}\right)^2}},$$

where

$$\Delta D = \sqrt{\sin^2 \alpha_2 (\Delta d_2)^2 + d_2{}^2 \cos^2 \alpha_2 (\Delta \alpha_2)^2 + \sin^2 \alpha_1 (\Delta d_1)^2 + d_1{}^2 \cos^2 \alpha_1 (\Delta \alpha_1)^2}.$$

These formulas are derived in the Appendix.

Values of

$d_1 = 30$ cm
$d_2 = 50$ cm
$\Delta d_1 = \Delta d_2 = 10\,\mu$m
$S = 2$ m
$\Delta S = 0.1$ mm
$\alpha_1 = \alpha_2 = \pi/2$
$\Delta \alpha_1 = \Delta \alpha_2 = 0.01$,

which would be typical for a nominal 4 GeV beam and 1 Tesla-m nominal field integral, give an absolute angular uncertainty of $\Delta\Theta/\Theta = 8.7 \times 10^{-5}$ and a relative angular uncertainty of $\Delta\Theta/\Theta = 7.1 \times 10^{-5}$. The relative calculation uses the same formulas but neglects the survey and alignment errors by setting $\Delta S = \Delta\alpha_1 = \Delta\alpha_2 = 0$, since the apparatus can be assumed to be stationary for short intervals. Presupposing a field integral measurement of approximately equal precision will allow a momentum determination to better than 10^{-4}.

DETECTOR GEOMETRY

The total length of the apparatus is about 6 m. This distance is required both to implement a full wiggler that produces no net effect on the beam trajectory and to allow sufficient path length for the SR beams to make an accurate angular measurement.

A full wiggler arrangement of the dipole magnets is proposed to make the detector transparent to the beam, that is, to allow the beam to emerge from the detector along the same trajectory it would have taken had the detector not been present. While such an arrangement requires at least three magnets, only the more powerful central magnet needs to be precisely characterized.

The synchrotron radiation detectors are required to achieve a spatial resolution sufficiently fine to do a 10^{-4} measurement in the limited space available for the detector. It is expected that available video cameras will allow $10\,\mu$m resolution. This resolution, together with a travel distance of about 2 m between detectors will allow sufficiently accurate angular measurement.

A nominal bend angle of at least $\Theta = 0.1$ rad is desirable. At 4 GeV this requires a field integral for the main dipole of approximately $\int B dl = 1$ Tesla-m. Because magnets are limited to about 1 Tesla due to emittance considerations, the main dipole must be about 1 m in length. A full wiggler thus requires 2 m of magnet.

The vertical kickers must be kept as compact as possible to avoid interfering with the SR beams. However, they must be sufficiently powerful to create a transverse SR stripe about 1 cm in length at the detectors, approximately 1 m downline. This implies a bend angle of about 0.01 rad, so $\int B dl$ for the kickers must be around 0.1 Tesla-m. A field of 0.4 Tesla would allow the kickers to be limited to 25 cm apiece in length. Four such magnets are necessary to make the detector transparent to the beam, so about 1 m of kicker magnet is required.

Increasing the distance between the detectors along each SR beam improves the precision of the absolute measurement, but does little to further improve the relative measurement. The value of such extension must be weighed against the use of very limited space in the accelerator enclosure.

CONCLUSIONS

The requirement of mean energy measurement with a precision of 10^{-4} can be met by a monitor using synchrotron radiation as described. The two most critical aspects of the monitor will be the spatial resolution of the SR detectors and the accuracy of the magnetic field measurement. Careful design should allow both of these parameters to fall well within the required values.

ACKNOWLEDGEMENTS

Grateful acknowledgement is made to P. Kloeppel and R. Rossmanith for their assistance in carrying out this analysis and to M. Fripp for assistance in preparing this manuscript.

APPENDIX

The uncertainty analysis for the angular measurement is carried out as a Taylor series expansion to first order. That is, measurement errors (uncertainties) are assumed to be propagated by the relation:

$$F = f(x) \quad \Longrightarrow \quad \Delta F = \Delta x \frac{\delta f}{\delta x}.$$

The total error in a derived quantity is assumed to be equal to the sum in quadrature of the partial errors in the derived quantity due to the error in each quantity from which the new quantity is derived. Applying these rules to

$$D = d_2 \sin\alpha_2 - d_1 \sin\alpha_1$$

gives

$$\Delta D = \sqrt{\sin^2\alpha_2(\Delta d_2)^2 + {d_2}^2\cos^2\alpha_2(\Delta\alpha_2)^2 + \sin^2\alpha_1(\Delta d_1)^2 + {d_1}^2\cos^2\alpha_1(\Delta\alpha_1)^2}.$$

Similarly, letting $T = D/S$ gives

$$\Delta T = \sqrt{\left(\frac{\Delta D}{S}\right)^2 + \left(\frac{D\Delta S}{S^2}\right)^2},$$

and since $\Theta = \sin^{-1}(T)$, it follows that

$$\Delta\Theta = \left(\frac{\Delta T}{\sqrt{1-T^2}}\right).$$

Therefore

$$\Delta\Theta = \sqrt{\frac{\left(\frac{\Delta D}{S}\right)^2 + \left(\frac{D\Delta S}{S^2}\right)^2}{1-\left(\frac{D}{S}\right)^2}}.$$

REFERENCES

1. Watson, S. *et al.*, *Precision Measurements of the SLC Reference Magnets*, SLAC-PUB-4908, March 1989.
2. Wollnik, Hermann, *Optics of Charged Particles*, p. 31, Academic Press, Inc., Orlando, Florida, 1987.

The use of Digital Signal Processors in LEP beam instrumentation

P.Castro, L.Knudsen, R.Schmidt
CERN-SL, GENEVA, Switzerland

1 Abstract

At LEP, DSPs are used in the systems measuring the polarization and luminosity. In both systems the DSP receives the digitized information from a Silicon strip calorimeter. With these calorimeters the position and the energy of electromagnetic showers are measured. The luminosity monitors measure the rate of Bhabha events which is proportional to the luminosity. The signature of such an event is one e- and one e+ measured in coincidence by 2 calorimeters left and right from the interaction point. After each bunch crossing signals proportional to the total energy from the calorimeters are recorded with the DSP to decide within 22 μs, whether the event is a Bhabha event or not. The polarimeter operates at the repetition frequency of 30 Hz. Input to the DSP are 128 ADC channels, 53 for the silicon strips and 55 from other sources such as diodes etc. The output are mean and sigma of the electromagnetic shower distributions. The use of fast processors operating in real time is of particular interest for beam instrumentation in large accelerators : the natural repetition frequency of any circular accelerator is the revolution frequency. The use of DSPs allows to analyze data from turn by turn beam observations and replaces complicated electronic circuits.

2 Introduction

DSP's are fast microprocessors used in various types of data acquisition systems, in particular in High Energy Physics experiments. However, in beam instrumentation they are not very common. To illustrate potential applications we discuss some examples for LEP, where DSPs are used in beam polarization and luminosity monitors [1] [2]. In this paper we do not intend to present an overview of the use of DSPs in High Energy Physics and Accelerator physics as it has been done elsewhere [3].

The monitors provide a large amount of data, between 10-400 kByte/s. The user is interested in the final result, the degree of polarization from the polarimeter and the luminosity from the luminosity monitors. DSP's are employed in the first level data processing to reduce the incoming data substantially. They are installed in an auxiliary building in a distance of some 100 m from the calorimeters.

Before discussing the data acquisition systems with the DSPs in greater detail, we describe the functionality of the two monitors.

Polarimeter For the polarization measurement polarized laser light is scattered off the electron beam. For each laser shot every 33 ms hundreds of high energetic photons are produced and reach a silicon / tungsten calorimeter 247 m downstreams. The calorimeter consists of 4 strip planes and 5 full planes. The full planes have a size of about 40mm x 40mm, the strip planes 16 strips each 2 mm wide. With the strip planes the vertical and the horizontal distributions of the scattered light are measured and the full planes measure the energy of the electromagnetic shower. The degree of polarization is evaluated from the distributions : by changing from left to right circular polarized light the mean of the vertical distribution moves. The amount of the mean shift is proportional to the degree of polarization.

The DSP is used for preprocessing the data from the silicon calorimeter. It reads out 128 ADC channels digitizing the signals from the silicon strips, full planes and some other analog signals. The mean and sigma of the electromagnetic shower distribution and other statistical values are calculated with the help of the DSP. The DSP also tags the data in case of saturation of the ADCs. The data is written into a memory which is read out by a microprocessor (Motorola 68030) (see fig.1). The DSP reduces the amount of data by a factor of one to two orders of magnitude.

The polarization monitor has also been used to measure the turn by turn beam profile and the center of the distribution of the circulating electron bunches. The first strip plane of the calorimeter is illuminated by the synchrotron light from a wiggler magnet. The ADC's digitize the signals from the 16 strips of the first plane once per LEP turn every 88 μs. It is possible to measure the whole profiles for 2000 turns or the mean of the distribution for 32000 turns.

Luminosity monitors At LEP the bunches collide at four interaction points (IP). For a fast measurement of the luminosity two monitors per IP are installed. One monitor is a set of two silicon calorimeters similar to the one used in the polarimeter. The calorimeters are installed at a distance of 8.5 m from the IPs, one calorimeter towards the inside of LEP and the other towards the outside (see fig.2). The luminosity is proportional to the number of Bhabha events. A Bhabha event is an elastic scattering of an electron and a positron. The probability for such an event is high for small angles (some mrad) and decreases strongly with increasing scattering angle.

The calorimeters are installed close to the beam at about 30 mm. The background from bremsstrahlung and other effects is high (between about hundred Hz and some kHz in each calorimeter. For accidental coincidences this yields a rate of up to 200 Hz, compared to about 20 Hz Bhabha event rate for a luminosity of $10^{-31} cm^{-2} s^{-1}$). With four bunches per beam the bunches collide every 22 μs. In the near future LEP will operate with eight bunches per beam and the time between two crossings will decrease to 11 μs.

After each crossing the DSP reads out from the ADCs a number proportional to the total energy deposited in each of the calorimeters. If the energy in both calorimeters is above a predefined threshold, a pair event is counted. If the energy in only one calorimeter is above the threshold, an event for the inside or outside

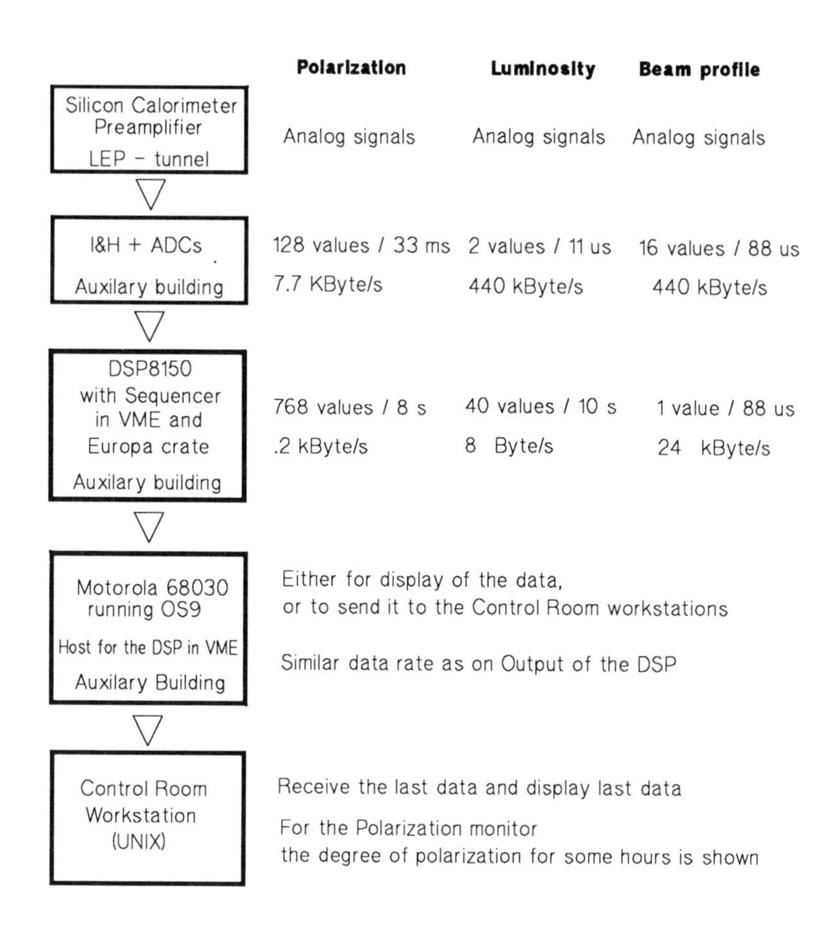

Figure 1: Data reduction by using the DSP

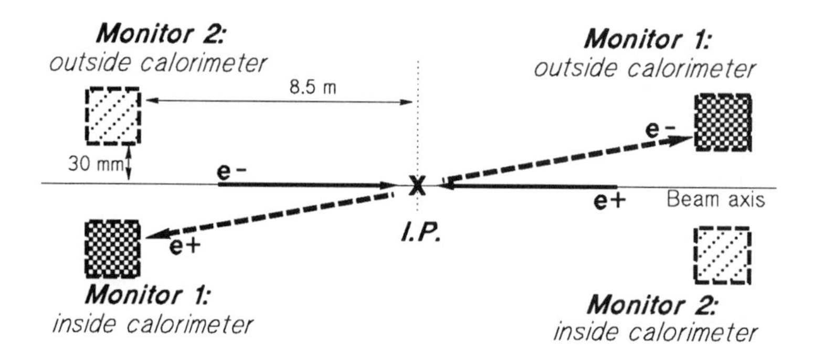

Figure 2: Schematic view of the Luminosity monitors

calorimeter is counted. To record the number of accidental coincidences the DSP measures coincidences between a particle in the inside calorimeter for one crossing and a particle in the outside calorimeter for the next crossing. In this case an accidental event is counted. Every 10 s the values of the counters is transferred to the microprocessors which averages for a time of 180 s to give the number of Bhabha events and therefore the luminosity.

3 The DSP in the data acquisition system

Overview : For the applications described above a commercially available VME-board is used, the DSP8150 from CES, Geneva, Switzerland [4]. It uses the Motorola DSP56001 and is designed to be used with an ADC board. In order to control the analog digital conversion the module contains a dedicated sequencer. The sequencer is a memory (RAM), its data is read out with a frequency of 5 MHz. One sequencer word is used as command for the ADC board, the sequencer executes a sequence of commands and works in parallel with the DSP. Therefore the entire performance of the DSP is available for data processing without having to wait for ADC conversions etc.

The DSP board is housed in a VME crate. In the same crate another processor (Motorola 68030) with a UNIX like operating system (Microware OS9) serves as a host for the DSP. The communication between host processor and DSP board is implemented by use of a memory buffer which can be accessed from either the DSP or from OS9 (dual port ram). This allows the read-out to be independent of the DSP processing. Fig.3 shows a general block diagram of the DSP8150 board and the host microprocessor. To carry out the synchronization of the different modules, several trigger pulses and control signals are required.

Example : The following example illustrates the operation of the data acqui-

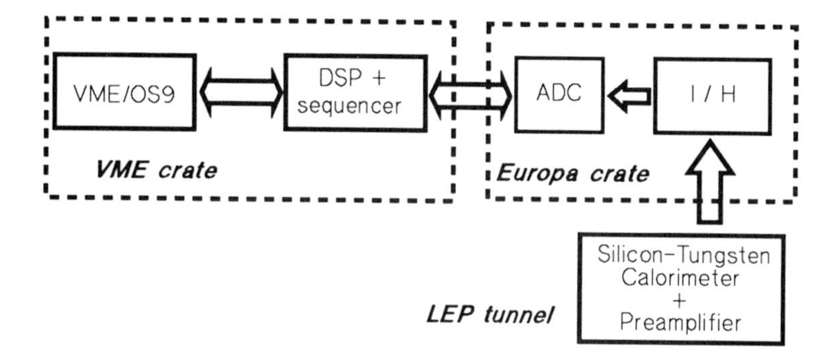

Figure 3: Block diagram of the data acquisition system

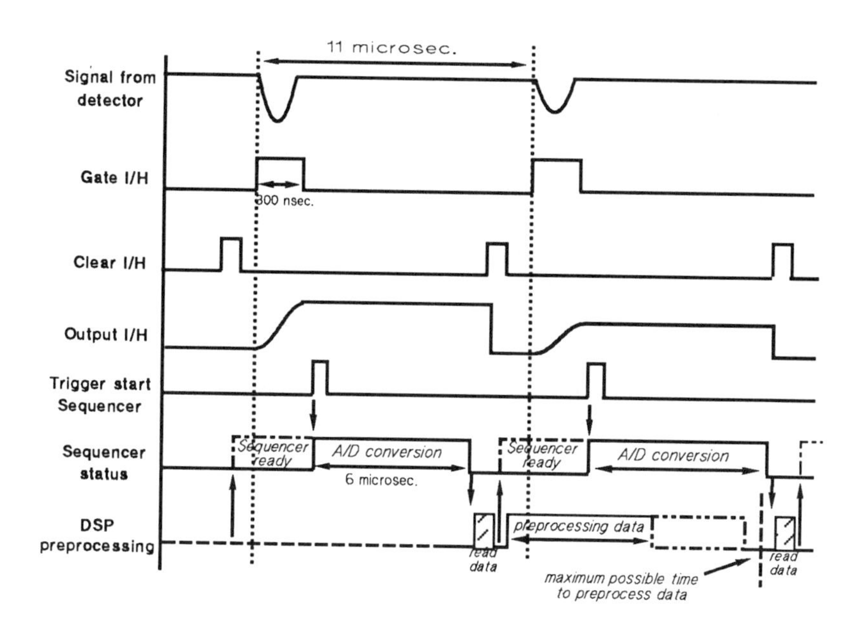

Figure 4: Timing diagram for the acquisition system

sition system for the luminosity monitors (see fig.4), dealing with an event every 11 microseconds (for a LEP operation with 8 e+ and 8 e- bunches). The analog signals from the calorimeter are integrated in an Integrate and Hold unit (I&H) during a gate pulse synchronized with the bunch crossings. Before the start of the integration the I&H requires a clear signal. A set of 32 I&H are multiplexed to one ADC. The ADC module is built up of 4 high speed CMOS 12 bits-ADC. The sequencer controls every function related to the analog/digital conversion. It requires an initialization from the DSP and a trigger pulse to start the command sequence for the A/D conversion. Once started it selects the I&H channels to be digitized, triggers the ADCs and transfers the data into a buffer (FIFO). The DSP waits for the sequencer to finish : it reads a sequencer status word and when the sequencer is ready it reads the data from the buffer. Then the DSP initializes the sequencer to prepare for the next event. The preprocessing of the data is done in parallel with the sequencer reading out the next event. The available time for the preprocessing is about 10 μs. The processing has to be finished before data from the next event is available in the FIFO.
In total three external signals are supplied by the LEP timing system together a specific simple electronics module :

- A gate for the integration.
- A signal to start the command sequence.
- A signal to clear the Integrator&Hold (this signal can also be provided by the DSP program, as in the polarimeter).

So far we discussed the functionality of the DSP card together with the ADCs and the I&H unit. In the next section the communication between the host microprocessor (Motorola 68030) and the DSP card is described.

In fig.5 the timing of the system for some successive events is shown. The DSP processes a sequence of events in real time. After each event the result is stored in the DSP memory. When the DSP has processed the total amount of events required, it writes the data into an output buffer with two pages. At the same time one page can be accessed by the DSP and the other page by the host microprocessor. When the DSP finished writing the data. e.g. on page 1, it changes to page 2. Then host microprocessor can access page 1 and read the data etc.

4 Software for the DSP

The DSP56000 is optimized to execute algorithms in as few operations as possible. The whole architecture has been designed to ease the interface and maximize the throughput. No overhead is caused by an operating system and interrupts can be executed within a delay of a few hundred ns. The DSP is provided with 62 CPU

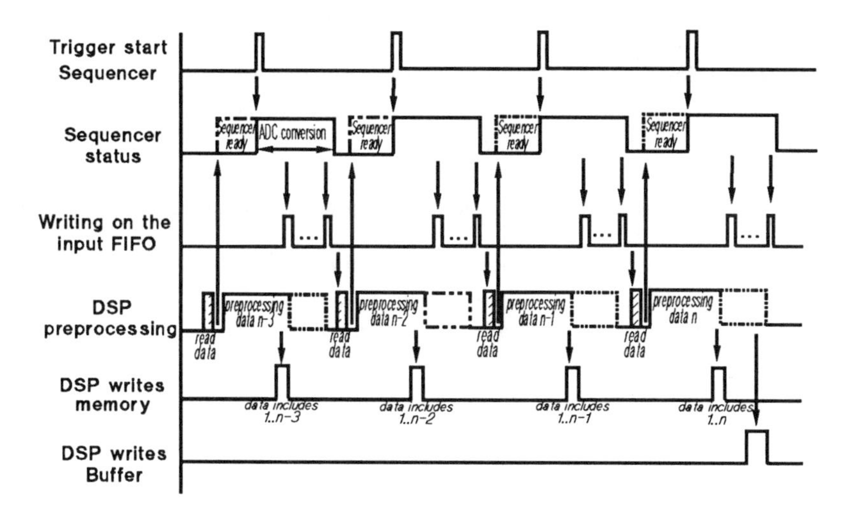

Figure 5: DSP and Sequencer processes and data transfer

like instructions [5] and contains on-chip two separate data buses together with the program bus. Certain instructions can be executed in parallel. Therefore with 10 MIPS the processor is a very fast tool in frontend data-acquisition systems. For using the DSP8150 board one has three levels of related software :

- The sequencer needs to be supplied with a sequence of commands to control the ADC card multiplexers, the analog-digital conversion and the data transmission to the DSP8150 input buffer (FIFO).

- The DSP5600 is programmed in assembler language. The program can have up to 512 instructions. As an example, within one instruction one multiplication and two data move commands are possible.

  Apart from setting up the environment of the DSP (e.g. loading the command sequencer into the sequencer) a typical program has three parts : 1) read event, 2) process event, 3) store results.

- A program in the host microprocessor deals with the DSP (in our case the language C is used). It loads the DSP program into the DSP memory as well as the command sequence for the sequencer. When data from the DSP is available, it reads the data, does further processing and if desired sends the data to the workstations in the LEP control room.

The development environment for the DSP56001 exist on many different platforms, e.g IBM-PC/DOS, UNIX, VAX/VMS and Macintosh. We use an IBM-PC.

A C-compiler to write DSP programs is available although we do not use it. The reason are the relative few instructions in one program. For an optimized use of the DSP performance the assembler language is more appropriate.

The development path can be illustrated as following:

- After a DSP program is written it is running through an assembler.
- Linking the relocatable file with libraries (if any) and thereby producing an absolute load file ready for loading into the DSP-memory. This file is in ASCII format and is transferred to the VME crate with the DSP card via TCP/IP-FTP.
- At this stage a simulator can test performance of the program.
- Transfer of the program to the DSP external bootstrap program memory. This part is specific for the host computer environment. We use a C-program which reads the program from the file and writes it into the bootstrap memory of the DSP, which is memory mapped to VME.
- The DSP is ready to receive the bootstrap command for loading the program to the on-chip program memory.

5 Experience with the DSP

The use of a DSP requires a certain expertise because three levels of software are involved. The documentation for the DSP8150 board is insufficient, i.e. examples are missing. Therefore it took some time to understand the detailed functioning of the board. Once this was accomplished, it was relatively easy to develop DSP programs. An essential part of the time is devoted to the test of the program. In particular the execution time must be smaller than the available time for data processing between two events. The DSP board works very reliably. Because of its architecture the execution time is determined and for one program always the same. No software maintenance is required because the DSP board is independent from the environment. No interrupts between the DSP board and the Motorola 68030/OS9 are used. The program in the DSP is written in a way to be independent from the Real Time operating system in the VME. If the host microprocessor or its operating system are changed, the DSP program does not have to be modified. However, it might be required to rewrite the program running in the host computer. A disadvantage of the DSP56000 is the lack of floating point arithmetic which complicates the program development.

For both, the polarimeter and the luminosity monitor it is important not to be limited by any dead-time due to the data acquisition system. This aim could be accomplished by the use of the DSP without building complicated electronics circuits.

For the luminosity monitors the processing time between two events is about 10 μs. A typical application program with about 30 to 50 instructions can be executed withing this time.

6 Conclusions

For beam instrumentation equipment most data acquisition systems are operating synchronously with the revolution frequency or their harmonics. A typical application is the digitization of a signal for each revolution. Two methods can be employed:

- The signals can be digitized each turn and stored in a large memory. This technique has been employed for the BOSC system [6]. The analysis of the data is done offline. This method is limited by the size of the available memory.
- A second possibility is the use of a DSP : the time between two successive signals is available to perform the data processing to reduce the amount of data. For instruments monitoring instruments this technique is very attractive.

The use of a DSP is very flexible. Similar programs can be used for different applications. The reliability of the DSP is high. We expect that with the development of the new generation of DSPs with floating point arithmetics the speed will further increase and the program development will become easier. Because of the increased level of complexity a programming in the language C will be essential. The first LEP instrument using a DSP of the next generation will be the Q-meter.

7 Acknowledgement

The authors would like to thank J.d.Vries, B.Dehning, G.P.Ferri and M.Placidi for their contributions to the use of the DSP. We also thank C.Bovet for his continuous support and W.Herr for reading this manuscript.

References

[1] J.Badier et al., Part.Acc.Conf.,San Francisco 6-9 May 1991

[2] G.P.Ferri, EPAC 1990, Nice, June 12-16

[3] D.Crosetto, Digital Signal Processing in High Energy Physics, CERN Computing School 1990, Ysermonde, Belgium, CERN Yellow Report

[4] DSP8150 User's Manual Creative Electronics System S.A., Geneva, Switzerland, 1990.

[5] DSP56000/DSP56001 Users's Manual, Motorola, MOTOROLA INC.,1990

[6] A.Burns et al., EPAC 1990, Nice, June 12-16

[7] H.Schmickler, private communication

Offset Calibration of the Beam Position Monitor Using External Means*

Y. Chung and G. Decker
Argonne National Laboratory, Argonne, IL 60439

ABSTRACT

Determination of the offset of the electrical center of the beam position monitor (BPM) relative to the mechanical center is required for the absolute beam position measurement in the injector synchrotron and the storage ring of the APS. Conventionally, RF signal is sent to an antenna or through a wire carefully aligned in the vacuum chamber and the signal is measured at each electrode. A new method using only external means which does not involve mechanical alignment inside the vacuum chamber was applied to the injector synchrotron BPM unit in this work. The result shows a good agreement with the wire method within 15 μm error. A similar measurement on the storage ring BPM will also be discussed.

INTRODUCTION

The Advanced Photon Source consists of an electron linac, a positron linac, a positron accumulator ring, an injector synchrotron, and a storage ring. The storage ring has a circumference of 1104 meters, along which will be placed 360 beam position monitoring stations, each composed of four button-type capacitive pickup electrodes. The injector synchrotron, which is exactly one third the circumference of the storage ring, has BPM's that are of the same type of button electrodes as the storage ring. A total of 80 stations are planned. Shown in Table 1 are accelerator parameters relevant to diagnostic instrumentation design.

A measurement accuracy of ±200 μm for the storage ring is required relative to the magnetic centerline of adjacent sextupole magnets, with a required resolution of better than 25 μm. The APS storage ring BPM specifications are listed in Table 2. The reason for the tight accuracy specification is that, at commissioning, the dynamic aperture of the machine approaches the physical aperture when the rms placement error of the sextupole magnets is equal to 200 μm. After commissioning, the insertion device vacuum chambers (1.2 cm vertical full aperture) will be installed and become the limiting aperture.

The storage ring vacuum chamber is an aluminum extrusion with a roughly elliptical inner bore near the positron beam, a photon exit slot, and an antechamber containing NeG pumping strips. The pickup electrodes will be

*Work supported by the U.S. Department of Energy, Office of Basic Energy Sciences, under Contract No. W-31-ENG-38.

Parameter	Storage Ring	Injector Synch.
Energy (GeV)	7	.45 – 7
RF Frequency (MHz)	351.93	351.93
Harmonic No.	1296	432
Minimum Bunch Spacing (ns)	20	1228
Revolution Period (μs)	3.68	1.228
Number of Bunches	1 – 60	1
Maximum Single Bunch Current (mA)	5	4.7
Bunch Length (2σ, ps)	35 – 100	61 – 122
Damping Times $\tau_{h,v}$ (ms)	9.46	2.7 @7GeV
Tunes $\nu_{h,v}$	35.22, 14.30	11.76, 9.80
Damping Time τ_s (ms)	4.73	1.35 @7GeV
Synchrotron Frequency f_s (kHz)	1.96	21.2

Table 1: Accelerator Parameters for Diagnostic Instrumentation

First Turn, 1 mA Resolution / Accuracy	200 μm / 500 μm
Stored Beam, Single or Multiple Bunches @ 5mA Total Resolution / Accuracy	25 μm / 200 μm
Stability, Long Term	±30 μm
Dynamic Range, Intensity	≥ 40 dB
Dynamic Range, Position	±20 mm

Table 2: APS Storage Ring Beam Position Monitor Specifications.

mounted on machined flats surrounding the positron side of the chamber. They will be located with a tolerance of ±0.004" (±100 μm) relative to the positron chamber center.

In case of the injector synchrotron, each BPM unit is machined out of stainless steel and welded between two vacuum chamber segments. The required accuracy of 1 mm is not so stringent, and therefore, a prototype unit will be calibrated on the test stand and the result will be applied to all others.

For the storage ring BPM, the tight accuracy requirement calls for calibrating each and every monitor for the mechanical offset and sensitivity in both the horizontal and the vertical directions. Since the BPM is an integral part of the vacuum chamber (≈ 5 m long), it will be very difficult to calibrate the BPM's using wire suspended inside the vacuum chamber. Therefore, a new method involving only external measurements is desired.

This paper describes the application of the external calibration method developed by G. Lambertson[1,2] to the offset calibration of the injector synchrotron BPM. It also serves as a feasibility study of the method for calibrating the storage ring BPM with an accuracy of less than 30 μm.

THEORY

Consider the schematic of the button configuration in Fig. 1. The position of the beam (x_0, y_0) is determined from

$$
\begin{aligned}
\Delta_x &= V_1 - V_2 - V_3 + V_4 , \\
\Delta_y &= V_1 + V_2 - V_3 - V_4 , \\
\Sigma &= V_1 + V_2 + V_3 + V_4 , \\
Q &= V_1 - V_2 + V_3 - V_4 , \\
X_0 &= \frac{\Delta_x}{\Sigma} \approx S_x x_0 + R_x , \\
Y_0 &= \frac{\Delta_y}{\Sigma} \approx S_y y_0 + R_y , \\
Q_0 &= \frac{Q}{\Sigma} .
\end{aligned}
\tag{1}
$$

S_x, S_y, R_x, and R_y are sensitivity and offset functions. Associated with each button is a gain factor g which causes the difference between the mechanical and the electrical centers. The electrical center is the wire or beam position where Δ_x and Δ_y vanishes. The larger the gain factor of a button, the farther the electrical center from that button. With these gain factors, we can obtain the electrical center (x^e, y^e) relative to the mechanical center, which is the coordinate origin. That is,

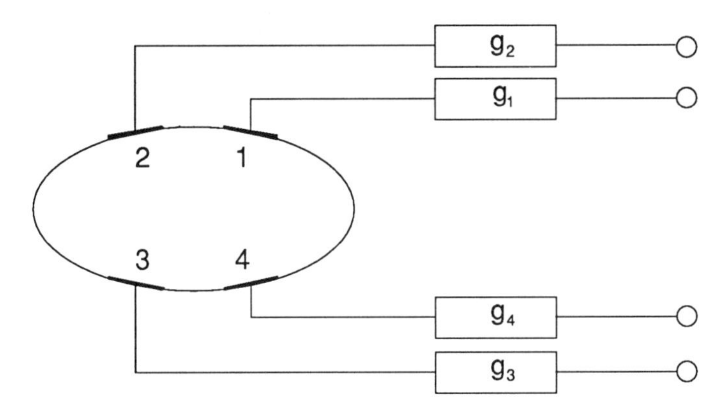

Fig. 1: The schematic of the button configuration. g's represent the gain associated with the buttons.

$$x^e = -\frac{1}{S_x}\frac{g_1 - g_2 - g_3 + g_4}{g_1 + g_2 + g_3 + g_4},$$
$$y^e = -\frac{1}{S_y}\frac{g_1 + g_2 - g_3 - g_4}{g_1 + g_2 + g_3 + g_4}. \quad (2)$$

The external measurement method described in Ref. 1 uses the coupling between the buttons to determine the gain factors. Here, we will briefly summarize the result. As shown in Fig. 2, the RF signal V_j is applied to the button j and then the signal V_i is detected at button i. The voltage on the button j is twice the applied signal times g_j, that is, $V_j^b = 2g_jV_j$. With the capacitive coupling coefficient G_{ij} (= G_{ji}), there appears current on the button i equal to $I_i = 2G_{ij}g_jV_j$. If the transmission line has characteristic impedance of 50Ω and if terminated with 50Ω, the detected signal will be $V_i = 2\cdot 50\cdot G_{ij}g_ig_jV_j$. Therefore, the normalized voltage V_{ij} is equal to

$$V_{ij} = \frac{V_i}{V_j} = 2\cdot 50\cdot G_{ij}g_ig_j. \quad (3)$$

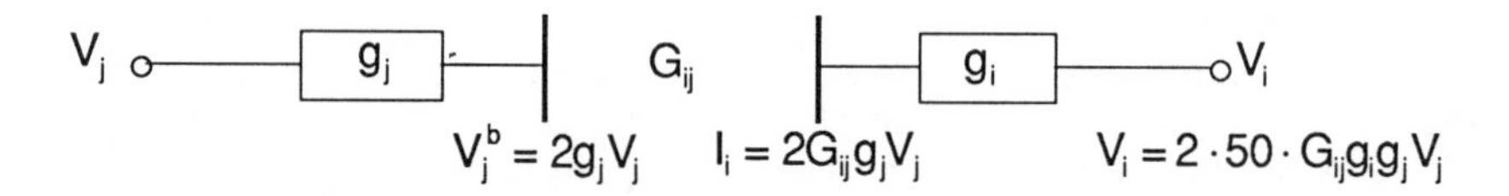

Fig. 2: The coupling between the buttons i and j.

If we assume 2-D symmetry of the button configuration, that is, $G_{12} = G_{34}$, $G_{14} = G_{23}$, and $G_{13} = G_{24}$, the gain factors g_i then can be obtained from three alternative combinations of the measured V_{ij} as shown in Eq. (4). Since we are interested in the ratios of the gain factors, the values of G's need not be known.

$$2\cdot 50\cdot g_1^2 = \frac{V_{21}V_{14}}{V_{42}}\frac{G_{13}}{G_{12}G_{23}} = \frac{V_{12}V_{31}}{V_{32}}\frac{G_{23}}{G_{12}G_{13}} = \frac{V_{41}V_{31}}{V_{43}}\frac{G_{12}}{G_{23}G_{13}},$$

$$2\cdot 50\cdot g_2^2 = \frac{V_{21}V_{32}}{V_{31}}\frac{G_{13}}{G_{12}G_{23}} = \frac{V_{21}V_{42}}{V_{14}}\frac{G_{23}}{G_{12}G_{13}} = \frac{V_{32}V_{42}}{V_{43}}\frac{G_{12}}{G_{23}G_{13}},$$

$$2\cdot 50\cdot g_3^2 = \frac{V_{32}V_{43}}{V_{42}}\frac{G_{13}}{G_{12}G_{23}} = \frac{V_{43}V_{31}}{V_{14}}\frac{G_{23}}{G_{12}G_{13}} = \frac{V_{32}V_{31}}{V_{21}}\frac{G_{12}}{G_{23}G_{13}}, \quad (4)$$

$$2\cdot 50\cdot g_4^2 = \frac{V_{43}V_{14}}{V_{31}}\frac{G_{13}}{G_{12}G_{23}} = \frac{V_{43}V_{42}}{V_{32}}\frac{G_{23}}{G_{12}G_{13}} = \frac{V_{14}V_{42}}{V_{21}}\frac{G_{12}}{G_{23}G_{13}}.$$

MEASUREMENT SETUP

In Fig. 3 is shown the measurement setup for the external calibration of the injector synchrotron BPM. Measurement of the RF characteristic of buttons is described in Ref. 3. The network analyzer was Hewlett-Packard Model 8753C with an S-parameter test set. To compensate for the weak button-to-button coupling, an RF amplifier with 45 dB gain (1.7 – 500 MHz) was used at port 1 of the S-parameter test set. A computer-controlled SP4T switch was used at the port 2 for multiplexing through the buttons. Its insertion loss was measured and corrections were made to the data. This helped minimize the number of times needed to disconnect and connect the cables and buttons.

The measurement was made at a fixed frequency of 351.93 MHz. In Eqs. (3) and (4), the normalized voltage V_{ij} is equal to the transmission coefficient s_{21} between buttons i and j. The network analyzer was put in CW mode and an IF bandwidth of 10 Hz was used to reduce the noise-to-signal ratio to a minimum. The data was transferred to a desktop PC for storage and analysis.

The buttons were selected from eight available buttons such that the measured capacitances were the closest to each other. Eqs. (3) and (4) can be applied to the offset measurement only when the characteristic impedances of buttons, cables, and connectors are carefully matched. Otherwise, the matrix V_{ij} will not be symmetric and the measurement error will be significant.

RESULTS

Table 3 shows data from a single measurement on the injector synchrotron BPM using the external calibration method. With the sensitivity functions S_x = 0.070 mm^{-1} and S_y = 0.057 mm^{-1}, we obtain $x_L{}^e$ = 155 μm and

Frequency = 351.93 MHz (in dB)

Out(i)↓ In(j)→	1	2	3	4
1		-39.3776	-45.4625	-36.1599
2	-39.4058		-35.7875	-45.8526
3	-45.4953	-35.7799		-39.6083
4	-36.1638	-45.8309	-39.5856	

g factors	1.0000	1.0009	1.0100	0.9678
X_0, Y_0, Q_0, S	-0.0108	0.0058	0.0104	3.9787

Table 3: The result of external calibration measurement on the injector synchrotron BPM. The upper table is the normalized voltage matrix V_{ij} while the lower one lists the gain factors, X_0, Y_0, Q_0 and S.

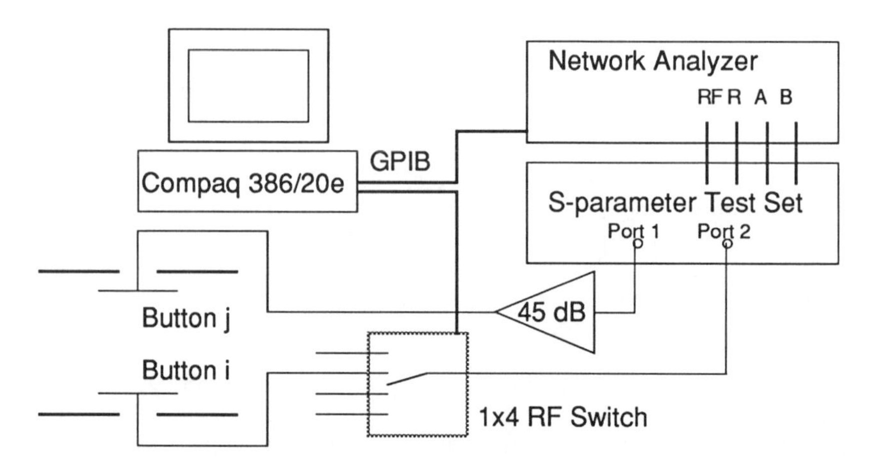

Fig. 3: Application of Lambertson's external calibration method to the offset calibration of the injector synchrotron BPM.

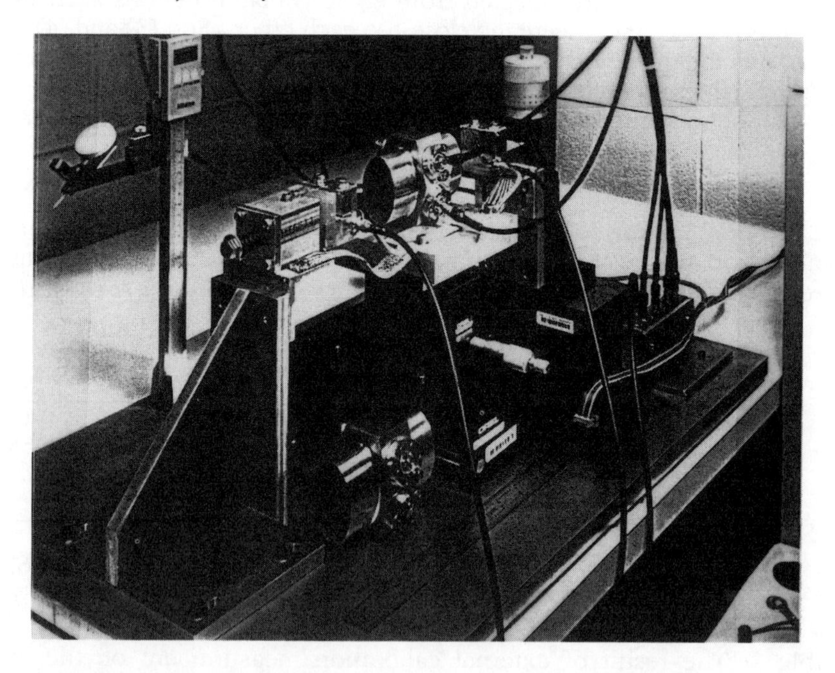

Fig. 4: Injector synchrotron BPM test stand.

y_L^e = -100 μm. A separate measurement using the wire method[3] gave x_w^e = 150 ± 6 μm and y_w^e = -103 ± 13 μm. The two results agree very well. It is to be noted that the asymmetry in the normalized voltage matrix V_{ij} is quite small. The largest error was 0.033 dB between buttons 1 and 3.

In Table 4 are listed results from measurements with different button configurations. In case 2, the buttons in case 1 were swapped about the x-plane (y = 0), and in case 3, the buttons in case 2 were swapped again about the y-plane (x = 0). The comparison of the results with those obtained using the wire method shows good agreement within a typical error of 15 μm.

(in μm)

	Wire Measurement		External Measurement†	
Case*	x_W^e	y_W^e	x_L^e	y_L^e
1	150 ± 6	-103 ± 13	159	-100
2	167 ± 4	-185 ± 4	152	-194
3	-81 ± 4	-189 ± 6	-73	-189

Table 4: Result of the offset measurement using the wire and the external methods on the injector synchrotron BPM. "w" denotes the wire method and "L" denotes the Lambertson method.

* Buttons were swapped symmetrically about the x- and y- planes for different configurations.

† Typical measurement error was ≤ 3 μm.

DISCUSSION ON THE STORAGE RING BPM CALIBRATION

A storage ring BPM calibration test stand is being built for measurements similar to those described in previous sections. In contrast to the injector synchrotron BPM, which is a separate machined unit to be welded to the vacuum chamber, the storage ring BPM has buttons directly mounted at an angle of 15.11° on the vacuum chamber.

The vacuum chamber has asymmetry about the y-plane due to the photon exit slot height of 0.426". Therefore, the wire alignment method used on the injector synchrotron BPM[3] is not applicable to the storage ring BPM. We plan to use tiny alignment holes approximately 500 μm in diameter with a laser beam passing through them. This will enable us to position the wire at the mechanical center within an error of 10 μm or better, excluding the machining error.

It is expected that the vacuum chamber will be deformed by as much as 500 μm at the photon exit slot once it is put under vacuum. The significance of the effect on the beam position measurement will be studied.

Details of the measurement setup and the results obtained from the wire and the external methods will be published separately in the near future.

REFERENCES

1. G. R. Lambertson, "Calibration of Position Electrodes Using External Measurements", LSAP Note-5, Lawrence Berkeley Laboratory, May 6, 1987
2. J. Hinkson, private communication.
3. G. Decker, Y. Chung and E. Kahana, "Progress on the Development of APS Beam Position Monitoring System", Proceedings of 1991 IEEE Particle Accelerator Conference.

The Diagnostics System for the Multiple Heavy Ion Beams Induction Linac Experiment, MBE-4[†]

S. Eylon for the MBE-4 Team*
Lawrence Berkeley Laboratory
1 Cyclotron Road
Berkeley, CA 94720

Abstract

MBE-4 is a four beam current amplifying induction linac experiment conducted at LBL as a part of the Heavy Ion Fusion Accelerator Research (HIFAR) program for studying accelerator physics issues of a heavy ion driver for inertial fusion. The four ion beams (Cs^+,10 mA) are focused by electrostatic quadrupoles and accelerated from about 200 keV to 900 keV through 24 induction gaps. Current amplification of up to nine times is achieved while the beam pulse duration is compressed from about 3 μs to 0.5 μs. The diagnostic system enables the complete time-resolved 2-D transverse phase space distribution of the beam to be measured. Reduction of the raw data yields the beam current, current profile, emittance, centroid position and angle as well as the beam envelope parameters. In addition, the longitudinal energy distribution is obtained from measurements using a calibrated electrostatic spectrometer. The diagnostic system hardware, as well as the the data acquisition and reduction routines, are controlled by an IBM pc-XT. We shall describe the diagnostic system and discuss its performance in view of the specific issues which result from the acceleration and amplification of multiple beams of heavy ions.

1. Introduction.

The Heavy Ion Fusion Accelerator Research Program (HIFAR) at LBL is assessing the multiple-beam induction linac as an inertial fusion driver.[1] In this concept multiple parallel beams of heavy ions are continually amplified in current and in voltage as they are accelerated to the parameters required to ignite an inertial fusion target (~ 10 Gev, 500 TW, 10ns). The MBE-4[2] is a four beam current amplifying ion induction linac experiment. This experiment models much of the accelerator physics of the electrostatically focused section (lower end) of a fusion driver. The four space-charge dominated, Cs^+, ion beams share the longitudinal acceleration-gap structures and the corresponding driving pulse-forming networks, while being transversely contained in four separate focusing quadrupole channels. The 5-10 mA beams are accelerated from approximately 200 keV to one MeV, amplified to final currents of about 20-90 mA and compressed from a duration of about 3 ms to about 0.3 ms by 24 accelerating gaps. Experiments[3] of particular interest on the MBE-4 are the possibilities of interactions between the multiple beams, longitudinal beam control i.e. acceleration, current amplification and pulse

† Work supported by the Director, Office of Energy Research, Office of Basic Energy Sciences, Advanced Energy Projects Division, U.S. Department of Energy under Contract No. DE-AC03-76SF00098.

* D. Keefe, T.J. Fessenden, A. Faltens, A.I. Warwick, C. Kim, H. meuth, T. Garvey, D.E. Gough, C. Lionberger, S. McCreight, and B. Ghiorso

ends, and the preservation of the longitudinal and transverse normalized beam emittance along the transport channel with and without acceleration for centered and displaced beams. This paper will describe the MBE-4 diagnostics and control system used in the course of the above studies.

2. Diagnostic requirements for the MBE-4.

Diagnostic access on the MBE-4 is possible on various axial locations (Fig. 1). Each location (diagnostic "box") permits, at four port holes, the attachment of diagnostic devices, most of which, at least in principle, can monitor the four beams independently. All diagnostic measurements derive from measuring the positive-ion current (or, by 50 W termination, from a voltage measurement), and are thus electromagnetic in nature. Essentially the following measurements can be performed:

Total current, via Faraday cup (Fig. 2a).
Transverse position and current density profile (Fig. 2b).
Transverse position and velocity (angular) distribution via a double slit & a Faraday cup (Fig. 2c).
Longitudinal velocity (energy) distribution, via an electrostatic bending beam energy analyzer (Fig. 2d).
Details of the coupling and biasing circuitry are given only in Fig. 2, and are omitted elsewhere.
Acceleration pulsers waveforms are monitored, via resistive voltage dividers placed across the acceleration gaps.

2.1. Time resolution

All measurements have to be fully time resolved to investigate in detail the beam pulse dynamics in the pulse frame, especially head and tail. The determination of position, velocity, and energy distribution require scans through the transverse, or longitudinal parameter space, respectively; for measuring the transverse position and velocity distribution, i. e. the position of the "radius" slit, and of the "angle" slit are repetitively, on a shot to shot basis, adjusted by computer controlled stepping motors. This is a time consuming process, where the current wave generated in each shot is recorded and stored by a digital storage oscilloscope and then transferred into a computer memory. While the determination of the machine total current pulses will require a signal bandwidth about 50 to 100 MHz, the requirement for position, velocity, and energy is more stringent. This is because these parameters may rapidly vary during a single pulse. For example, the already mentioned pulse compression along the MBE-4 leads to a 30% head to tail velocity difference along the beam bunch. The variation of the corresponding longitudinal particle energies within the current pulse is therefore even more pronounced. Here, a signal bandwith of 200 MHz, or even higher is needed. For a scan of the longitudinal energy, the energy analyzer adjusts (Fig. 5), again on a shot to shot basis, the voltages (+/- HV). This procedure is also computer controlled. The high-voltage potential selects,

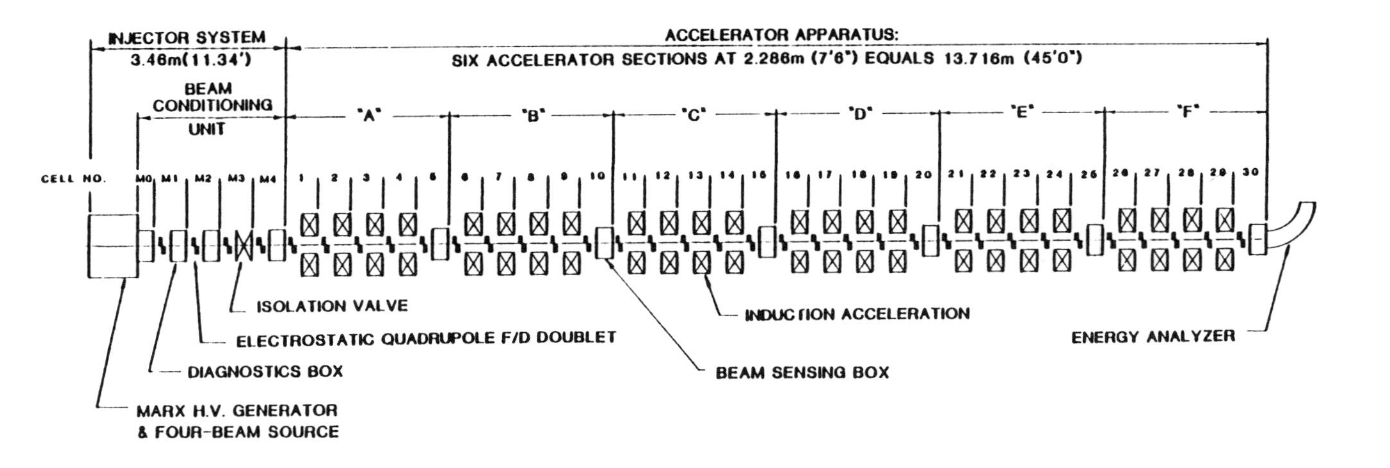

Fig. 1. The Multiple Beam Experiment system layout.

Fig. 2. The diagnostics system measuring setups:

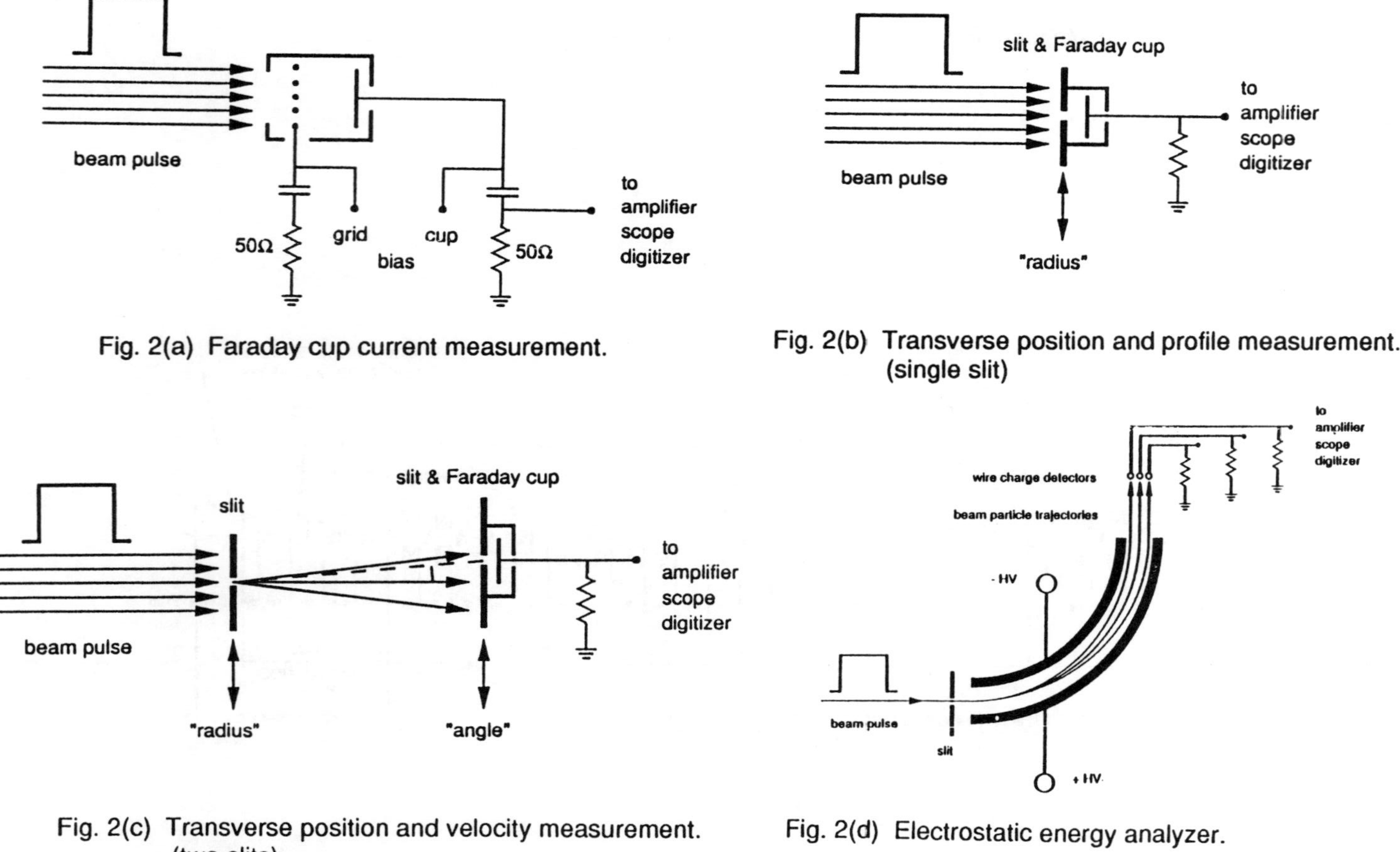

Fig. 2(a) Faraday cup current measurement.

Fig. 2(b) Transverse position and profile measurement. (single slit)

Fig. 2(c) Transverse position and velocity measurement. (two slits)

Fig. 2(d) Electrostatic energy analyzer.

from the beam pulse, those particles with a specific energy by forcing them onto a circular-sigmented trajectory. Those particles impinging on a wire or a the slit-Faraday cup will generate a current waveform which is again stored in the storage oscilloscope and then transferred into the computer. The composite energy distribution along the beam can be obtained by a complete scan of the analyzer high voltage settings.

2.2. Dynamic range

The voltage waveforms of the emittance (double slit) and the energy measurements are amplified by about 30 db, or thirty fold, using commercial amplifier modules by Comlinear Corporation. The amplifiers are placed as close as possible to the measuring devices to reduce noise effects during the the transfer of the signal to the oscilloscope. The signal strength is further enhanced about tenfold in the Faraday cup itself by employing a specific biasing polarity. For an absolute current cup measurement, the current collector is biased positive, and the grid negative. This insures that secondary electrons freed by the impinging ions are retained and do not distort the current reading. On the other hand, by reversing the bias polarity the secondary electrons may be used for increasing the cup signal. The raw signals for the energy analysis are of the order of 10 to 100mV. For the measurements of the beam transverse properties, particularly for the position and velocity distributions, the signal strength after the 30 db amplification is of the order of 1 to mV. This signal is low when compared with the energy signal because, here, the beam pulse is resolved into two characteristic quantities, namely radius (1st slit) and angle (2nd slit) leaving much less flux at a given position and transverse velocity.

3. Data Acquisition and processing

Data from the diagnostic probes are acquired using digital oscilloscopes controlled by IBM PC's. A set of programs[4] written in Pascal language performs various acquisition and analysis tasks. These programs are graphically-oriented and allow the experimenter to set up, perform, and examine the results of a variety of measurements. They automate the time-consuming multiple-pulse measurements and the tedious recording of diagnostic configurations.

The program READ reads whatever waveform is currently stored in the oscilloscope, displays it on the computer monitor, and stores it in a file. The programs WRITE and WAVEDISP read files produced by READ and display them both on the monitor and the oscilloscope, or on the monitor only, respectively. The program CONFIG allows the user to maintain a database of probe characteristics and to select the probe and location to be used in the current measurement. The program SETUP allows the user to control various aspects of the emittance scans and energy analyzer scans, such as the number of steps in the measurement. The program

VOLTS allows the user to interactively set voltages on an energy analyzer, pulse the machine, and observe the results. The program ANALYZER acquires data from energy analyzers by pulsing the machine while the analyzer voltage is adjusted to a series of values. SHOW is a data analysis and display program for files produced by ANALYZER. The program MOVE allows the user to interactively move harps, slit cups, and Faraday cups, and to fire the machine, as he observes the results. The PHASE program allows one to interactively move slit cups, pulse the machine and observe the resulting slit-cup signals plotted in phase-space coordinates; it is also used to specify the boundary of the region that will be scanned during an emittance measurement. The SCAN program performs an actual emittance measurement, pulsing the machine while the dual-slit diagnostic probe is set at a series of positions; it displays an ongoing plot of the results and stores the results in a file. The DISPLAY program provides various displays and calculated results based on files produced by SCAN. A typical DISPLAY output following an emittance (double slit) measurement contains the calculated time varying beam parameters like current, current profile, radius, divergence angle, centroid position and angle, emittance and the beam phase space distribution at any given time. Figure 3a show a beam current density profile and Fig. 3b the beam phase space distribution i.e. the angle X' (transverse velocity) distribution at a given position X and time, were the current measured at that point is given by the line length.

4. The diagnostics system performance.

The diagnostics system (including the accelerator) performance can demonstrated through measurements taken when performing evaluation experiments and other typical accelerator physics studies experiments.

Faraday-cups total current waveforms (Fig. 4) of the MBE-4 four beams measured at injection in and at end of the accelerator. The beam current pulses which are compressed from about 3ms to about o.3 ms show a current amplification of about 9. The current waveforms shown presents an overlay of waveforms taken in 25 different machine shots showing a low overall system jitter.

Faraday-cups total current waveforms (Fig. 4) of the MBE-4 four beams measured at injection in and at end of the accelerator. The beam current pulses which are compressed from about 3ms to about o.3 ms show a current amplification of about 9. The current waveforms shown presents an overlay of waveforms taken in 25 different machine shots showing a low overall system jitter.

Faraday-cups total current waveforms (Fig. 4) of the MBE-4 four beams measured at injection in and at end of the accelerator. The beam current pulses which are compressed from about 3ms to about o.3 ms show a current amplification of about 9. The current waveforms shown presents an overlay of waveforms taken in 25 different machine shots showing a low overall system jitter.

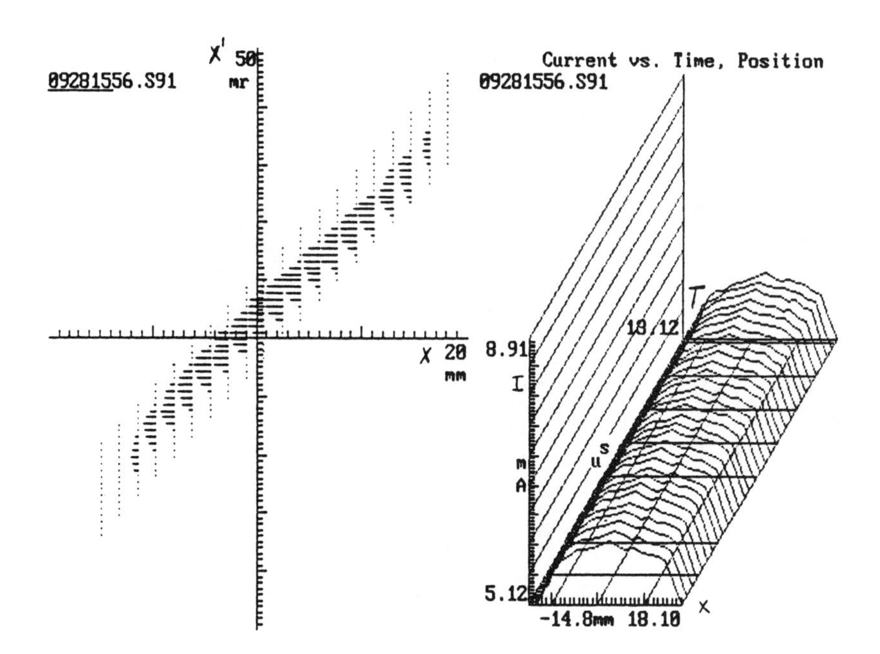

Fig. 3 DISPLAY output :
(a) Beam distribution in the phase space, angle (transverse velocity) X' at a position X.
(b) Beam profile, current I at a position X at a given time T.

Faraday-cups total current waveforms (Fig. 4) of the MBE-4 four beams measured at injection in and at end of the accelerator. The beam current pulses which are compressed from about 3ms to about o.3 ms show a current amplification of about 9. The current waveforms shown presents an overlay of waveforms taken in 25 different machine shots showing a low overall system jitter.

Emittance measurements were taken along the MBE-4 transport channel over a period of about two months to asses the emittance measurement and the system reproducibility. The scanners in each diagnostic were replaced several times while other beam parameters were kept constant. Fig. 5a shows the measured beam minimum and maximum emittance at each diagnostics station along the MBE-4. One can see that the emittance measurement reproducibility is within 5%. The emittance variations observed along the transport section were found to be due to the beam being displaced off the machine center. Figure 5b shows the beam un-normalized emittance along the MBE-4 for a drifting beam and an accelerated beam. One can see that the accelerated beam un-normalized emittance decreases along the MBE-4, i.e. the normalized emittance is conserved.[5]

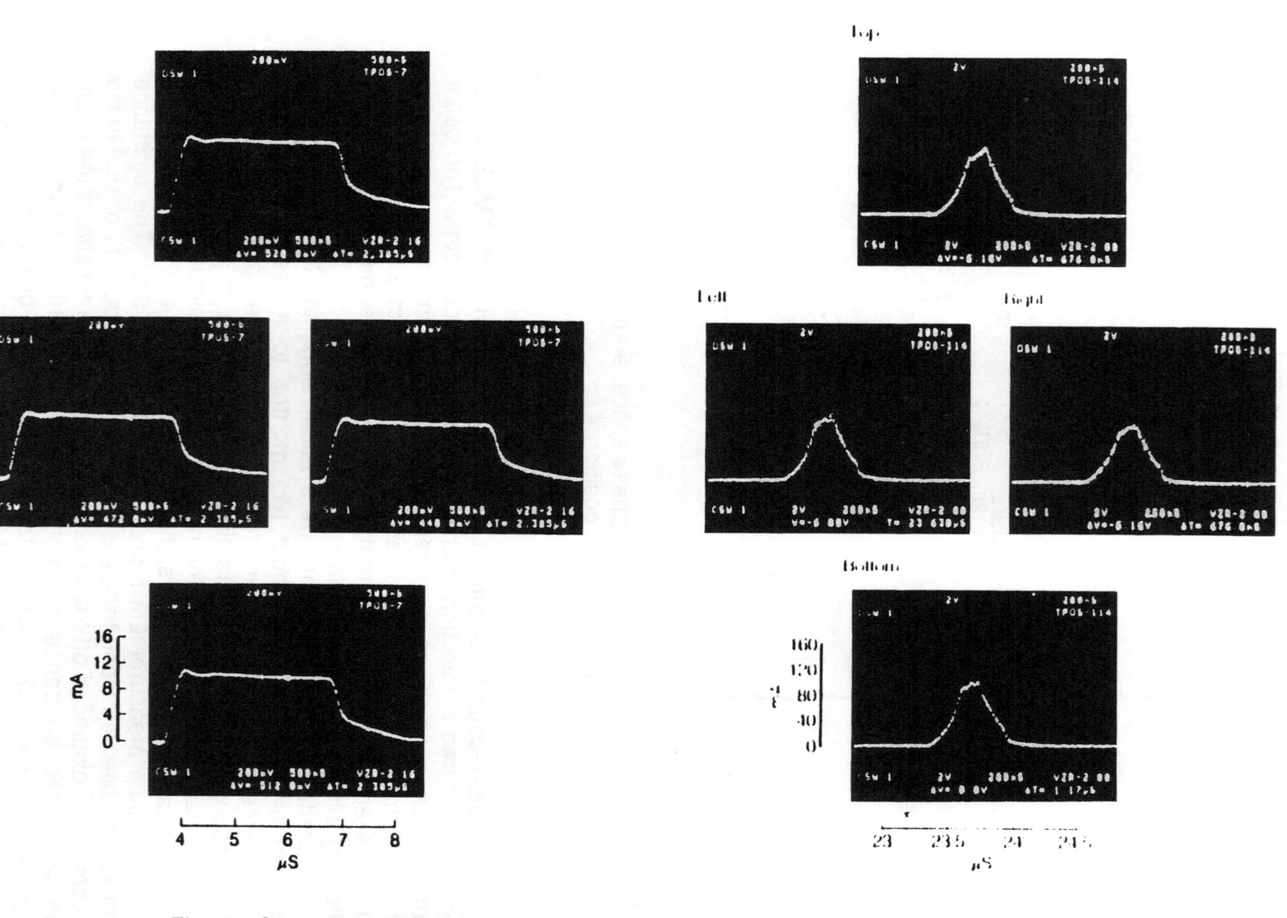

Fig. 4. Current amplification in MBE-4 (a) current at injection in accelerator (b) current at end of accelerator.

variations observed along the transport section were found to be due tc the beam being displaced off the machine center. Figure 5b shows the beam un-normalized emittance along the MBE-4 for a drifting beam and an accelerated beam. One can see that the accelerated beam un-normalized emittance decreases along the MBE-4, i.e. the normalized emittance is conserved.[5]

Energy measurements along the injected beam taken at the input to the transport section are shown in Fig. 5a. The beam energy was calibrated using time of flight measurements along the MBE-4. The measured spread in the energy at a given time point (position along the beam) is given by the error bar, about 0.3%, is consistent with the system designed resolution. Figure 5b shows energy measurements taken along a drift compressed beam bunch at the diagnostics station 25. One can see the compressing energy difference along the beam, furthermore that the beam reached a maximum in the compression and the beginning of an expansion at the bunch center followed by a change in the energy difference (slop) sign.

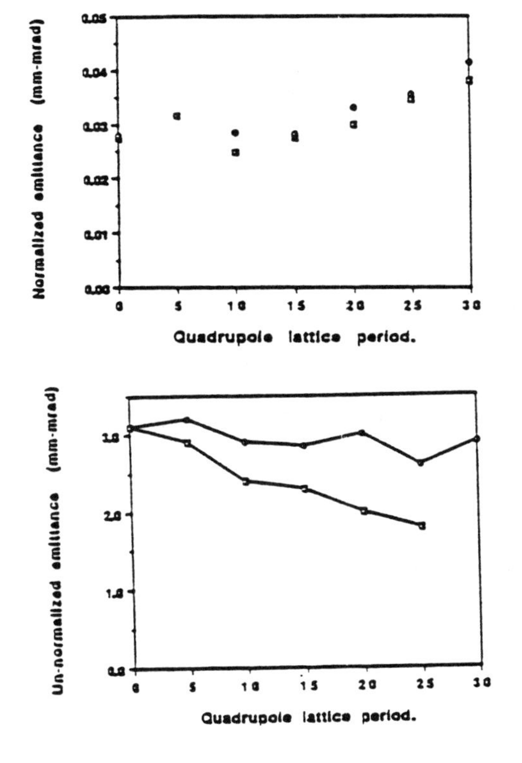

Fig. 5 Emittance measurements along the MBE-4:
(a) Evaluation of the measurements reproducibility.
(b) Normalized emittance conservation in an accelerated beam.

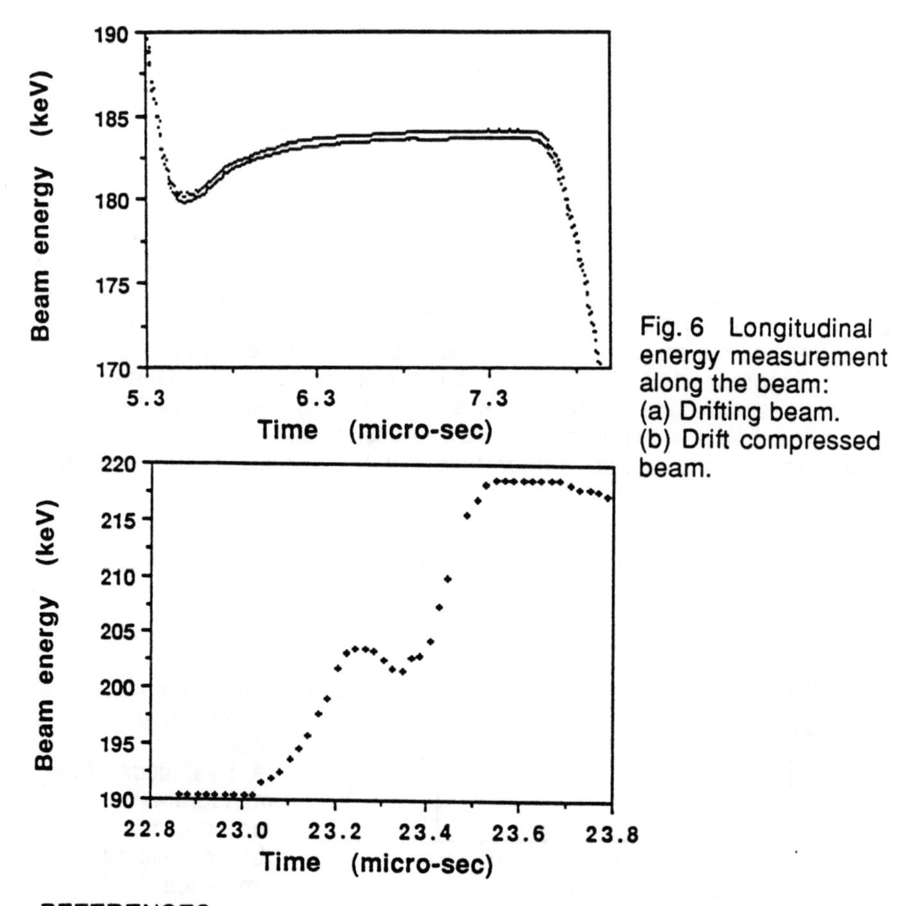

Fig. 6 Longitudinal energy measurement along the beam:
(a) Drifting beam.
(b) Drift compressed beam.

REFERENCES

1. D. Keefe, "Inertial Confinement Fusion," Ann. Rev. Nucl. Part. Sci., **32**, 391 (1982).
2. A. I. Warwick, et al., "Preliminary Report on MBE-4, An experimental Multiple-Beam Induction Linear Accelerator For Heavy Ions. L.B.L. Report 28529, November 1988.
3. T. J. Fessenden, "Emittance Variations in Current-Amplifying Ion Induction Linacs", to appear in the proceedings of the 1991 IEEE-APS Particle Accelerator Conference, May, 1991, San Francisco, CA.
4. M. Gross, "Diagnostic System Program Listings", L.B.L., HIFAR Note-280, Sep. 1990.
5. T. Garvey, et al., "Transverse Emittance Studies of An Induction Accelerator of Heavy Ions", to appear in the proceedings of the 1991 IEEE-APS Particle Accelerator Conference, May, 1991, San Francisco, CA.

Design of Beam Position Monitor Electronics for the APS Diagnostics.

E. Kahana
Argonne National Laboratories
Argonne, Illinois 60439

1. Introduction.

The BPM system must accurately measure the position of the positron bunches in the Storage ring. The sensors are a set of four capacitive pick-ups called "buttons", arranged symmetrically relative to the X and Y axis (Reference 1). There are 360 sets, approximately every one degree around the ring. The accuracy and resolution requirements for the single turn condition are 0.5 mm and 0.2 mm respectively, and for the multiple turns condition are 0.2 mm and 0.02 mm respectively. In the multiple turns condition the BPM system shall be capable of measuring each bunch once per revolution, and the data will be integrated by a signal processor for up to 256 revolutions.

The selected approach is the use of the AM/PM Conversion Monopulse method for measuring the beam position in the X and Y axis, as well as measuring the beam intensity. This selection was made due to the high accuracy achievable by this method and the capability to obtain measurements on a single turn.

The circuit includes three units: the Filter-Comparator unit, the Monopulse Receiver unit and the Signal Conditioning and Digitizing Unit (SCDU). The Filter-Comparator unit is located in the accelerator ring tunnel, while the other two units are located on a VXI board in the instrumentation cabinet. In the following paper each unit will be described, the technical specifications will be given and design considerations will be presented.

2. Filter-Comparator.

The functions of the Filter-Comparator unit are:

a. To convert the voltage impulse from the "buttons" to a pulse modulated at 351.93 MHz, for each of the four capacitive buttons used to sense the beam.

b. To compare the four RF signals and to create a beam intensity signal (Σ) and two deviation signals, one for the X axis (Δx) and one for the Y axis (Δy).

c. To provide a trigger for the timing circuit.

A block diagram of the Filter -Comparator is shown in Figure 1.

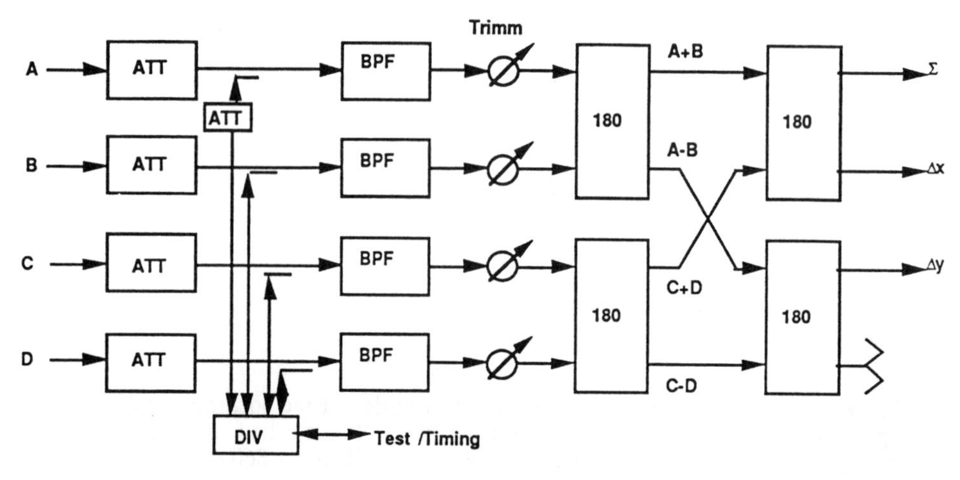

Fig. 1: Filter-Comparator Unit

Table 1 is a summary of the Filter-Comparator specification.

Table 1. Filter-Comparator Specifications.

Center frequency: 351.93 MHz
Bandwidth (3 db): 10 MHz nominal
Filter type: Bessel or Gauss
No. of sections: 3 to 5
Impedance: 50 ohm
Cancellation ratio: 50 db min.
Power handling: 5 W average, 1 Kw peak
Insertion loss: 4 db max., plus 6 db for the front attenuator
Test input coupling: 10 db.nominal
Time domain sidebands rejection: 60 db min.
Input leakage and high order signals rejection: 60 db min.

The selection of the ringing frequency for the filters is related to the accelerator design and harmonic number. The bandwidth is 10 MHz, in order to ensure a 100 nanosecond pulsewidth. The filters shall have no overshoot or undershoot in the dynamic range of interest, therefore a Gauss or Bessel characteristic is required. The four filters shall be very well matched, in order to ensure the cancellation by the 180 degrees hybrids. Phase and amplitude trimmers are used to obtain the maximum cancellation. 6 db pads are used at each filter input, to attenuate standing waves due to filter reflections. The 4-way power divider and the four couplers have two functions:

a. Couple and sum the filters' reflected signals, to be used for timing purposes in the BPM, as will be described further.

b. Distribute a test signal, injected for BPM testing. A small pad is added on one of the test lines, to create a calibrated imbalance.

The Filter-Comparator is located inside the accelerator tunnel, as close as possible to the buttons, in order to minimize matching problems. During operation, the unit may be subjected to Gamma radiation of up to 50 Krad/year. It is known that Teflon type materials are damaged by radiation, therefore they shall not be used on this unit. Rexolite type materials have been proven to withstand radiation much better, therefore it is recommended to use Rexolite dielectric for connectors.

3. Monopulse Receiver

The functions of the Monopulse Receiver are:
a. To amplify the signals to meet the required dynamic range.
b. To convert the amplitude differences between Σ and Δx and Δy into phase differences.
c. Measure the phase differences between the signals, which are proportional to the displacement errors in the X and Y axis, on a turn to turn basis.
d. Measure the Σ signal level.
A block diagram of the Monopulse Receiver is shown in Fig. 2.

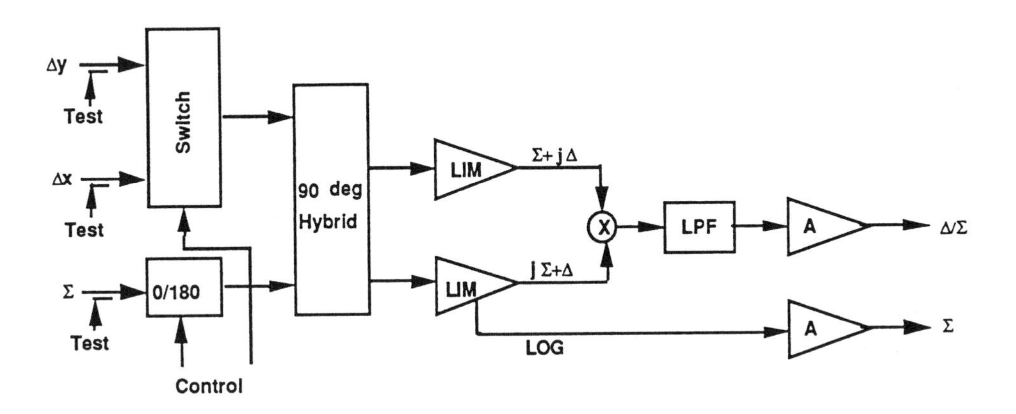

Fig.2: Monopulse Receiver.

Table 2 is a summary of the Monopulse Receiver specifications.

Table 2: Monopulse Receiver Specifications.

Input frequency: 351.93 MHz
Bandwidth: Optimized for settling time and noise
Impedance: 50 ohm
Input VSWR: < 1.2:1
Logarithmic amplifier linearity: +/- 1 db
Logarithmic amplifier output: 0 to 2 V
Input Dynamic range: -70 to -10 dbm
Phase detector output: +/- 1 V
Output noise: < 10 mV for -70 to -50 dbm, < 1 mV for -50 to -10 dbm
Settling time: < 20 nanosec. to 0.1%
Ratio accuracy: 50 db
Slope Accuracy: +/- 0.5% max. for phase detector from -45 deg to +45 deg.
Test input coupling: 10 db.
Channel switching time: <1 microsecond to 0.1% accuracy
DC Voltages: +/- 12 V

The critical part of the receiver is the conversion from amplitude to phase and the accurate measurement of the phase. The amplitude to phase conversion is achieved by the 90 degrees hybrids, which create the signals $\Sigma+j\Delta$ and $j\Sigma+\Delta$. These two signals are then limited by phase matched limiting amplifiers. The phase detectors will produce a bipolar video signal which is proportional to the phase difference between the two signals. The receiver will process the X and Y errors on alternate turns, by means of the switch on the Δ channel. In order to improve accuracy, a 0/180 degrees switched phase shifter is used on the Σ channel. By changing the sign of the Σ signal every revolution, bias errors are averaged out. Phase trimmers may also be used in order to improve the accuracy.

The Monopulse Receiver is located on a VXI board in the Instrumentation Cabinet. It shall be packaged in a box measuring no more than 5"x5"x0.5".

4. Signal Conditioning and Digitizing Unit.

The functions of the SCDU are as follows:

a. Detect the bunch signal in a time frame defined by the control system

b. Generate timing gates for the integration of the X and Y error signals and the detection of the intensity signal.

c. Integrate and measure the X and Y error signals

d. Measure the peak of the intensity signal.

e. Generate the commands to the X/Y switch and the 0/180 degrees phase shifter in the receiver.

f. Digitize the X and Y error signals as well as the intensity signal, by using 12 bit A/D converters.

g. Latch the data to be transferred to the FIFO memory board via a VXI interface.

Each BPM is "armed" once per revolution, in synchronization with the time of arrival of the selected bunch at the specific BPM. The "arm" gate is derived from the main clock, in a timing circuit which is not part of the BPM. The BPM internal timing is derived from the "buttons" signal, which is reflected by the ringing filters and passed to the timing circuit.

The SCDU block diagram is shown in Fig. 3.

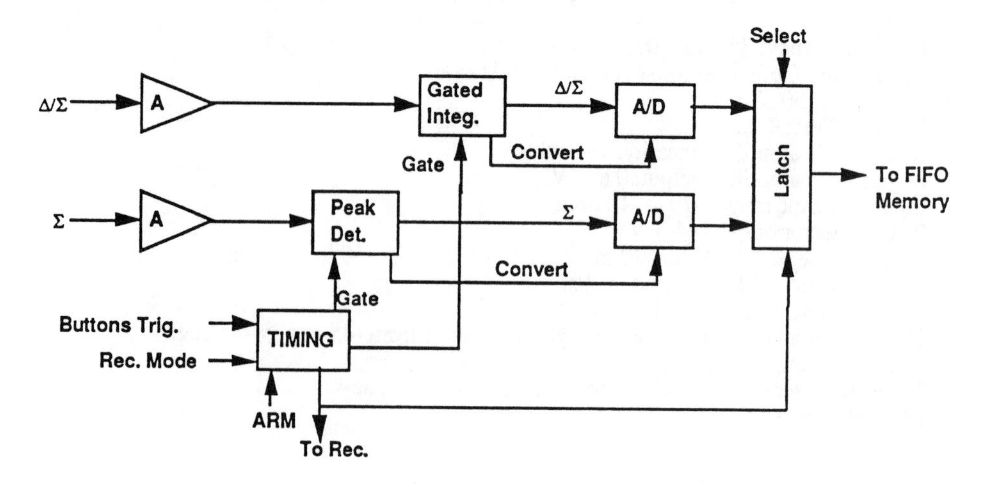

Fig. 3: SCDU

The signals from the Monopulse Receiver are first conditioned. Then, using the "buttons" signal, a gate is created for the Gated Integrator and the Peak Detector. The Gated Integrator will integrate the X or Y error signal during the gate time, in order to improve the S/N, and the value will be hold for digitizing. The Peak Detector will hold the peak of the signal in order to be digitized, while issuing the "convert" command to the A/D.

The timing circuit creates the gates for the Peak Detector and the Gated integrator, as well as the control for the receiver Δ channel switch and the 0/180 degrees switch. An external control will define one of three modes: X only, Y only or X/Y alternate. In this last case, we have four conditions, alternating as follows: 0/X, 0/Y, 180/X and 180/Y. During the switching time, the data is masked as invalid. Since the switching time is approximately 1 microsecond, the switching shall occur at the end of the turn period, after the signals were hold for conversion.

The A/D's are of the sampling type, with 12 bits resolution and 1 microsecond total conversion time. The encoder will add to the difference digital data 2 bits for the receiver status. The digital output is latched and transferred, upon receiving a select command, to a FIFO memory board, to be used for feedback, interlock and off-line data analysis.

The timing diagram is shown in figure 4. The following is a description of the timing diagrams:

1. The "Arm" signal enables the BPM to activate its measurement capability, when the bunch is at its location. The "Arm" signal is derived from the main clock by a separate circuit, not part of the BPM.

2. Using the "buttons" signal, a gate of about 50 nsec. is generated, which enables a gated integrator for the difference channel and a peak detector for the sum channel. The "buttons" signal is used in order to provide a high accuracy self triggering capability.

3. The gated integrator is used for the difference channel, in order to improve the measurement accuracy. The integrated signal is held, digitized and buffered for further processing

4. The gated peak detector will detect the peak of the sum signal and will hold, digitize and buffer it for further processing.

5. Once the signals are sampled and transferred to the A/D's, the BPM is reset, the receiver is switched to a new condition and the BPM is ready to take a new measurement.

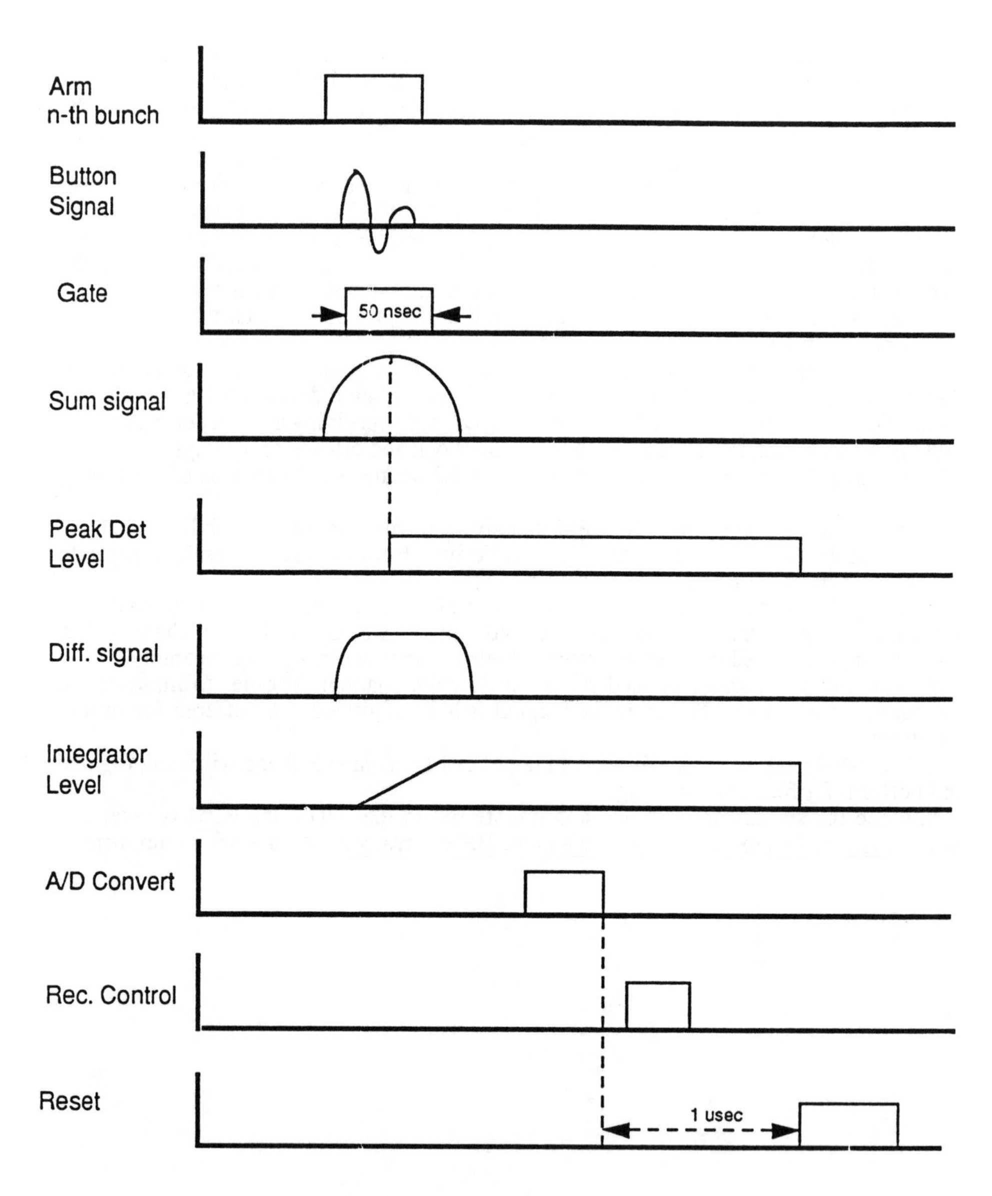

Fig. 4: Timing Diagram

5. References.

1. G. Decker, Y. Chung and E. Kahana: "Progress on the Development of APS Beam Position Monitoring System", Proceedings of the Workshop on Advanced Beam Instrumentation, April 1991, Vol. 1, pg. 254-261.

Comparative Study of the Proposed Absolute Energy Measurements for CEBAF

I. P. Karabekov
(Yerevan Physics Institute)
(Yerevan, Alikhanian Brothers St. 2, Republic of Armenia)

Abstract

The need to measure the incident beam energy close to the target absolutely with precision no worse than one part in 10^4 is an important part of the CEBAF Hall A physics program. Since the appearance of the Conceptual Design Report of 13 April 1990, where the requirement was established, the problem has been considered at both CEBAF and the Yerevan Physics Institute.

The following methods were considered for solving the problem:

- Backscattering of a plane electromagnetic wave by the relativistic electron beam. Calculation shows that the intensity of the backscattered radiation in a bandwith of 10^{-4} near the maximum frequency is about 1 photon per second at 4 GeV and 0.3 mA. (To be published.)
- Three- and four-magnet chicanes [1,2] with appropriate detector systems. Such a system was used at SLAC for absolute measurement of the SLC beam energy with precision 3×10^{-4}. Calculations show that similar accuracy can be achieved for CEBAF in both proposed systems. The tolerances for each type are presented in this paper.
- Angular distribution of synchrotron radiation[3,4,5]. Calculations have shown that precision of about 2.5×10^{-5} is achievable for CEBAF. The tolerances are also presented in this paper [3,4,5].

The high precision energy measurement for CEBAF is discussed.

1.0 Introduction

The precision of measurement of the beam mean energy by use of magnetic spectrometers depends on the accuracy with which the field intensity B and the bending radius ρ are measured. But because we cannot measure ρ directly for bending angle $\Theta_B < 2\pi$, it will be determined using other measurable quantities: deflection of beam trajectory H, bending angle Θ, length of trajectory δs.

All these parameters may be measured with limited accuracy. The main problem is to produce a design of the beam deflection with magnetic field where the error of the measurements are minimal.

The guiding formula for energy determination in the case of an extended magnetic field and used for the bending angle measurement may be written as:

$$E = \left(\frac{\rho}{\delta s}\right) \int_{s_1}^{s_2} B ds. \tag{1}$$

An alternative form adopted for deflection measurement is:

$$E = \left(\frac{300}{\delta s}\right) \int_{s_1}^{s_2} B ds \frac{H}{1 - \cos\Theta_B}. \tag{2}$$

Formulas (1) and (2) are applied to the three and four magnet chicanes and are the basis for the calculations of their tolerances.

Tables of the Main Parameters of the Chicanes Proposed for CEBAF

Table 1. Main parameters of 3-magnet chicane

Parameter	Symbol	Size
1. Bending angle	Θ	10^{-1} rad
2. Magnet length	l_m	1.0 meter
3. SR path length	l_{SR}	7.0 meters
4. Vertical bend angle	Θ_v	$2 \cdot 10^{-3}$
5. Number of spectrometer magnets	N_B	1
6. Number of kicker magnets	N_k	4
7. Number of position detectors	n_{det}	2
8. Total length of chicane	L_T	8 meters

Table 2. Main parameters of 4-magnet chicane

Parameter	Symbol	Size
1. Bending angle	Θ_B	$16° \times 4$
2. Dispersion	D_B	1.73 meters
3. Bending radius	ρ_B	11.1 meters
4. Number of magnets	n_B	4
5. Length of magnets	l_m	3.0 meters
6. Length of long straight section	l_{SL}	3.0 meters
7. Length of short straight section	l_{SS}	2.0 meters
8. Beam position monitor	BPM	
- resolution		10^{-4} rad
- number		4
9. Beam entrance angle resolution	BEA	10^{-5} rad
10. Total length of chicane		20 meters

Layouts of the chicanes are shown in Figure 1.

Proposed Layouts of Spectrometer Chicanes

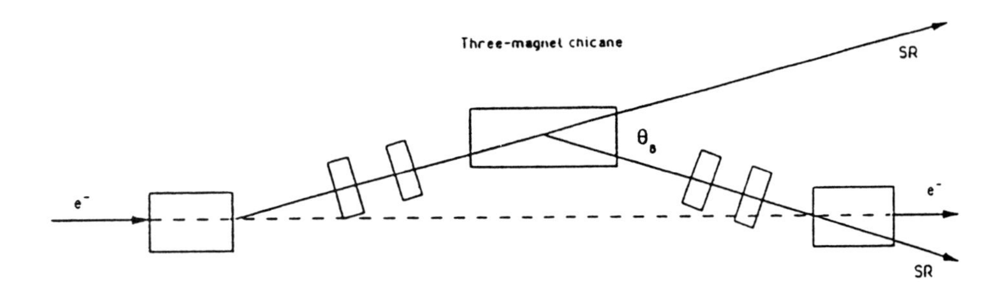

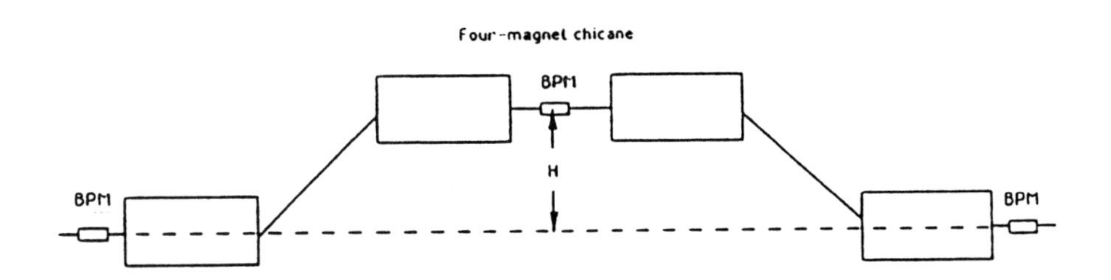

Figure 1

Tables of the Errors Limiting the Resolution

Table 3. Systematic error for 3-magnet chicane

Source of error	formula	size of error	contribution to energy error
Magnetic measurement	$\int B \cdot ds$ + monitoring	10^{-4}	$1. \times 10^{-4}$
Detector position	$\frac{<\Delta x>}{L_s} \sqrt{n_d}$	$0.86 \cdot 10^{-5}$	0.6×10^{-4}
Kicker rotation	$\Delta\phi \cdot \Theta_{v_k} \sqrt{n_k}$ $\Delta\phi = 2 \times 10^{-3}$	$0.8 \cdot 10^{-5}$	$0.8. \times 10^{-4}$
Survey error	$\frac{\lvert\Delta x\rvert}{L_s}$ $\lvert\Delta x\rvert = 30\mu m$	10^{-5}	$1. \times 10^{-4}$

Total $\Delta E = 1.73 \times 10^{-4} E_0$

Table 4. Systematic error for 4-magnet chicane

Source of error	formula	size of error	contribution to energy error
Magnetic measurement	$\frac{1}{\sqrt{n_m}} \int B \cdot ds$ + monitoring	0.5×10^{-7}	0.5×10^{-4}
BPM Measurement	$\sqrt{n_{BPM}} \lvert\Delta x\rvert$	1.73×10^{-4}	$1. \times 10^{-4}$
Beam entrance angle	$\lvert\delta x\rvert = \lvert\delta\phi\rvert\rho \sin 32^\circ$ $+ \lvert\delta\phi\rvert l_{SL}$	$1.6 \cdot 10^{-4}$	$2.06 \cdot 10^{-4}$
Survey error	$\lvert\Delta x\rvert \cdot \sqrt{n_{BPM}}$	5.2×10^{-5}	0.3×10^{-4}

Total $\Delta E = 2.29 \times 10^{-4} E_0$

2.0 Absolute Measurement of the Beam Mean Energy Using Synchrotron Radiation

The principle of the method described in [3] is to measure the variation in the angular distribution of the synchrotron radiation (SR) in the vertical plane. This distribution, which depends upon $\gamma = E_0/mc^2$, is shown in Figure 2. If one can measure and compare the SR beam intensity at two angles, $\psi = 0$ and $\psi = \psi_1$ (see Figure 3), the energy can be determined absolutely, since the relation between the photon fluxes at the two angles can be calculated accurately for the given γ. Consider two horizontal slits of equal width, d, but different lengths ℓ_1, ℓ_2, so that the numbers of photons per unit time through the slits are equal:

$$\Delta\theta_1 \; N_K(\psi = 0) = \Delta\theta_2 \; N_K(\psi = \psi_1), \tag{1}$$

where N_K is the flux density of the radiation in the bandwidth $\Delta\lambda/\lambda = K$. The aperture angles are $\Delta\theta_1 = \ell_1/L$, $\Delta\theta_2 = \ell_2/L$, where L is the distance between the slits and the point on the trajectory where the SR beam originates.

The mean energy E_0 of the beam is determined absolutely when the two photon fluxes are equal, which is measured as the zero difference.

3.0 Synchrotron Radiation Vertical Distribution

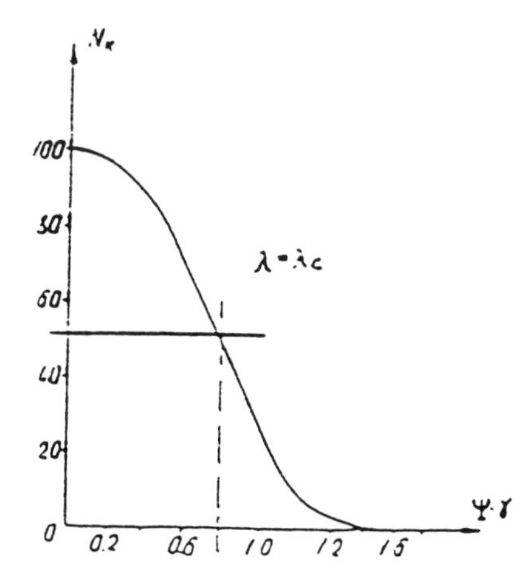

Figure 2. SR photon distribution along the axis vertical to the orbit plane; ψ is the vertical radiation angle and γ is the Lorentz factor of the electron.

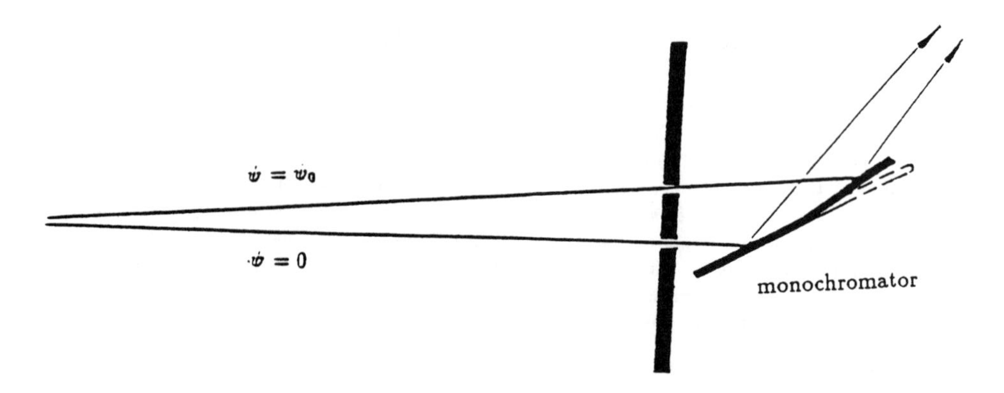

Figure 3. Separation of Photon Fluxes at the Angles $\psi = 0$ and $\psi = \psi_0$

Tolerances and Resolution of the Synchrotron Radiation Analysis Method

The working point is $\gamma\psi = 0.801084$, where the second derivative of the function of photon vertical distribution is equal to zero.

Table 5. Main parameters for AEM-4GeV

Source of Error	Symbol or Formula	Tolerance	Maximum Error $\Delta E/E_0$
angle between orbit plane and the slit system symmetry plane	$\Theta\psi$	$0.945 \cdot 10^{-5}$ rad	10^{-5}
quantum fluctuations	ΔN_k	$2.140 \cdot 10^{5}\,s^{-1}$	$0.5 \cdot 10^{-5}$
the arc in bending	$\Delta\theta$	0.148 rad	10^{-5}
wavelength	$\Delta\lambda/\lambda$	$2.700 \cdot 10^{-3}$	10^{-5}
accuracy of measurement of field intensity	$\Delta B/B$	$0.600 \cdot 10^{-4}$	10^{-5}
accuracy of measurement of the monochromator angles	$\delta\theta$	$2.720 \cdot 10^{-5}$	10^{-5}
emittance parameters	$\delta\psi = y' + (y/L)$	$1.250 \cdot 10^{-5}$ rad at $L = 5 \cdot 10^{3}$ mm	10^{-5}

Total $\Delta E = 2.5 \times 10^{-5} E_0$

4.0 An Optimal Variant of CEBAF High Precision Energy Measurement

The above results of the investigations and possibilities at CEBAF permit us to propose the following solution for the beam mean energy measurement with the precision of 10^{-4} and higher. It is evident that this goal may be achieved only by the proposed method of measuring the synchrotron radiation vertical distribution, with the tolerances presented in Table 5.[3,4] On the other hand, for the creation of SR, a high precision magnetic field is required. For this purpose, a special short magnet will be inserted in the extracted beam transport system as close as possible to the target area (see Figure 4). The best place for this magnet is the position of the next-to-last dipole, which for this case will be removed.

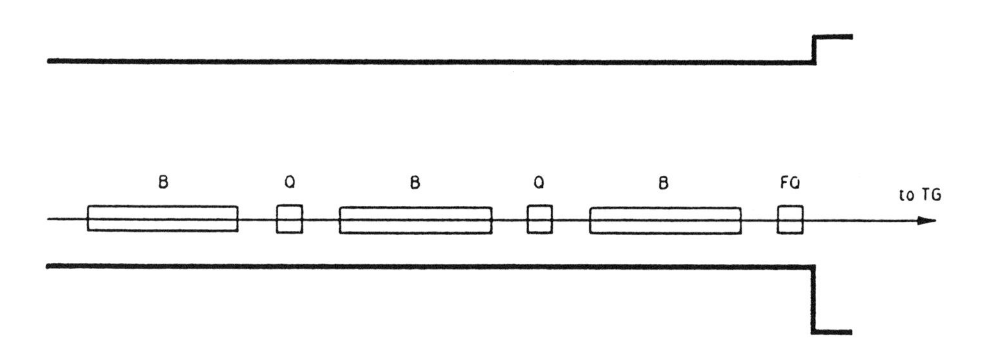

Figure 4

To compensate for the distortion of the electron beam path caused by the inserted magnet,an optical system like the three-magnet chicane is required. Such a kind of optics may be performed using neighboring ordinary magnets by increasing the bend angle (see Figure 5). The additional angle $\delta\Theta$ may be achieved by increasing the current in magnet conductors. The tolerable increase in the current is about 20%, which creates accordingly $\delta\Theta = 0.2\,\text{rad}$.

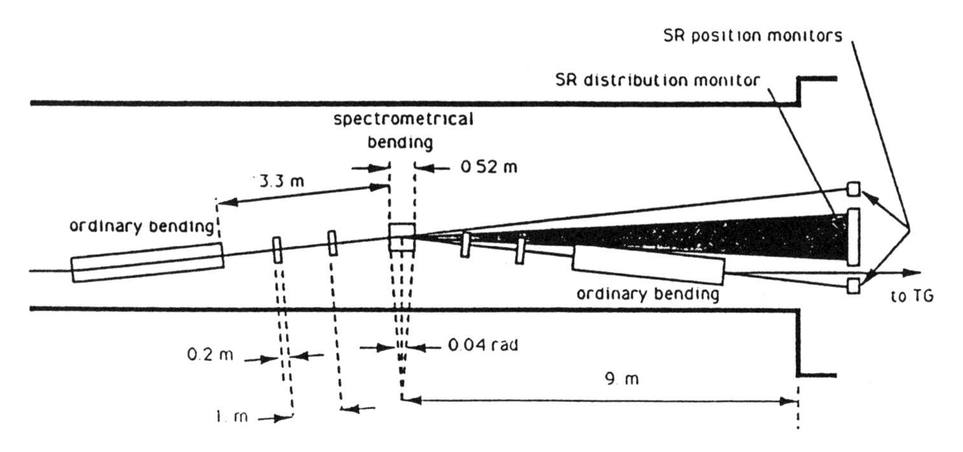

Figure 5

The length of the high precision magnet is about 0.5 m; straight sections of more than 2 m each are sufficient to install four 0.2 m long kicker magnets. If the SR path length is about 7 m and the distance between them is measured with accuracy 3×10^{-5} m, the preliminary beam absolute energy determination will have accuracy of three units of 10^{-4} and ultimately with accuracy up to $2.5 \cdot 10^{-5}$ using the SR analysis method.

The main parameters of the inserted optics and the main error sources limiting the resolution are presented in Tables 6 and 7.

Table 6. Main parameters of the proposed chicane

Parameter	Symbol	Size
1. Bending angle	Θ	4×10^{-2}
2. Magnet length	L_m	0.52 m
3. SR path length	L_{SR}	7.0 meters
4. Vertical bend angle	Θ_v	$2 \cdot 10^{-3}$ rad
5. Number of spectrometer magnets	N_B	1
6. Number of kicker magnets	N_k	4
7. Number of position detectors	n_{det}	2
8. Total length of chicane	L_T	no additional space needed

Table 7. Systematic errors for the proposed chicane

Source of Error	Size of Error	Contribution to energy error
1. Magnetic measurement	$1. \times 10^{-4}$	$1. \times 10^{-4}$
2. Detector position resolution	5.76×10^{-6} rad	1.44×10^{-4}
3. Kicker rotation	8×10^{-6} rad	2.0×10^{-4}
4. Survey error	0.5×10^{-5} rad	1.1×10^{-4}

Total $\Delta E = 2.25 \times 10^{-4} E_0$

5.0 References

[1] J. Kent *et al.*, "Precision measurement of the SLC beam energy," SLAC-PUB-4922, LBL-26977, March 1989.

[2] B. Bevins and P. Kloeppel, "Proposed high accuracy beam energy measurement at CEBAF," (presented to this workshop).

[3] I. P. Karabekov *et al.*, Nucl. Instrum. Methods **A286**, 37-40(1990).

[4] I. P. Karabekov, R. Rossmanith, "Measurements of the absolute value of the beam energy and its deviation at CEBAF," CEBAF TN 90-0224, April 23, 1990.

[5] I. P. Karabekov, "Specification for construction of absolute energy monitor for CEBAF," CEBAF TN 91-045, July 8, 1991.

DESIGN AND OPERATION OF THE AGS BOOSTER IONIZATION PROFILE MONITOR*

A.N. Stillman, R.E. Thern, W.H. Van Zwienen, R.L. Witkover
AGS Department, Brookhaven National Laboratory
Upton, NY 11976

Abstract

The AGS Booster Ionization Profile Monitor (IPM) must operate in a vacuum of about 3 x 10^{-11} Torr. The ultra-high vacuum imposes certain requirements on detector gain and restrictions on construction techniques. Each detector is a two-stage micro-channel plate with an integral substrate containing sixty-four printed anodes. Formed electrodes provide uniform collection fields without the use of resistors, which would be unacceptable in these vacuum conditions. An ultra-violet light calibrates the detector in its permanent mounting. An extra set of electrodes performs a first order correction to the perturbations imposed by the horizontal and vertical collection electrodes. This paper will present details of the design of the profile monitor and recent operational results.

INTRODUCTION

The AGS Booster is a synchrotron whose purpose is three fold. It will increase the AGS proton intensity by allowing the injection of four 1.5 GeV pulses rather than one 200 MeV pulse. It will accumulate twenty pulses of polarized protons to boost their intensity as well. Lastly, it will accelerate heavy ions to a momentum suitable for injection into the AGS and RHIC, the Relativistic Heavy Ion Collider. It is an intensifier for AGS beams and a necessary pre-accelerator for RHIC.

The profile monitor in the AGS Booster is a residual gas ionization monitor. The Booster beam ionizes gas molecules in the volume over two sets of collecting wires, one vertical, the other horizontal. The ions would naturally drift radially outward from the space charge of the beam, but electrical fields that are strong enough to redirect their natural radial motion send them to the wires. For strong enough collection fields, the ions travel in rather straight trajectories from their point of generation to the collection wires. The pattern on the wires is thus either a horizontal or vertical projection of the intensity of ion production, which is directly proportional to the beam intensity. Standard, high sensitivity integrating electronics and a real time computer interface acquire the signals, display them, and allow control of the collecting fields and other variables. Figure 1 is a block diagram of the profile monitor and its electronics.

* Work performed under the auspices of the U.S. Dept. of Energy.

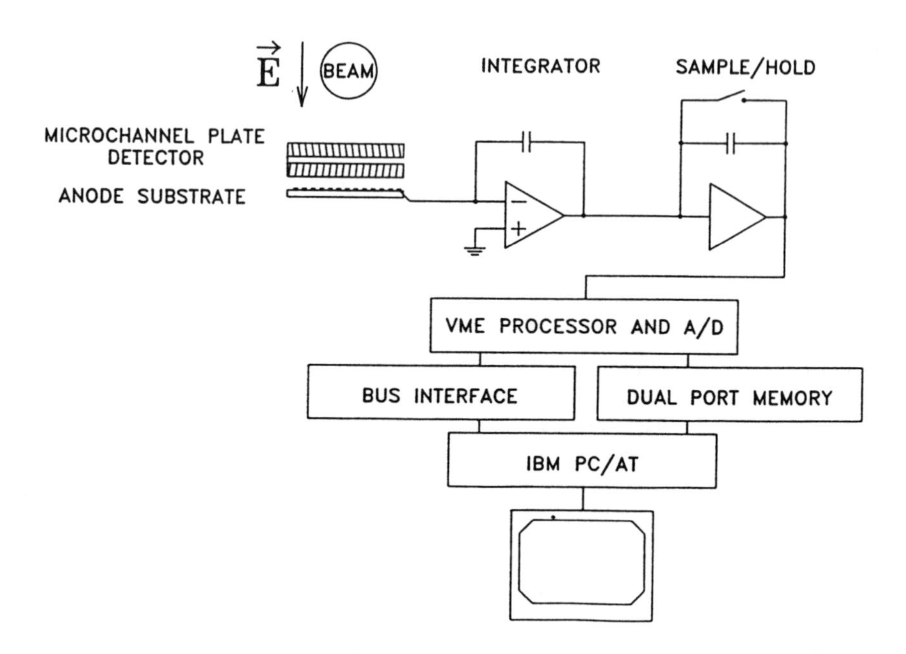

Fig. 1. A block diagram of the Ionization Profile Monitor and its readout electronics. AGS Booster timing signals control the integration and sampling times.

The most severe constraint on the ionization profile monitor design is the ultra-high vacuum of the AGS Booster, nominally 3 x 10^{-11} Torr. This pressure corresponds to a residual gas density of approximately 10^6 gas molecules per cm^3. At these rarefactions, the beam produces ionization signals on the order of picoamps. Furthermore, there is an extremely limited range of construction materials and techniques that one can use in vacuum of this quality. Resistors, for instance, cannot grade potentials in the vacuum since they outgas too much. Aluminum softens at the bake out temperatures in use, so metal parts must be of stainless steel. An appropriate choice of detector, therefore, is a microchannel plate. The microchannel plate has high gain, operates only in a vacuum, and can tolerate bake out temperatures of 300°C.

MCP DETECTOR

A typical microchannel plate has a gain of 10^4 electrons per input electron. Cascaded plates provide higher gain, though not strictly the product of the gains of the individual plates, since saturation effects in the last plate dominate very quickly. In the AGS Booster IPM, the detector is a dual microchannel plate[1] with a specified gain of 10^7 electrons per electron. Ionic inputs to microchannel plates cause slightly different gains from the electron inputs. In fact, the gain of a typical plate can vary with ion species and energy.[2] In order to generate beam inten-sity profiles that are not merely qualitative, the profile monitor design must make these gain variations insignificant. Careful shaping of the electrical collecting fields minimizes these gain effects.

The heart of the profile monitor is the dual microchannel plate detector, which comprises the microchannel plates and their associated anode array. A small distance from the exit end of the plates is a ceramic substrate with sixty-four linear anodes. Each anode is a collecting wire, and runs the length of the plate. The whole detector assembly attaches to four supporting posts on the vacuum side of a flange. Ceramic insulation covers the wires from the detector to two thirty-five pin instrumentation feed-throughs. Ceramic also insulates the wires that provide the bias voltage to the front and back sides of the plates. The unity of the detector/anode assembly allows the anode spacing to be at a rather fine pitch. In fact, the anodes are 1.1 mm wide and are 1.47 mm apart. The length of the anodes is 75 mm, the length of the detector active area.

The gain of the detector is set by the voltage across the input and output faces of the dual plate. There is no electrical separation between the two plates. Generally, the voltage is about 1 kV per plate for the specified gain of 10^7. However, the second plate can easily saturate in the presence of large input signals. This saturation is very deleterious, since microchannel plates have a finite life determined only by the charge extracted from them. Accordingly, the power supply that provides bias volt-age also contains a protection circuit. This circuit reduces the bias voltage to half the full scale voltage, i.e. 500 V per plate. The circuit trips when the current flowing through the detector is .7% of the current that flows with no signal. This is the recom-mended value for the fraction of strip current at which to trip. This power supply circuit uses an optically isolated operational amplifier to compare the signal current to a set current. It also provides a small differential voltage between the exit side of the last microchannel plate and the collecting anodes.

FIELD SHAPING

The collecting fields in this profile monitor merit special consideration. There are constraints on them as well. The primary constraint is that they direct ions to the face of the detector in trajectories that accurately project the beam inten-sity distribution onto the anode wires. This voltage gradient, sufficient to overcome beam space charge effects, seems to be subject to a law of diminishing returns. There always seems to be a further reduction of beam profile width as the collection volt-age increases.[3] However, the ability to generate large fields in confined vacuum chambers falls off rapidly at the higher voltages. The historical value of the voltage gradient at the AGS is somewhere in the range of 1 kV/cm.

A second constraint on the collecting field value, not often noted in devices like these, is that the microchannel plate is an energy analyzer for ions below about 20 keV. Thus, to form accurate profiles, the minimum ionic energy must be above this threshold. The ions of minimum energy form in the tail of the beam closest to the face of the detector, so the collecting field at this distance must accelerate these ions to 20 keV. With careful electrode design, it is possible to maintain a potential of -20 kV at this distance.

The ultra-high vacuum conditions also cause problems in generating graded potentials. Since resistors are forbidden and external resistors would require prohibitively many feedthroughs, shaping of the collection field electrodes is the best way to shape the field itself. Figure 2 shows the final shape of the electrodes and a plot of the equipotential lines. The size of the walls of the box-shaped electrodes should be one quarter the length of the electrode.[4] A close inspection of the equipotentials shows that they are certainly flat enough not to cause any distortion of the beam profile image. A four inch clear aperture being necessary for the beam, the electrodes must form part of the perimeter of an eight inch cube. They are of stainless steel, polished and with no sharp corners.

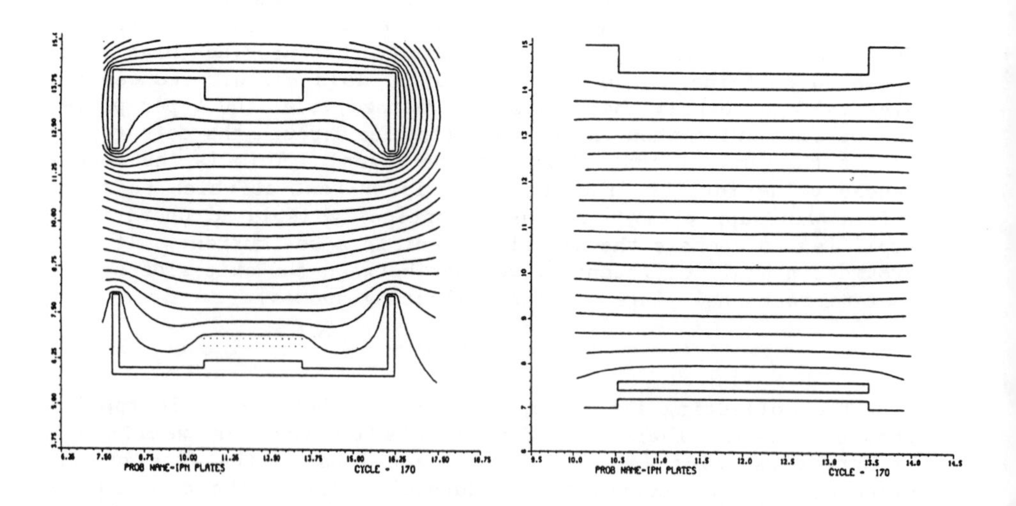

Fig. 2. Poisson calculations of equipotential lines in the ionization profile monitor collection volume. These views are cross-sections of the collection volume and show the characteristic U-shape of the electrodes. Each electrode lip is one quarter the length of the electrode. In both views the -70 kV electrode is at the top and the ground electrode at the bottom;

the voltage between potential lines is 3.5 kV. Scale is in cm. In (a), the input face of the microchannel plate detector is the upper dotted line. This face is held at -2 kV by the bias power supply. The characteristic dual plate capacitor potential lines have given way to flat potential lines due to the lips on the electrodes. In (b), a magnified view of the collecting volume shows that its electric field is essentially constant. Here the microchannel plate detector is the rectangle above the lower electrode.

Each plane has a set of electrodes and the working voltage on each set is nominally 70 kV. To correct the kick on the beam from these fields, a third set of electrodes is between the two collection electrodes. These correction electrodes are at 45° to the horizontal and vertical collection electrodes. The corrector voltage, 99 kV, generates an electric field that cancels, to first order, the effect of the vector sum of the collector fields. Figure 3 indicates the electrodes and their electric fields.

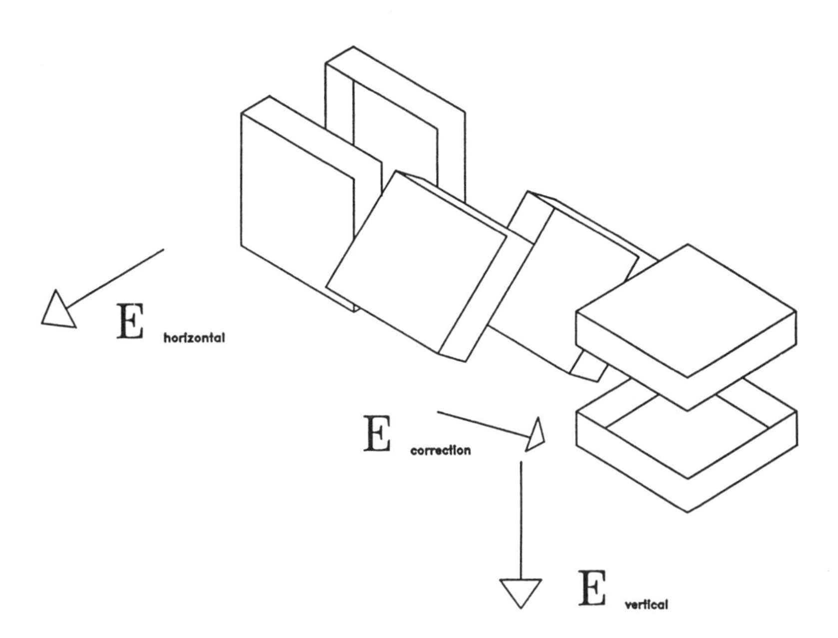

Fig. 3. Schematic diagram of the orientation of the collection electrodes and the corrector electrode. The corrector electrode generates an electric field opposite to the vector sum of the collector fields. The beam direction is from lower right to upper left.

READOUT ELECTRONICS

Each detector plane has sixty-four channels. Each of these channels is input to a low leakage integrator. A commercial VME sample/hold multiplexer and A/D converter with an imbedded microprocessor stores the integrator outputs in a local memory. An IBM PC/AT then reads the memory locations via a commercial bus interface that connects the PC bus to the VME bus. The A/D can scan through all channels and load them into memory in about 1 msec. Timing for the data acquisition is by real time interrupts to the PC, which controls the scanner as if it were the VME host. The timing signals themselves are either standard AGS Booster time line codes or the Booster T0 signal. Both are available for selection by the PC.

A simple calibration system is also available under PC control. An ultra-violet spectroscopy lamp of 180 nm wavelength illuminates the detector briefly in response to a calibration pulse from the computer. The light shines on the detector face through a sapphire window in the vacuum chamber and then through the hole in the top of the collection electrode, since the lamp is physically outside the vacuum. The quantum efficiency of the microchannel plates at this wavelength is about .5%. This calibration system can provide a roughly uniform input to the detector and will monitor the inevitable gain decline of the detectors' gains.

SPECIAL CONSIDERATIONS

High vacuum systems and high gain microchannel plates come with their own special set of precautions. All electrode and mechanical support materials are either stainless steel or ceramic. They must be bankable to 350°C and in the case of metal parts, vacuum fired. Gas evolution is a serious problem in high vacuum systems and devices not essential to acceleration or control should require as little pumping as possible. Microchannel plates in particular may take days to pump down.

The microchannel plates also merit special handling. They store well under a nitrogen atmosphere or in vacuum, but they do not like air for extended periods of a year or so. During assembly into a detector, clean room practices are necessary, including gloves and preferable filtered air. Testing the detector before inserting it into the vacuum chamber consists of verifying a high resistance (20 to 200 megohms) between the faces of each microchannel plate. Any application of bias voltage to the plates will start the electron multiplication. If this bias voltage is present while the plates are in a poor vacuum or in air, the plate will destroy itself due to saturation currents. Essentially, its useful life is spent in the test.

OPERATIONS AND RESULTS

The first use of the IPM came during the AGS Booster commissioning period. Before the Booster had any beam, the calibration system generated a set of profiles. Commands from the remote computer turned on the calibration lamps, which provided the signal for the detectors. With a bias of about 600V per microchannel plate (1200V per detector), small signals appeared in the central positions. These were on the order of the quantization error of the A/D electronics. Increasing the bias voltage to about 800V per plate caused the abrupt appearance of sharp profiles. Thereafter, as the profiles grew with bias, the gain seemed to follow the standard gain versus voltage relationship for photon multipliers. To avoid saturating the output current in the second plate, the bias never went above 1000V. Figure 4 shows typical profiles after using the lamp. The illumination passes through a circular aperture in the high voltage electrode and causes the steep shoulders and non-gaussian shape.

When Booster beam was available, the procedure for obtaining profiles differed slightly. Capturing profiles of real beam requires a collection voltage to be present on the electrodes. Since the power supplies that generate the collection voltages reach their terminal voltage rather slowly, they are turned on first. The AGS Booster control system generates the timing signal for the sample/hold gate and the PC then captures profile data. At this point, the only operator requirements are the setting of voltages for bias and collection, and the setting of gain for the initial stage of integrating electronics. The timing of profile acquisition is preset through the AGS Booster controls system. This timing scheme allows a timing resolution of microseconds throughout the Booster acceleration cycle. Figure 5 is an example of some of the first profiles of the Booster beam. Notice that the horizontal and vertical profiles have different heights and widths. Further analysis showed that the total counts in each profile were the same to within 5%, as they should be.

Presently, the software that controls the IPM also massages the display. This is new since the acquisition of the very first profiles and completes the basic design of the IPM. Figure 6 shows the most recent type of display. The program computes the first three moments of the profile and does a Gaussian fit to the data, superimposing the fit on the profile. These software improvements give a polished look to the data and also allow the rapid evaluation of peak position and total counts.

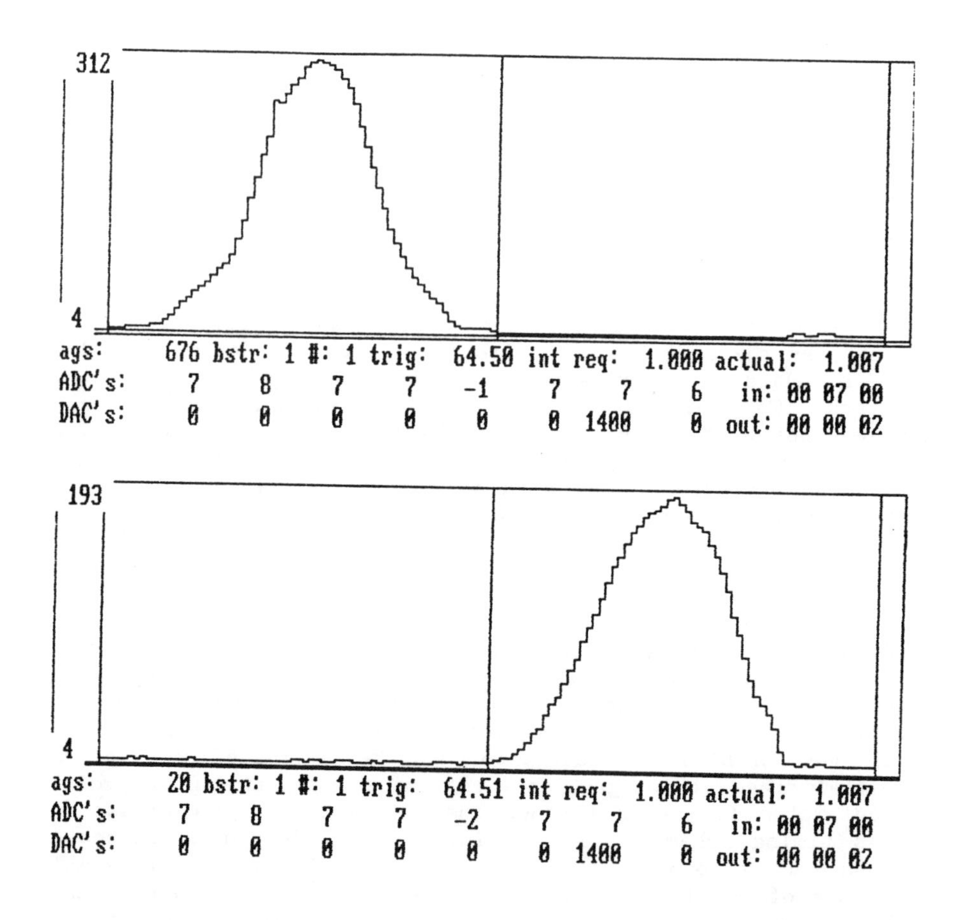

Fig. 4 Calibration system profiles using an ultraviolet lamp. The lamp generates 180 nm photons which the microchannel plate detectors amplify with a quantum efficiency of about .5%. The irregular shapes of the profiles are due to the circular aperture in the electrode through which the light shines, and the casual placement of the lamp. The table at the bottom of the display indicates various readbacks from the electronics and timing signals for the integrator gate. (a) Horizontal detector. (b) Vertical detector.

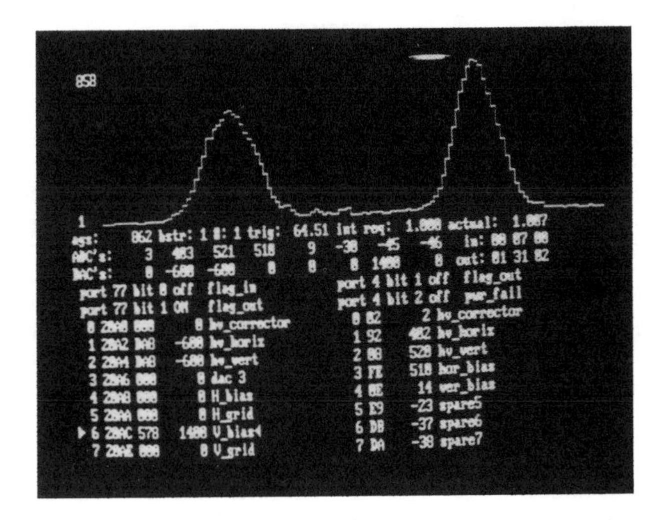

Fig. 5. The first profiles of actual beam in the Booster. The horizontal and vertical planes have different shapes, but the area under each profile is the same. The table below the display is a list of electronic and timing setpoints used in IPM trouble-shooting during the first attempts at operation.

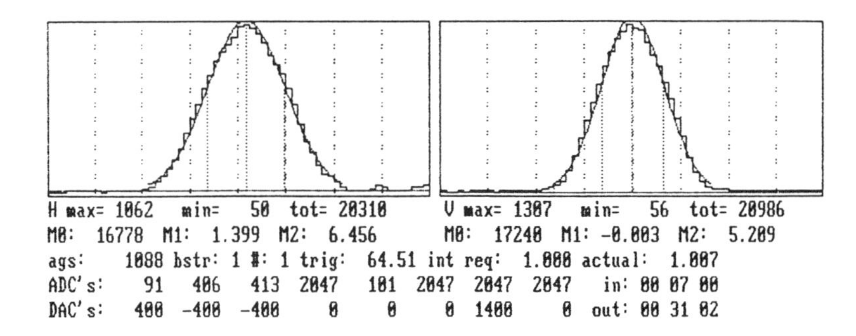

Fig. 6. The present form of display for the profile data. The numerical data give the first three moments of the profiles and readbacks described in Fig. 5. The other data are channel statistics. The software autoscales the display, thus both horizontal and vertical planes have the same height, though the numerical maxima differ. Note the relative constancy of M0, the area under the profiles. Since the same beam generates both profiles, identical detectors would give identical areas.

SUMMARY

The AGS Booster Ionization Profile Monitor has the ability to generate beam profiles even in the tenuous residual gas of a 3 x 10^{-11} Torr vacuum. It uses a microchannel plate detector with an integral anode assembly to generate profiles with a 1.47 mm wire pitch. A computer based scanning ADC system allows for profile acquisition on AGS cycle time scales. Formed electrodes generate fields without the use of resistors and the high voltage design minimizes the distorting effects of beam space charge and differential detector response. A calibration system is integral to the design and first results are very promising. They indicate that the AGS Booster IPM will work in a vacuum range where ionization devices have never seen use.

REFERENCES

1. Galileo Model 3810 with 64-element anode. Manufactured by Galileo Electro-Optics Corp., Galileo Park, P.O. Box 550, Sturbridge MA, 01566.

2. Hellsing, M., et al., "Performance of a Micro-channel Plate Ion Detector in the Energy Range 3-25 keV," J. Phys. E: Sci. Instrum., 18, 920 (1985).

3. Thern, R.E., "Space-Charge Distortion in the Brookhaven Ionization Profile Monitor," Proc. 1987 IEEE Particle Accelerator Conference, Washington, D.C.

4. Leal-Quiros, E., and Prelas, M.A., "New Tilted Poles Wien Filter with Enhanced Performance," Rev. Sci. Instr., 60, No. 3, 350 (1989).

TYPICAL SPECIFICATIONS OF VARIOUS BEAM DIAGNOSTIC DEVICES

P. Strehl[1]

GSI, D6100 Darmstadt, FRG

ABSTRACT

The characteristics of some important beam diagnostic elements, including the mechanical design parameters and electronic signal processing are discussed. Special reference is given to the limitations, determined by the electrical noise at low intensities and arising from thermal effects for high intense beams.

INTRODUCTION

The main quantities which are important for all kind of particle accelerators are : beam current i(t), beam position, beam profile i(x,y), beam emittance in the transverse and the longitudinal phase spaces. In the following we will discuss typical specifications of representative beam diagnostic devices, provided for the measurement of these quantities.

BEAM CURRENT

A often used device is the Faraday cup. For uncooled cups the low current detection limit is in the order of some pA. This may be shifted to even some nA for cooled Faraday cups due to the formation of galvanic elements. For the suppression of secondary electrons a combination of an electric field in front of a magnetic field is very efficient. Magnetic fields should be in the order of some 100 Gauss, which can be achieved using small permanent magnets. Beyond the current limits a well designed Faraday cup measures the absolute current with an accuracy of some percent. The design of Faraday cups becomes more complicated if thermal problems due to intense beams arise. In such cases the necessary energy to melt the range volume $V(R,t)=\pi R^2 \cdot t$ should be calculated. Here R is the beam radius and t the range of the particles within the material under discussion. The energy needed to melt or even vaporise a given volume can be very roughly estimated from the energies for :

- heating up V(R,T) to the melting point, melting the material,
- heating up to the boiling temperature, vaporising the material.

---

1. Consulting physicist of Scientific International Inc., Princeton, N.J. 08542

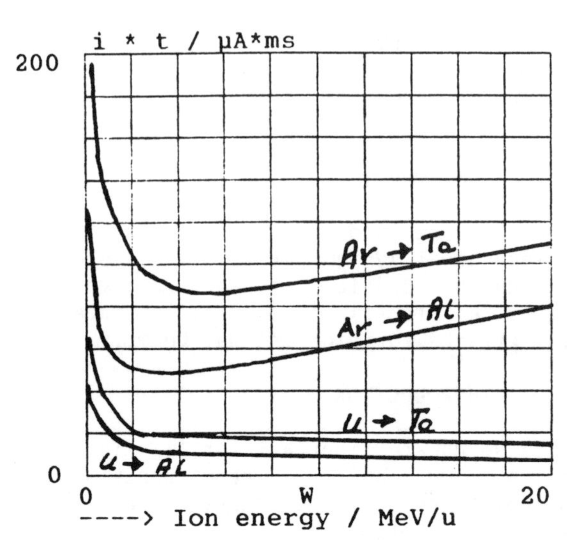

Fig. 1 : Necessary energy for melting

Fig. 1 gives the current·time product for Argon and Uranium beams to melt the range volume of Aluminium and Tantalum in dependence of energy. The energy loss data were taken from [1]. Such an estimation is helpful for designing Faraday cups and other beam stopping devices. For example : if for a given beam current the allowed pulse lengths is less than the actual beam pulse length, cooling the device will not help since the material will be melted or even vaporised by only one single beam pulse.

The situation can be improved by selection of an other stopping material or, if possible by increasing the beam radius.

No problems due to thermal heating will arise if a toroidal beam transformer can be used for current measurement. By designing a passive beam transformer the following arguments should be considered :

- The drop of the signal is determined by L/R and requires large L and small R, where L is the transformer conductance which is proportional to N^2 and R is the load resistance.
- The signal itself is proportional to R/N (see [2]) and demands large R and small N, which is completely in contrast to the argument given above.
- The thermal noise is proportional to R and therefore requires also small R.

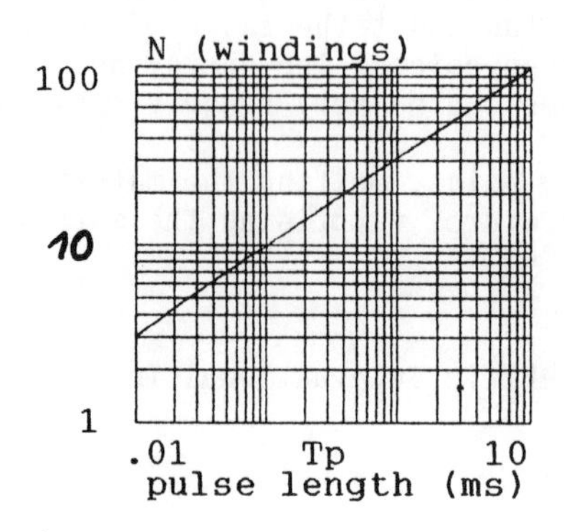

Fig. 2 : N = f(T_p)

For the following reasonable mechanical and electrical parameters the required number of windings has been calculated :

inner core radius : 40 mm
outer core radius : 60 mm
core height : 40 mm
rel. Permeability : 100.000
Vitrovac, Ultraperm (see [3])
load resistance R : 50 Ohms
maximum drop : 1%

Fig. 2 gives the necessary number of windings in dependence of the beam pulse length T_p.

Assuming that the bandwidth of the system is determined by the relation BW = FAC· $1/T_p$ the required beam current for a given signal to noise ratio can be calculated easily in dependence of the selected bandwidth. Fig. 3 shows the result for 10 < FAC < 1000. Obviously, this holds only if the number of windings is taken from fig. 2.

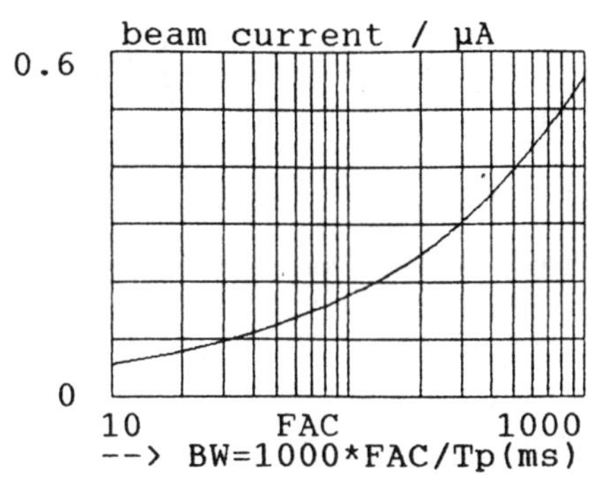

Fig. 3 : Required beam current

BEAM PROFILE

Mostly used devices for profile measurements are profile grids, viewing screens and residual gas monitors. The important parameters of a profile grid measuring system are given in Table 1.

Table 1. Important parameters of a profile grid measuring system

Parameter	Value
Diameter of the wires	0.05 ... 0.5 mm
Spacing	0.5 ... 5 mm
Length	40 ... 100 mm
Material	W-Re alloy (typically)
Insulation (frame)	glass ceramics
Number of wires	15 ... 127
Maximal power rating	0.5 ... 1 W/mm
Sensitivity (lowest range)	2 ... 5 nA/V
Dynamic range	$1:10^3 \ldots 1:10^6$
Number of ranges	4 ... 16
Integration time	1 ... 20 ms (typically)
Maximal output voltage	10 V

Besides the data given in the table the accelerator physicist is interested on the minimum beam current required for a profile measurement and the maximum allowed beam current.

Minimum beam current : Taking also multiple charged ions into account there can be a big difference in the detection limits when particles are stopped within the wires or the particles pass through the wires.

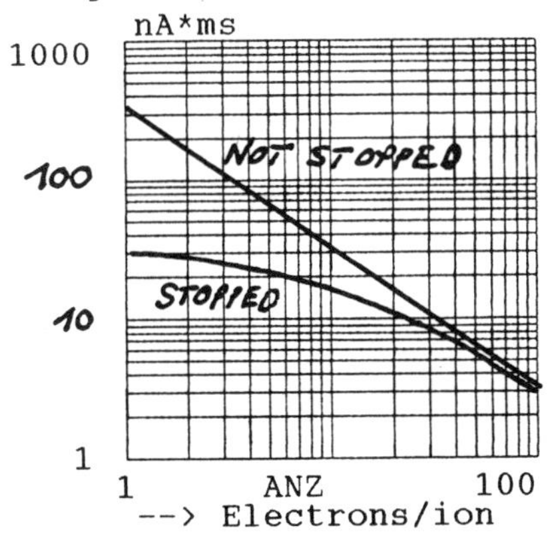

Fig. 4 : Required i·t, see text.

In the first case the current from the wires to the input amplifiers results from the sum of stopped charged ions and emitted electrons. In the second case only the emitted secondary electrons can be detected, which can shift the minimum required beam current to much higher values. This is demonstrated in fig. 4. The calculations hold for positive ions in the charge state 10^+ with a parabolic intensity distribution of 5 mm radius.

Taking the specifications of a typical grid electronic into account (see table 1) a minimum charge of 100 pA•5 ms has been assumed for the wire in the beam centre.

Maximum beam current : In the case of passing particles the deposited power per wire depends on much more parameters than in case of stopped particles. Therefore we will give here only an estimation for the simpler case that the particles are stopped completely within the wires. For a test grid with 40 mm long W-Re wires, and a diameter of 0.1 mm having a resistance of 0.64 Ohms the maximum allowed power were determined by electrical heating in vacuum [4]. The wire was broken at a temperature of 1600 °C (current of 2.8 A), which corresponds to 5 Watts.

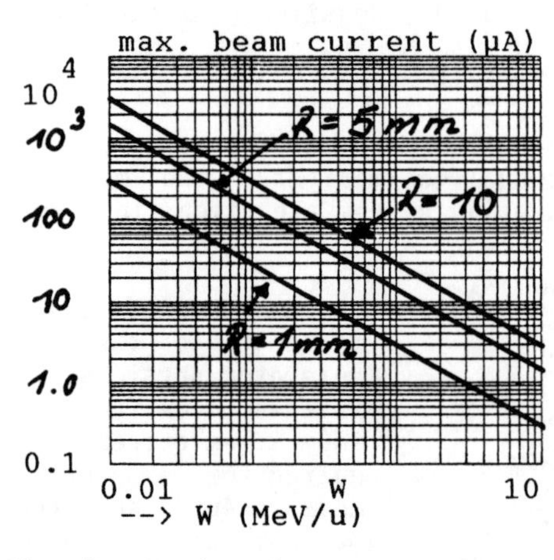

Fig. 5 : Maximum beam current

Fig. 5 gives the maximum allowed beam current in dependence of beam energy for beam radii of 1, 5, 10 mm. The data, which hold for a charge over mass ratio of 0.05 and a maximum of 5 watts on the centre wire can be easily scaled to other parameters.

The intensity ranges for profile measurements can be extended by use of gas-filled profile grids at the low intensity side down to some pps (see [5]) and residual gas monitors in case of high intensities, (see [6]).

BEAM EMITTANCE

Destructive measurements of the transverse emittance are mostly based on the stepwise motion of a slit in front of a profile detector. The detector may be a sandwich of collecting strips, a profile grid, a second scanning slit in front of a small Faraday cup or a single scanning wire.

Width of the slit	: $w = 0.1\,mm$
Maximum of divergence	: $x'_{max} = \pm 50\,[mrad]$
Number of detectors	: $N = 32$
Width of the detectors	: $d = 0.2\,mm$
Space between detectors	: $s = 0.2\,mm$
Distance between slit and detectors	: $L = 128\,mm$
Resolution in x'	: $\delta x' = 1.6\,mrad$
Width of the detector set	: $\pm 6.4\,mm$

Table 2 : System parameter

Table 2 gives typical parameters of a slit - detector sandwich combination. Before designing a emittance measuring system one should keep in mind that there will be a considerable reduction in current behind the slit.

For the parameters given in table 2 the fraction of currents on the detector channels has been calculated assuming again a parabolic intensity distribution. The curve shown in fig. 6 holds for a position of the first slit where the intensity is reduced to 10% of the maximum in the beam centre.

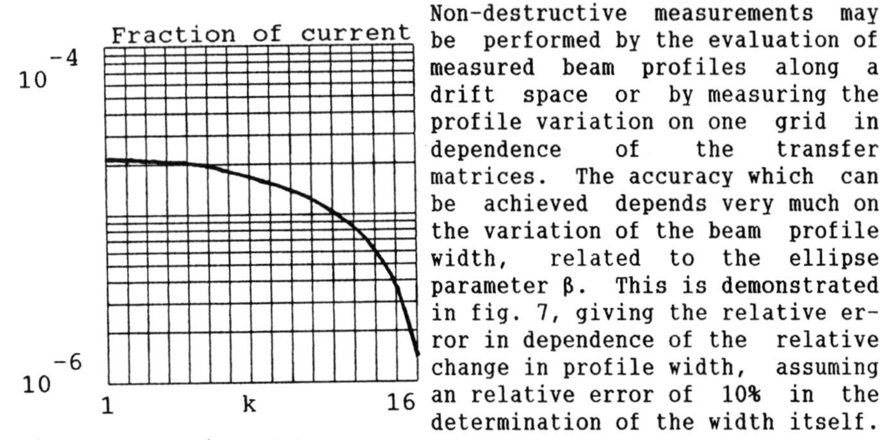

Fig. 6 : Fraction of beam current on the channels (k).

Non-destructive measurements may be performed by the evaluation of measured beam profiles along a drift space or by measuring the profile variation on one grid in dependence of the transfer matrices. The accuracy which can be achieved depends very much on the variation of the beam profile width, related to the ellipse parameter β. This is demonstrated in fig. 7, giving the relative error in dependence of the relative change in profile width, assuming an relative error of 10% in the determination of the width itself.

Although the estimation shown in fig. 7 has been performed for an emittance measurement in the longitudinal phase space (see [7]) the trend holds also for the algorithm under discussion : the accuracy for the determination of the ellipse parameters can be improved considerably if the variation of the envelops goes through a waist at the location of the profile measuring device.

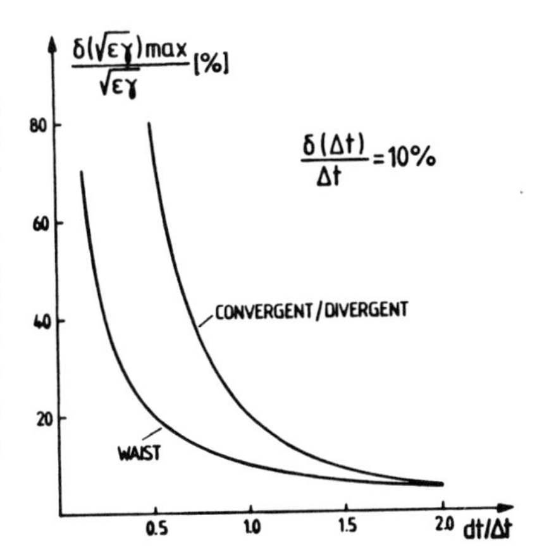

Fig. 7 : Error estimation

MEASUREMENTS IN THE LONGITUDINAL PHASE SPACE

Besides the determination of longitudinal emittance (see for example [8]) monitoring of bunches can be very helpful for determination of bunch length and shape, energy measurements by time of flight techniques, observation of phase relations between the particles and the accelerating rf, position measurements. Mostly used devices are ring-shaped capacitive pick ups. For such a pick up the necessary number of single charged particles within one bunch has been calculated assuming the following parameters :

ring diameter: 50 mm
probe length : 10 mm
impedance : 50 Ohms
bandwidth : 1.5 GHz
signal:noise : 1
bunch length : 0.2-2ns

The results are shown in fig. 8. Although relativistic effects enhance the signal for higher β-values the necessary number of charges within one bunch increases for Δt = 1 ns and Δt = 2 ns due to the increasing geometrical bunch length.

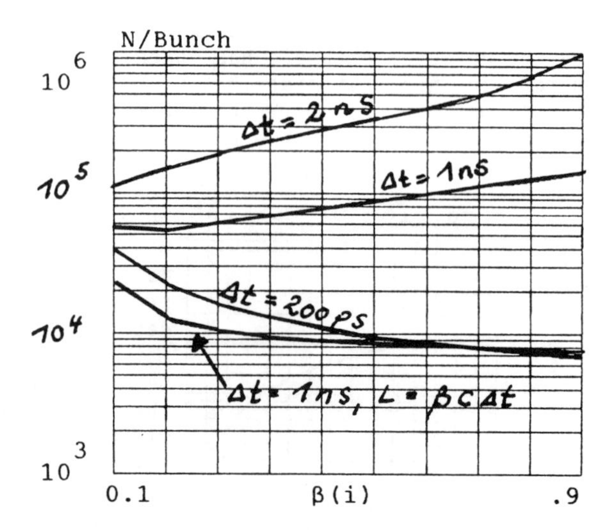

Fig. 8 : Required charges/bunch.

This is demonstrated in fig. 8 by the lowest curve where the probe length has been chosen to be proportional to β and Δt.

The measurement of beam position by capacitive pick ups may be based on the two different principles shown in fig. 9 and fig. 10.

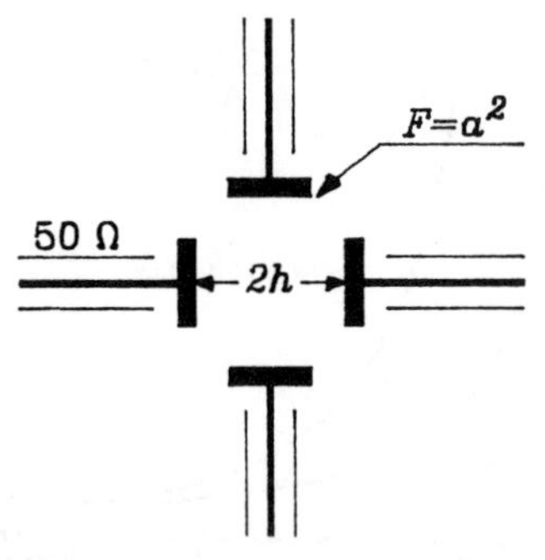

Fig. 9 : Scheme for position measurement (curve $\Delta 1$ in fig. 11).

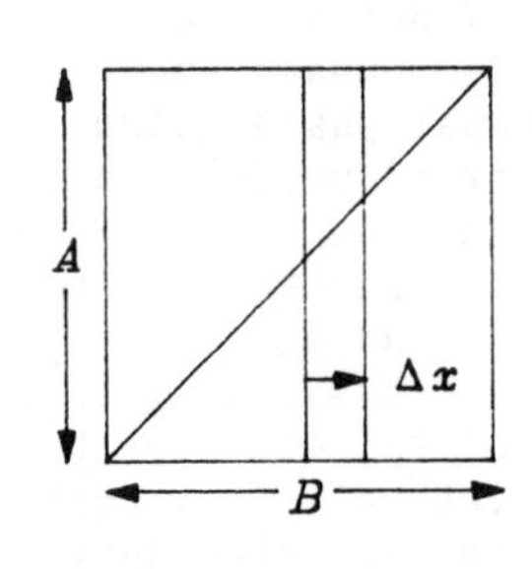

Fig 10 : see fig. 9, (curve $\Delta 2$ in fig. 11)

According to the scheme of fig. 9 the beam position is determined by the difference signal from opposing plates. This is completely different from the scheme of fig. 10 where the position results from the difference signal of the two parts of one divided plate.

Taking β=0.1, Δt=1ns and a plate area of 10mm•10mm the required number of charges within a bunch has been calculated for both systems. Fig. 11 shows the results in dependence of h. Again a bandwidth of 1.5 GHz and a signal to noise ratio of 1 in a 50 Ohm system has been assumed. The data for $\Delta 1$ (see fig. 9) and $\Delta 2$

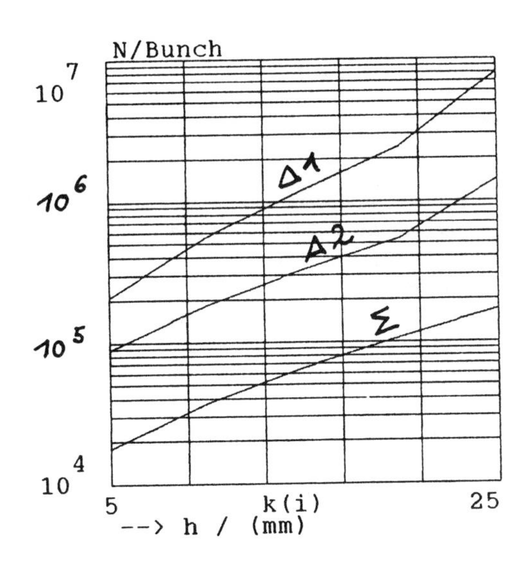

Fig. 11 : Necessary number of charges for beam position measurements. The geometric schemes are shown in fig. 9 and fig. 10.

(see fig. 10) hold for a displacement of 1mm. For both systems two opposing plates have been taken into account. is calculated for the sum signal of two plates.

REFERENCES

1) Hubert, F., Fleury, A., Bimbot, R., Gardes, D.: Range and stopping power tables, Annales de Physique, Vol. 5, Suppl., p. 1-214.

2) Unser, K.: beam current transformer with DC to 200 MHz range. IEEE Trans. on Nucl. Sci., Vol. 16/3, Washington 1969.

3) Vacuumschmelze A.G., Hanau, Germany : private communication

4) Kaufmann, W., Walter, H.: GSI-Darmstadt : private communication.

5) Breskin, A., Charpak, G., Majewski, S., Melchart, G., Petersen, G., Sauli, F.: Nucl. Instr. and Meth., 161 (1979), p. 19.

6) DeLuca, W. H.: Beam detection using residual Gas Ionization. IEEE Nucl. Sci. NS-16, 813 (1969).

7) Strehl, P.: A new method for longitudinal emittance measurements. Particle Accelerator Conference, Santa Fe, New Mexieo, 1983, IEEE Trans. on Nucl. Sci., Vol. NS-30, Nr. 4, p. 2198.

The Parametric Current Transformer, a beam current monitor developed for LEP

K. B. Unser, CERN, CH-1211 Geneva 23 (Switzerland)

Abstract: Toroidal transformers are used to measure the beam current in beam lines and accelerators. Placing such a transformer in the feedback loop of an operational amplifier will increase the useful frequency range (active current transformer). A magnetic modulator can be added to extend the response to DC current, maintaining with a control loop the transformer core at a zero flux state. The magnetic modulator in the parametric current transformer gives not only the DC response but provides parametric signal amplification up to a transition frequency of about 500 Hz. The low frequency channel (magnetic modulator) and the high frequency channel (active current transformer) are linked together in a common feedback loop. A large dynamic range together with good linearity and low distortion is obtained. This arrangement protects the magnetic modulator from dynamic errors in case of a sudden beam loss, which could impair its zero stability. Dynamic overload protection is an important condition to obtain high resolution and good zero stability, even in applications which require in principle only a very limited frequency response.

Introduction

Beam current transformers are among the oldest examples of beam instrumentation. Their development has followed the evolution of particle accelerators. Two important milestones of this development should be mentioned here:

The current transformer was placed in the feedback loop of an operational amplifier (H. Hereward and J. Sharp[1]). This extended the low frequency range by a factor approximately equal to the gain of this amplifier. The differentiation time constant L/R of the "Active Current Transformer" could exceed 1000 seconds, making it possible to measure the circulating beam in the proton synchrotron during several seconds with a negligible shift of the baseline.

A magnetic modulator [2] and a control loop was added to prevent any magnetic flux change in the core of the active beam current transformer. This "zero flux DC current transformer" was originally developed for beam current measurements in the ISR [1], a storage ring, where the proton beams would circulate for days and weeks. It is an example of a technology developed for particle accelerators which has found many industrial applications [3] for precision DC and AC current measurements.

A new generation of beam current monitors.[5] was developed for the LEP project. This gave the opportunity to introduce a number of new ideas to improve the performance and to reduce the influence of environmental factors like stray magnetic fields, electromagnetic interference and mechanical vibrations (microphony). The new instrument is called the Parametric Current Transformer (PCT), because the magnetic modulator provides parametric amplification in the low frequency channel, up to a transition frequency of about 500 Hz.

The development work was done in collaboration[7] with an industrial company in France (technology transfer) who intended to produce this instrument commercially. This meant that a number of economical factors had to be considered which were of lower importance in earlier projects. The priorities for a commercial product are cost, reliability and performance - in that order! The new design goal was to reconcile these requirements without sacrificing the performance. This was achieved by reducing the number of components and their cost (cables, connectors, electronic components and circuit boards) and by cutting down on the volume, the weight and the power consumption.

This paper gives a summary of the new techniques which are now available for DC beam current measurements. It does not necessarily imply that all of them are required in every practical application.

System description

The simplified block diagram (Fig. 1) of the PCT system shows 3 distinct transformers and their associated circuits:

- the zero flux transformer (T_5) together with the L/R integrator circuit.
- the magnetic modulator (T_1, T_2, T_3) with excitation generator and demodulator.
- the ripple feedback transformer (T_4) for the ripple compensation circuits.

The transformers are surrounded by electrostatic screens and some of the windings are screened from each other to eliminate unwanted coupling. Current feedback and calibration windings are common to all transformers. Inductive coupling with the beam is symbolically indicated with a one turn beam coupling winding.

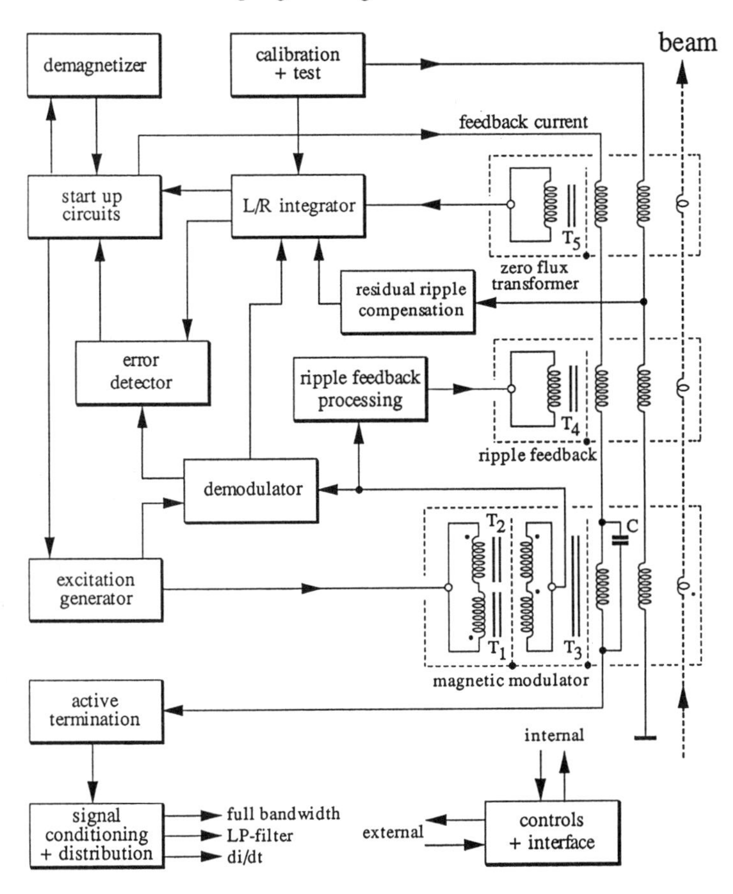

Fig. 1. Simplified block diagram of the PCT

The magnetic cores are demagnetized (depolarized) automatically each time the mains power is applied. The demagnetizer generates a sinusoidal 50 Hz current (>12 A_{pp}) in the feedback windings and this current decays exponentially with a time constant of a few seconds. Demagnetization of the modulator cores is enhanced by programming the excitation generator simultaneously to the highest amplitude before bringing it progressively down to the normal excitation level.

Demagnetizing is important for the zero flux transformer to define the working point close to the center of the B/H loop. This helps to reduce microphony effects, where mechanical vibrations produce a modulation of the residual (remanent) flux and generate parasitic signals. The microphony effects, without this precaution, are very disturbing and could limit the resolution of the monitor in a practical application (vicinity of vacuum pumps etc.).

The magnetic modulator has a memory of previous exposure to a large current. This is probably due to a residual remanence effect. Zero readings may change by more than 1 mA after measuring a current of 1 A, which was, for some reason, not compensated by feedback. This is not only a static offset error, but it is followed by a tendency to drift back during days in the direction of the original zero state. Demagnetization at low frequency permits erasure of this memory effect with an residual error of less than ± 2 μA.

The zero remanence state of the magnetic cores has to be maintained under all operating conditions. This is the task of the start-up circuits, which apply the feedback current after the demagnetizing cycle is completed, on condition that there is no error signal from the circuits in the feedback loop. Error signals are generated if an excessive external current is applied. This is also transmitted as an error message to the control interface. The error detector has the additional function to supervise the positive and the negative power supplies. A drop in power causes an immediate controlled shut-down followed by a demagnetizing cycle when the power is restored again.

The calibration circuit applies a precision current source to the calibration windings. This is useful as a system test and permits the calibration of the entire data acquisition chain (for both polarities) in a typical application. There is also another function of this circuit: in the control state "test", a known current is added to the current in the feedback windings. The feedback current will try to compensate the error caused by this current source. The change in the zero reading of the PCT can be used to calculate the internal d.c. loop gain of the PCT.

Fast current changes (beam or feedback current) are shorted out with capacitor C, which is decoupled[4] from the modulator with the help of an additional transformer core T3. This capacitor both protects the magnetic modulator from fast transients and attenuates at the same time high frequency components in the modulator output signal, which are coupled into the feedback current loop. This coupling, an undesirable effect, is the origin of modulator ripple in the PCT output signal. A processed modulator output signal is returned back via the ripple feedback transformer to compensate this unwanted signal at the source (reduction up to 98%).

Earlier instruments[4] of this type required a complete 19"- crate with 8 plug-in modules to house the electronics. The new design, in spite of many additional circuit functions, requires only 2 Eurocards (100×160 mm) with 4 micro modules in surface mount technology. The total power consumption was reduced by 94% and is now only 3 watts (at zero input current). The electronics is placed in a sealed box without ventilation holes ($185 \times 130 \times 70$ mm).

The interconnection between the front-end electronic box and the back-end chassis is a single cable with 3 shielded wire pairs. The first carries the analog signals, the second the power supply and the third the multiplexed bidirectional controls and the power for the demagnetizer. The back-end chassis contains only the analog signal conditioning and distribution circuits, the control interface and the power supply.

The Magnetic Beam Sensor

The magnetic beam sensor consists of 5 separate magnetic cores, packed together in the toroid assembly (Fig. 2). The cores are strip wound toroids having a useful cross-section between 5 and 25 mm^2 depending on the application. Small cross-sections of the cores were possible thanks to the choice of a high modulation and transition frequency of the system. The soft magnetic material is a thin ribbon (5 mm wide, 23 μm thick) of Vitrovac® 6025*, an amorphous magnetic alloy with the composition $(CoFe)_{70}(MoSiB)_{30}$. This material features higher values of permeability and can be used at higher frequencies than conventional (crystalline) nickel/iron alloys.

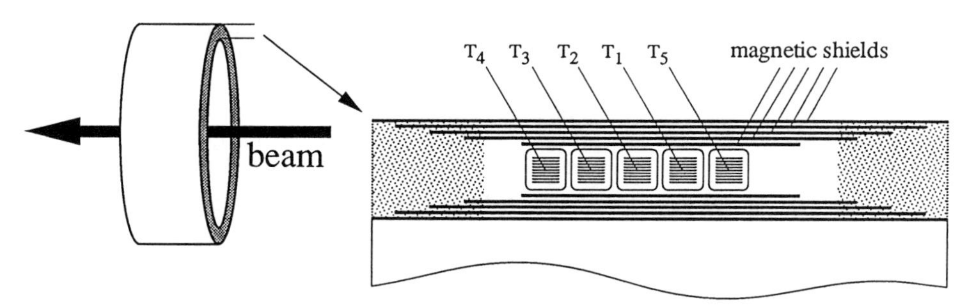

Fig. 2. Toroid assembly, simplified cross-sectional view (windings not shown)

The cores for the 2nd-harmonic magnetic modulator (T_1 and T_2) are the most critical components of the system and the magnetic properties of these cores determine the resolution and the zero stability of the instrument. Vitrovac® 6025 is now produced in quantity, but the normal commercial grade has a very large spread of magnetic characteristics. A special quality is selected by the manufacturer using a detailed set of specifications containing among others the following selection criteria:

- low value of magnetostriction ($\lambda s < 0.2 \times 10^{-6}$)
- low value of saturation flux density ($Bs < 0.5$ Tesla)
- good surface quality
- no brittleness

The selected material is submitted to a series of tests to determine the specific annealing conditions[5] for each production batch and the important parameters for the modulator application, i. e. the modulator gain and the magnetic modulator noise. The magnetic noise (Barkhausen noise) depends essentially on the number and the structure of the magnetic domains in the material, which can change with the composition and the annealing treatment of the material. Less than 5% of the material received will pass these tests, but the rest can be used for all other applications, where these specific characteristics are not relevant.

Certain aspects of the fabrication of the cores have been treated in an earlier publication[5] and will not be repeated here. The winding of the modulator cores is a very critical operation. The ribbon has to be continuously controlled with the microscope for mechanical defects (micro fractures and surface defects). The correct winding tension has to be carefully maintained. The insulation between the layers, a mylar foil of 2 μm thickness, is very delicate and difficult to handle. It has to be placed with great care to maintain a minimum and equal spacing between

---

* Vitrovac® 6025 is a trade name of Vacuumschmelze GMBH, D-6450 Hanau, Germany

the layers. It is not only necessary to wind all cores with exactly the same number of layers, but also to position the start and the finish of the ribbon in a well defined position in respect of each other. The magnetic ribbon is not simply cut at 90° to the longitudinal axis of the tape but at a very narrow angle in order to distribute the discontinuity in the cross-section over a larger circumference. The finished cores are vacuum impregnated and cross field annealed. The toroidal excitation winding is applied and all magnetic parameters are measured and recorded. Core pairs are selected by matching the dynamic hysteresis loop to better than 1% (defined by the factor of attenuation of the modulation frequency in the common output winding).

The magnetic modulator, in the center of the assembly, is a very sensitive magnetometer to external magnetic fields. This is an undesirable feature which can only be attenuated by extensive magnetic shielding. The magnetic shield of the PCT consists of a number of concentric magnetic cylinders of different length, inside and outside the magnetic cores (Fig. 2). The shields which are closest to the cores consist of several layers of Vitrovac 6025 and provide the best shielding factor, but this material is only available with a maximum width of 50 mm. All other shields are Mumetal. Seen in this context, the small cross section of the magnetic cores is also an important advantage for efficient magnetic shielding. It helps to bring the inner and the outer shields closer together and reduces the volume of the magnetic beam sensor.

This shielding attenuates the external field by a factor between 50 to 500, depending on the number of shields in use. This is not enough in many applications. One has also to consider that high permeability shields are easily saturated by a strong external magnetic field.

The Excitation Generator

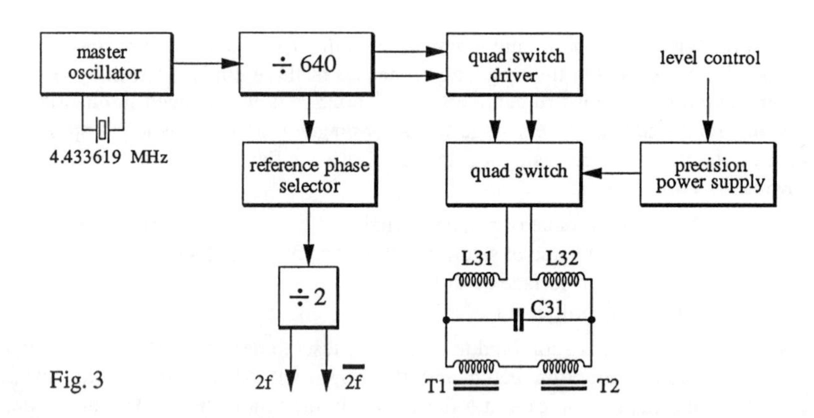

Fig. 3

Basic design consideration: The resolution and zero drift of the PCT should only be limited by the magnetic noise of the magnetic modulator. The contribution of noise from the electronic circuits should ideally be considerably lower. As a design limit, the tolerance for these contributions have been arbitrarily set to $\leq$ 10 nA rms of equivalent beam current.

This translates into the following specifications for the excitation generator (assuming matching errors $\leq$ 1% for the modulator core pair):

- variation of 2nd harmonic distortion: $\leq$ 5 ppm
- variation of frequency: $\leq$ 10 ppm
- variation of amplitude: $\leq$ 50 ppm (parts per million)

The excitation frequency of the magnetic modulator should be as high as possible, but eddy currents in the core material impose an upper limit which is in our case around 7 kHz.

A crystal controlled master oscillator (Fig 3) with a stability $\leq$ 3 ppm and a synchronous divider generate the excitation frequency (f = 6927.3 Hz). The tolerance is less than 1 ns for the differential timing error (difference in duration of the positive and the negative half period) and less than 100 μV for the differential amplitude error. The difference of rise and fall times and the corresponding transmission delays of the digital control signals have to be taken into account. A perfect symmetry of all pulse forming elements is required and symmetric transmission lines for the timing signals are used. The circuit board lay-out is critical. A quad DMOS transistor array (on a single chip) in a symmetrical H-bridge configuration[5] is used to switch the output of a precision regulated power supply. A passive low pass filter (L31; L32 and C31) eliminates the higher frequency components. The capacitor C31 supplies high current peaks (Fig. 4) in an avalanche discharge to drive the cores hard into saturation and recuperates a large part of the stored energy on the return swing. The optimum value of this capacitor and the optimum value of peak excitation current as a function of resolution are individually determined for each magnetic sensor in a semi-automatic test set up.

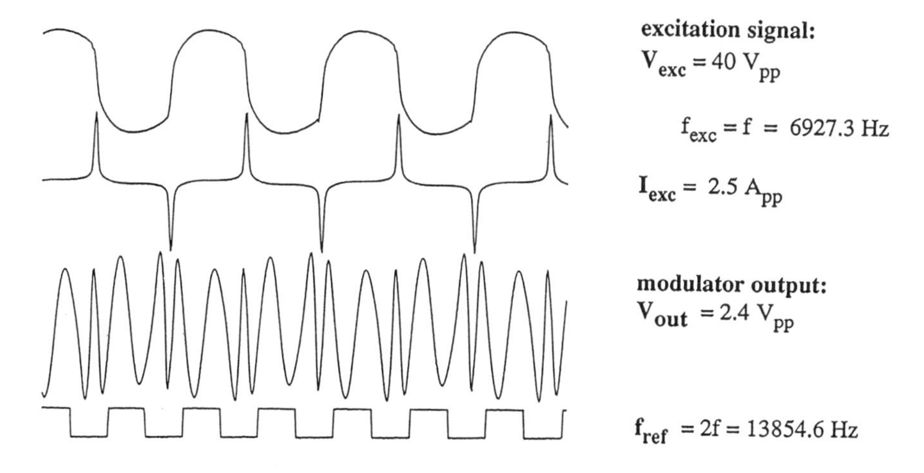

Fig. 4. Typical signal waveforms observed on a magnetic modulator (plot of display, averaged signals, on LeCroy 9410 oscilloscope). The core matching in this example is better than 0.5%. The exact waveform of the output signal is an individual "signature" for every magnetic modulator.

The Demodulator

The demodulator has to detect and to amplify the 2nd harmonic component in the output signal of the modulator. It has to satisfy the following specifications, taking into account the parametric amplification[5] in the magnetic modulator:

Resolution: 0.1 μV rms (for a bandwidth of 1 Hz)
10 μV rms (for a bandwidth of 500 Hz)

This 2nd-harmonic signal is completely masked by a parasitic output (see Fig. 4) of the magnetic modulator, resulting from the core matching error of the magnetic modulator. This parasitic signal is composed of the modulation frequency f and a spectrum of odd harmonics (3f; 5f; 7f; 9f; 11f etc.). It has an amplitude of several volts, more than 150 dB higher than the required resolution for the 2nd harmonic signal.

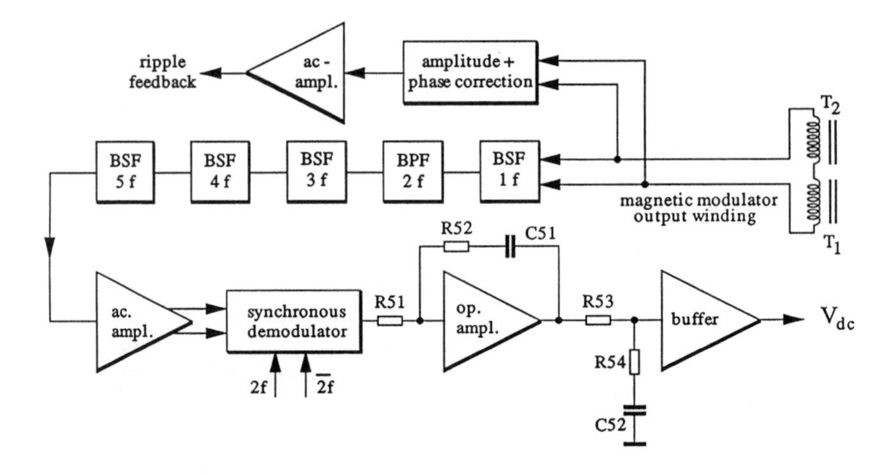

Fig. 5. Simplified block diagram of demodulator

To obtain the specified resolution, it is necessary to attenuate, with a filter, the parasitic signals by more than 50 dB and to amplify the 2nd harmonic component at least 30 dB before demodulation. The bandwidth of the filter should be 2.5 kHz above and below 2f in order to accommodate the upper and lower sidebands of the modulated signal with an acceptable phase error. This is one of the conditions which has to be satisfied to make the overall feed-back loop stable, considering a transition frequency of 500 Hz for the low frequency channel.

The filter is a passive LC-network and consist of 1 band pass (BPF) and 4 band stop (BSF) sections. The signal, after demodulation in a synchronous detector, is integrated with the time constant R51 × C51 for a 6 dB/octave (frequency) roll-off. This determines the transition (cross-over) frequency between the modulator channel and the active current transformer channel.

The Active Current Transformer

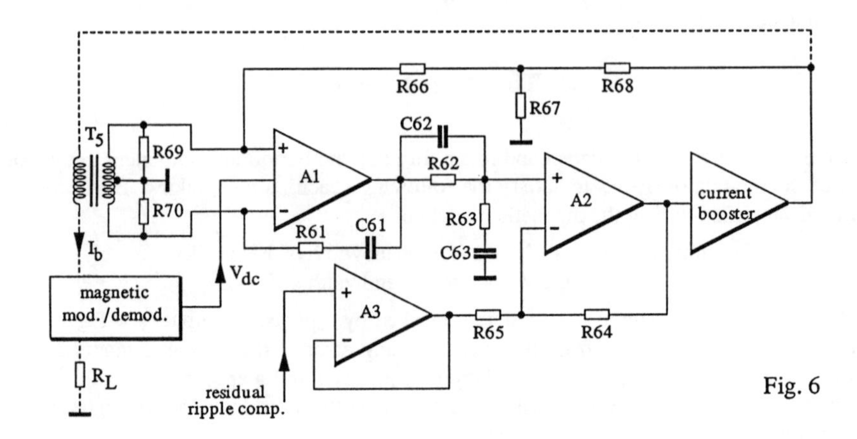

Fig. 6

The active current transformer with the zero flux transformer (T_5) and the overall feedback loop of the PCT is shown in Fig. 6. The signal gain for medium and high frequencies (up to 1 MHz) is provided by a composite amplifier (the L/R integrator), consisting of A1; A2 and a current booster for max. 100 mA. The DC and low frequency gain comes from the magnetic modulator/demodulator in cascade with a part of the high frequency channel. The 2 channels have therefore 2 independent inputs, but one common output and one common feedback loop, which defines the (closed loop) signal gain of the system. The signal path is always via the channel with the highest open loop gain at any particular input frequency.

Considerations of loop stability impose an upper frequency limit (< 1/10 f) for the transition from the low to the high frequency channel. A high transition frequency has many advantages. It reduces the required core cross-section for the zero flux transformer T_5, limits the noise contribution of amplifier (A1) and reduces the microphony effect of core T_5. All these effects increase rapidly at lower frequencies.

The open loop gain of both cascaded channels is very high (> 150 dB at DC). This is how good linearity, low distortion and the large dynamic range of the PCT is obtained. It requires a carefully tailored roll off (gain and phase) in the direction of the unity gain cross over frequency (1 MHz) of the system. Phase correction elements in the active current transformer (R61, C61 - R62, C62 - R63, C63) and in the demodulator (R52, C51 - R53, R54, C52) have to be set for optimum loop stability and a clean step response. Range switching does not influence the dynamics of the feedback loop, because the (virtual) load impedance R_L is constant and small.

Damping resistors (R69 and R70) eliminate undesirable high frequency resonances of the zero flux transformer T5. They cause a small gain error (2 to 3 %) in this channel, which is compensated by a corresponding amount of positive feedback (via R66, R67 and R68).

The parasitic output signal of the magnetic modulator, coupled into feedback loop, causes an unwanted error current (modulator ripple) in this loop. This effect is unfortunately enhanced by the low output impedance of the current buffer and the low value of (virtual) load impedance (R_L= 50 ohms). The ripple feedback via T_4, mentioned earlier, reduces this effect already by a large factor. The remaining ripple signal is measured at the calibration winding and added via A3 and A2 to the signal of the current booster (compensation by "bootstrapping").

The Active Termination

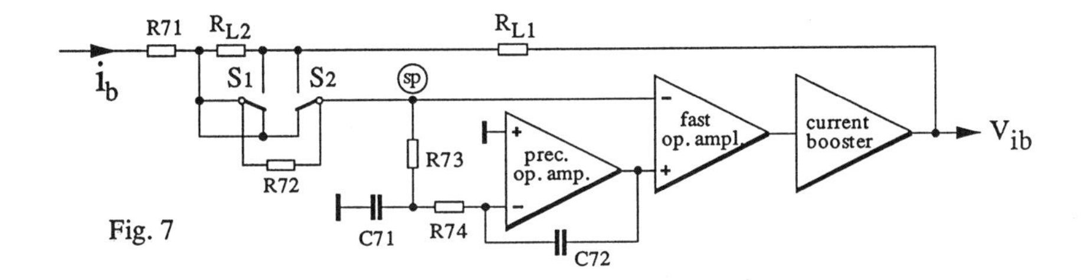

Fig. 7

The active termination converts the feedback current i_b into an output voltage V_{ib}. Two precision load resistors (R_{L1} and R_{L2}) in the feedback loop of an operational amplifier provide a virtual ground reference at the summing point (sp). The operational amplifier is in reality a composite amplifier with a separate high and low frequency channel and a current booster (100 mA max.). This arrangements permits an accurate measurement of the average beam current, even if the input signal consists of very short pulses, separated by a long time interval.

The switches S1 and S2 select the current ranges A and B without interrupting the feedback path and without adding the contact resistance to the load resistor values. The load resistors R_{L1} and R_{L2} are composed of several precision resistors in parallel in order to keep the power dissipation in each of them at a low level. Resistor R71 (50 Ω) defines the actual impedance of this active termination in the main feedback loop and keeps it at a constant and low value to reduce the effects of parasitic capacitance of the different elements in this loop, a condition for loop stability, independent of the selected current range

Signal conditioning and distribution

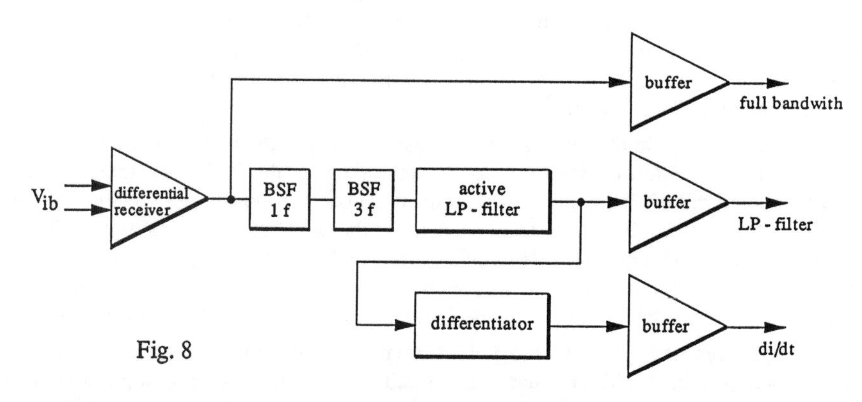

Fig. 8

The analog signal V_{ib} from the front end is transmitted over a symmetrical transmission line, up to a distance of several hundred meters. A differential line receiver rejects all common mode noise which may be present on the line. The signal path is either direct or via a low pass filter and optional band stop filters[6] (BSF) for f and 3f to reject spurious modulator noise. A differentiated signal, proportional to beam loss, is also provided. There is a buffer amplifier for every signal output.

Results

The following specifications can be obtained with a selected sensor:

Sensor dimension	225 mm o.d. 175 mm i.d. 100 mm length
Range A, full scale	any range from 10 mA to 100 A (both polarities)
Range (B)	1 to 20 % of range A
Linearity error *	± 0.001 % ± zero error
Resolution *	± 0.3 ppm of range (A) ± 0.4 µA rms (± 1 µA rms typical)
Zero drift *	± 1 µA/°C (± 5 µA/°C typical)
Zero drift (24 h) *	± 2 µA rms (at constant temperature)
Bandwidth	DC to 100 kHz
Accuracy (calibration source)	± 0.05 %

*) measurements with a 1 sec. integration window

Resolution and zero drift are not at all limited by the electronics, but depend only on the quality of the magnetic sensor. The quality of the Vitrovac 6025 material is in this respect of crucial importance. Material purchased 5 years ago gave in general better results than that currently produced at present. One observes very different zero drift behavior among sensors which are built with exactly the same batch of material. This is an indication, that all the factors which influence these characteristics are not as yet clearly identified.

The temperature drift is not so much caused by the temperature coefficient of the material itself than by the uneven mechanical constraints in the two modulator cores. The temperature dependent zero drift is generally reduced after a period of artificial ageing (temperature cycling between 20°C and 80°C) and the temperature coefficient of the sensor becomes in any case more reproducible. It is a good idea to incorporate a temperature gauge in the beam sensor, if the temperature drift is critical in a particular application.

During long term zero drift test one can sometimes observe in intervals of several hours or days a fairly sudden change of up to 2 μA. The cause of these phenomena are not known.

Acknowledgments

Many people have made their contributions: C. Bovet and R. Jung gave their support to the project and many useful discussions are gratefully acknowledged. The computer controlled test bench for dynamic testing of core samples was designed and built by P. Buksh. The mechanical design of the core winding machine and the different tooling required in the project was first the responsibility of A. Maurer and at a later date of G. Burtin.

J. Bergoz, A. Charvet, R. Lubès and P. Pruvost (BERGOZ, F-01170 Crozet. France) designed circuit lay-outs and built the different prototypes. They made experiments with the different construction methods for the magnetic cores and the toroid assembly and performed an incredibly large number of tests to find the optimum annealing procedures. Their practical experience in building a large number of PCT's is a valuable help for any future improvements.

G.Herzer, R. Hilzinger, W. Kunz and R. Wengerter (Vacuumschmelze GMBH, D-6450 Hanau, Germany) contributed with their knowledge of amorphous magnetic alloys and helped with the selection of a suitable quality of Vitrovac material.

References

1. K.B. Unser, "Beam current transformer with DC to 200 MHz range", **IEEE Trans. Nucl. Sci.**, NS-16, June 1969, pp. 934-938.
2. F.C. Williams, S.W. Nobel, "The fundamental limitations of the second-harmonic type of magnetic modulator as applied to the amplification of small DC signals", **Journal. IEE (London)**, vol. 97, 1950, pp. 445-459.
3. H.C. Appelo, M. Groenenboom, J. Lisser, "The zero flux DC current transformer, a high precision wide-band device", **IEEE Trans. Nucl.. Sci.**, Vol. N.S.-24, No.3, June 1977, pp. 1810-1811.
4. K.B. Unser, "A toroidal DC beam current transformer with high resolution", **IEEE Trans. Nucl. Sci.**, NS-28, No. 3, June 1981, pp. 2344-2346.
5. K.B. Unser, "Design and preliminary tests of a beam intensity monitor for LEP" **Proc. IEEE Particle Accelerator Conference**, March 1989, Chicago,Vol. 1, pp. 71-73.
6. R.L. Witkover, "New beam instrumentation in the AGS booster", April 22-24, 1991, KEK, Tsukuba, Japan: **Proceedings of the Workshop on Advanced Beam Instrumentation**, Vol. 1, pp. 50-59.
7. Collaboration contract K 017/LEP (CERN/BERGOZ) Geneva, December 1986.

A Pseudo Real Time Tune Meter for the Fermilab Booster

Guan Hong Wu, V. Bharadwaj, J. Lackey
Fermi National Accelerator Laboratory
P.O. Box 500, Batavia, Illinois 60510

Oct. 1991

1 Abstract

A tune meter has been developed and installed for the Fermilab Booster. It is capable of measuring the tunes in two planes over the energy ramping cycle with an accuracy of 0.001 in pseudo real time. For each plane this system uses one stripline pick-up type BPM and one single turn ferrite kicker. Data acquisition, processing and control is implemented in a VME-bus based system as an integrated part of the Fermilab Accelerator Controls Network (ACNET)[1] system. The architecture enables the tune measurement and control to be contained in one intelligent system. Here we will present architecture, software, results and future development of the system.

2 Introduction

The Fermilab Booster[2] presently operates over an energy range of 200 *Mev* to 8 *Gev* at 15 Hz cycle. The nominal betatron frequencies in the horizontal and vertical planes are $\nu_x = 6.7$ and $\nu_y = 6.8$ respectively but vary substantially over the cycle because of changes in the synchrotron lattice functions. A correction-magnet assembly consisting of a horizontal and vertical dipole, a quadrupole and a skew quadrupole is placed in each short and each long straight section. The quadrupoles and skew quadrupoles are designed to accommodate the space-charge tune shift at injection and to control the tune against inherent resonances and the coupling resonance of the horizontal and vertical oscillations over the entire cycle. Historically tune control in the Booster has been a difficult job because the tune measurement is slow and unreliable. Also, the tune measurement and tune control was not implemented in an integrated system. The new tune meter described here measures both the horizontal and vertical tunes over the entire energy ramping cycle in pseudo real time. The new architecture provides tune measurement and control in an integrated system with the control system.

3 Architecture

Fig. 1 is a block diagram showing architecture of the tune meter. A CAMAC controlled sub-system pulses the kicker at a programed frequency, duration and amplitude to keep the beam excited throughout the cycle. The turn-by-turn beam position analog signals for the entire cycle, after passing a high-pass filter, are captured and digitized in the ADC board[3]. The data for a whole cycle is moved to a DSP memory board and is Fourier transformed into frequency spectrum of consecutive windows. The DSP board also finds amplitudes and positions of peaks in the frequency spectrum. After this is done the results are buffered and the system is enabled again for next beam cycle. All these activities are controlled by the VME host processor, a Motorola VME133XT, which is in turn commanded by an interactive application program running on one of ACNET consoles. Communications (commands, status informations and data) between the console and the VME crate are done by the standard ACNET system.

4 Software

Two user programs make the system function as desired. One is a microprocessor program which resides in the VME133XT and the other is a application program in a VAX console. Fig.2 shows the basic flow chat of the micro-processor program. The software environment of the tune meter consists of a standard set of tasks for data acquisition running under the MTOS operating system. The interface between these standard tasks and the micro-processor program written by the user is protocol called Object Oriented Communications (OOC)[4]. The console program acts as a master while the micro-processor program is the slave. The micro-processor program performs sequential operations coordinating actions of all the VME boards;

acquiring, processing and moving data according to commands received from the console. The menu driven console program sends command settings and reads status or data from the micro-processor program. Thus, the console program has access control devices so it can control the kickers for tune measurements and change correction quadrupoles strength setting to excise tune control over the cycle.

5 Results and Future Development

The tune meter is being used for measuring tune in a test mode. Shown in Fig. 3 are on-line displays of the fractional tune in horizontal plane. In Fig. 3-a about 50 beam cycles were measured at the same beam conditions. For Fig. 3-b, as the beam intensity changed so did the tune. Fig. 3-c shows the tune responding to the changes in ramping current in the correction quadrupoles.

For these measurements, the beam was kicked every 1 millisecond or every 550 turns, the high voltages applied to the single turn kicker magnet was about .5 KV at the beginning of cycle and 1.5 KV at the end. With such kicker strength (about .87 Gause at the beginning and 2.57 at the end) the effect on the beam transmission throughout the cycle is negligible. For each cycle of data (about 20000 turns depending on where is considered to be the end of the cycle), the DSP did the Fourier Transformations 40 times, each time 512 words to give one tune measurement.

The system's dead time, when operated under these conditions, is about one third of a second during which the DSP is processing data and the ADC is prohibited data taking for next beam cycle. While there are rooms for further optimizations, the quality of tune measured and speed of the measurements as shown are practically good enough for both tune measurements and controls. The architecture design has proven powerful and reliable.

We are encouraged by the results to carry on further development. We have already started to implement interactive tune control part of the project. The next is to measure chromaticity. The architecture is being adopted for other instrumentation applications.

6 Acknowledgements

The Authors give thanks to many people who h
this system, especially among them: Charlie Brieg
communications and Mike Shea for providing us
designed.

REFERENCES

1. Dixon Bogert, "The Fermilab Accelerator Co
   24.
2. E. L. Hubbard(editor), "Booster Synchrotr
   port, TM405
3. Mike Shea and Al Jones, "Fairly Quick VM
   internal report.
4. Lee J. Chapman, "An Object-Oriented Comn
   A293(1990) 347-351.

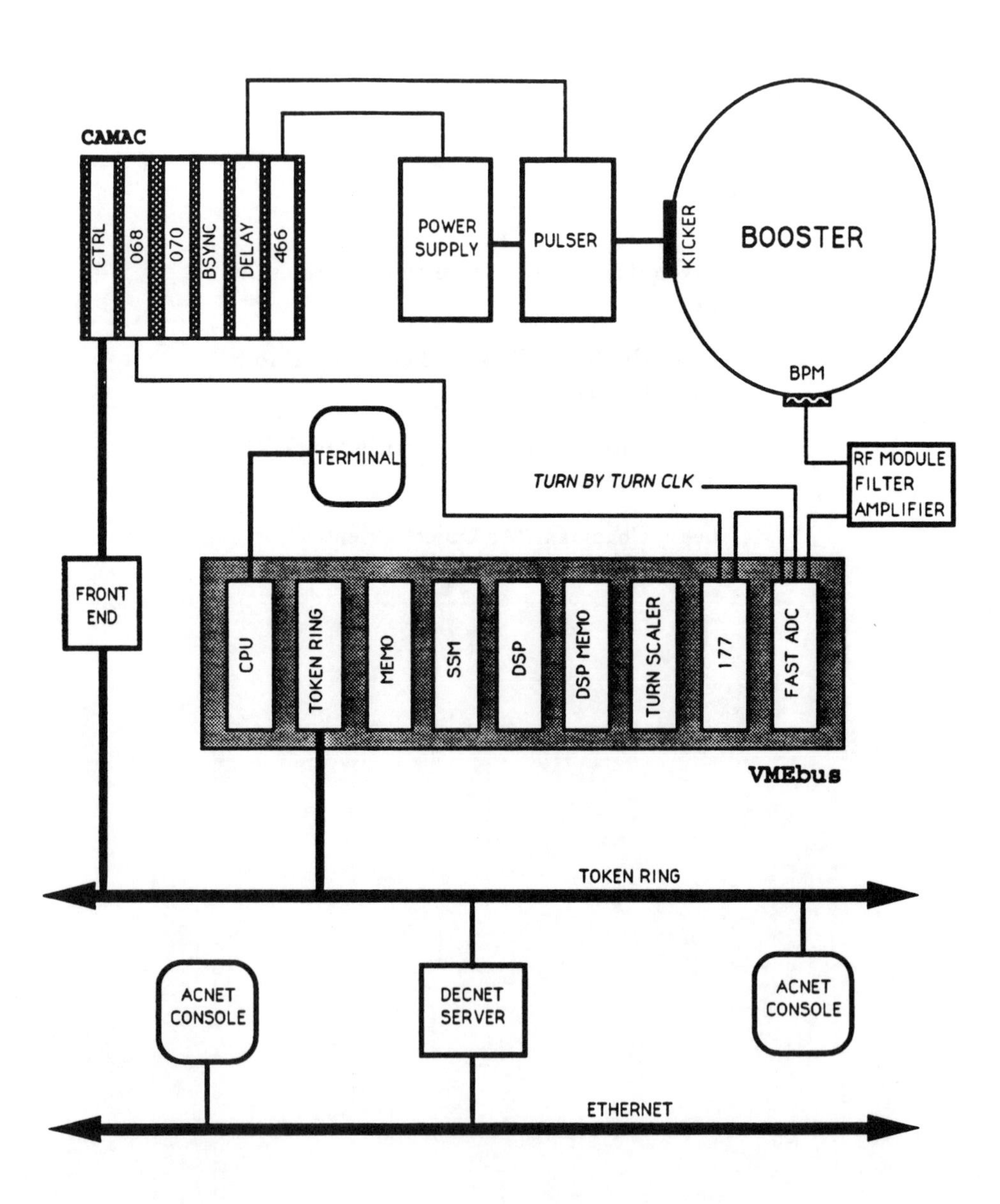

Fig. 1 The tune meter system block diagram.

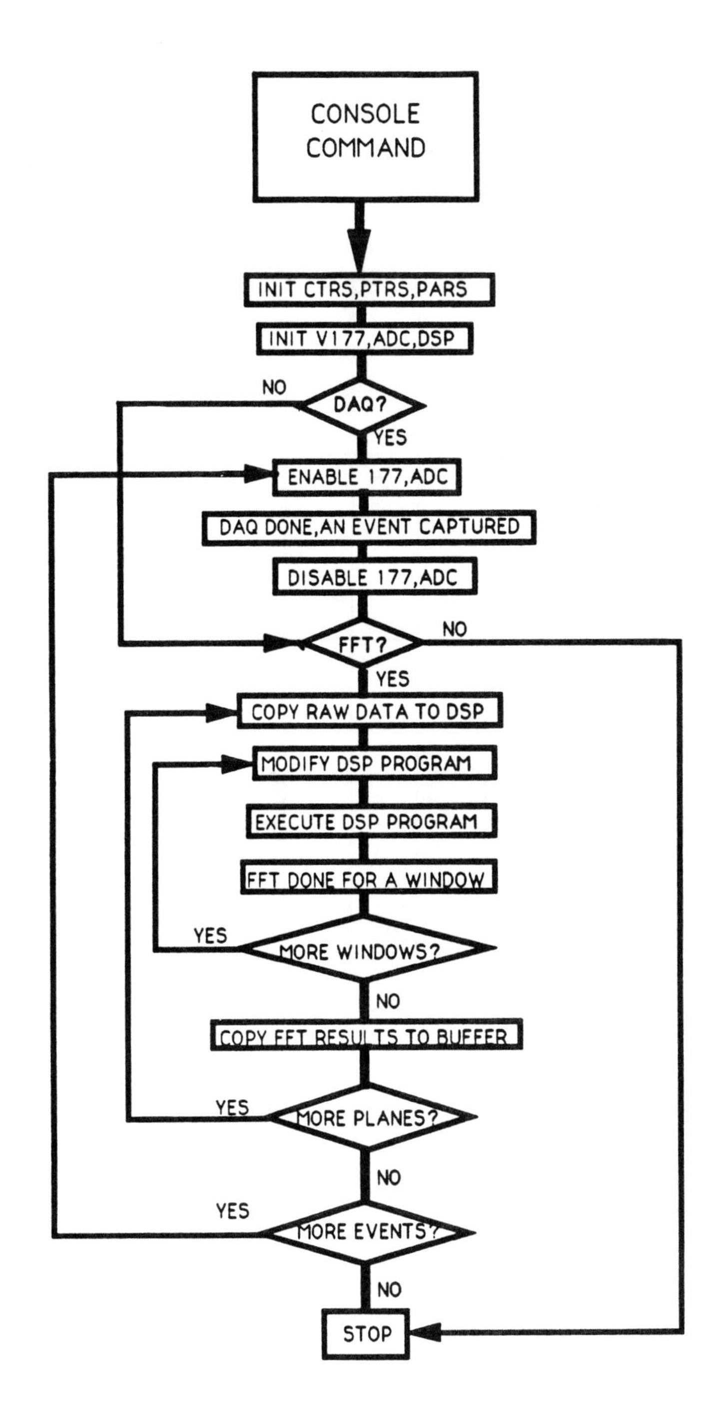

Fig. 2 The tune meter micro program flow chart.

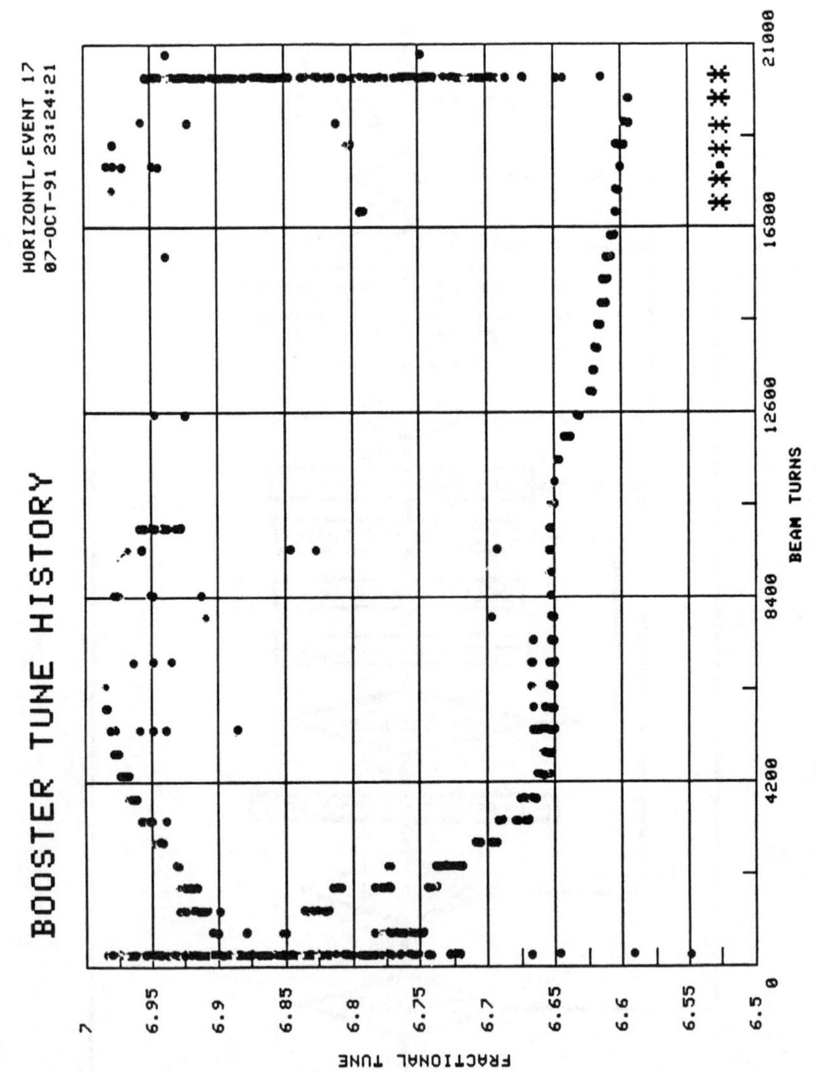

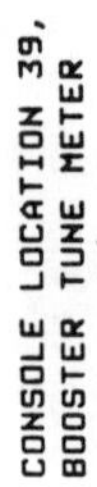

Fig. 3a *On-line display of the fractional tune in horizontal plane for about 50 beam cycles at the same beam conditions.*

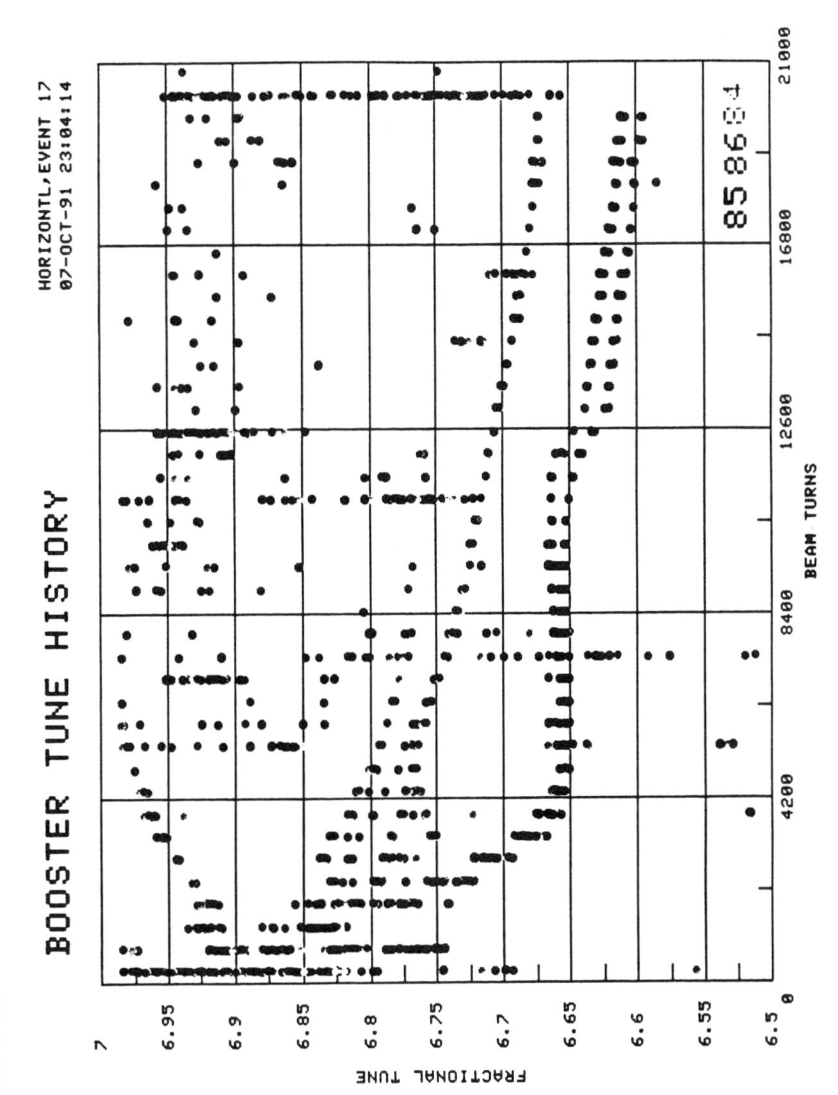

Fig. 3b *On-line display of the fractional tune in horizontal plane as the beam intensity changed.*

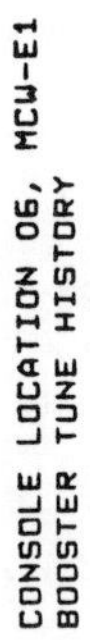

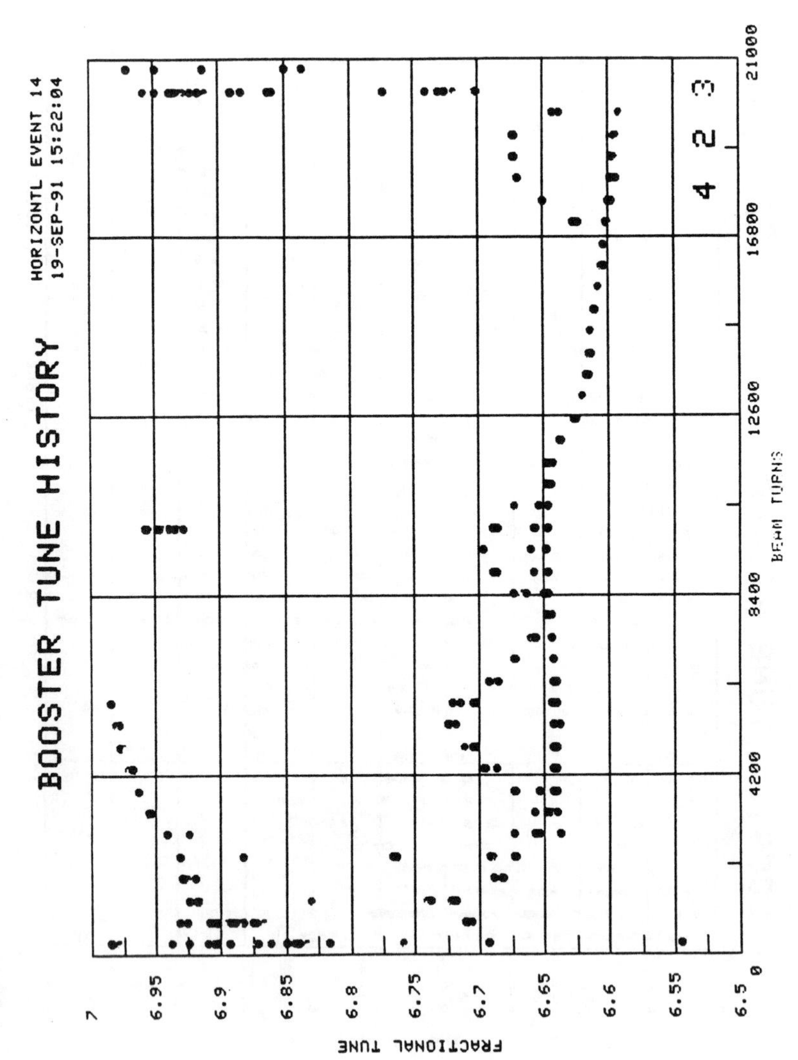

Fig. 3c *On-line display of the fractional tune responding to the changes in ramping current in the correction quadrupoles.*

List of Participants

Phil Adderley
CEBAF, MS 34A
12000 Jefferson Avenue
Newport News, VA 23606
804-249-7200 ADDERLEY@CEBAF

Advanced Technology Laboratories
1111 Street Road
Southampton, PA 18966
215-355-8111

Roberto Aiello
SSC Laboratory, MS 1046
2550 Beckleymeade Avenue
Dallas, TX 75237
214-708-3295 SSCVX::AIELLO

Walter Barry
Lawrence Berkeley Laboratory
1 Cyclotron Road
Berkeley, CA 94720

Edward Barsotti
Fermilab, MS 222
P.O. Box 500
Batavia, IL 60510
708-840-8104
EBARSOTTI@ADCALC.FNAL.GOV

David Beechy
SSC Laboratory, MS 1046
2550 Beckleymeade Avenue
Dallas, TX 75237
214-708-3450

Julien Bergoz
Bergoz, Inc.
01170 GEX, Crozet
France
33 50 41 00 89

Willem Blokland
Fermilab, MS 308
P.O. Box 500
Batavia, IL 60510
708-840-2681
BLOKLAND@ADCALC.FNAL.GOV

Paul Boccard
Old Dominion University
2631 Roadside Lane
Chesapeake, VA 23325
804-543-9176

Bruce Bowling
CEBAF, MS 85A
12000 Jefferson Avenue
Newport News, VA 23606
804-249-7240 BOWLING@CEBAF

David Brown
Los Alamos National Laboratory
MS H838
Los Alamos, NM 87545
505-667-3277

Ernest Buchanan
Fermilab, MS 341
P.O. Box 500
Batavia, IL 60510
708-840-4699

Karel Capek
CEBAF, MS 34A
12000 Jefferson Avenue
Newport News, VA 23606
804-249-7197 CAPEK@CEBAF

Kevin Cassidy
Tektronix
17052 Jamboree Blvd.
Irvine, CA 92714-5897
714-660-8080, Ext. 306

Vincent Castillo
Brookhaven National Laboratory
Bldg. 911B
Upton, NY 11973
516-282-3772

Ceramaseal
P.O. Box 260
New Lebanon, NY 12125
518-794-7800

Yu-Chiu Chao
SLAC, Bin 12
P.O. Box 4349
Stanford, CA 94309
CHAO@SLACVM

Young Choi
SSC Laboratory, MS 4005
2550 Beckleymeade Avenue
Dallas, TX 75237
214-708-3450

Youngjoo Chung
Argonne National Laboratory
Bldg. 360, C229
9700 S. Cass Avenue
Argonne, IL 60439
708-927-4601

Commonwealth Controls
7630 Whitepine Road
Richmond, VA 23237
804-271-7700

Jim Crisp
Fermilab, MS 308
P.O. Box 500
Batavia, IL 60510
708-840-4460

Mark Curtin
Rockwell International
MS FA38
6633 Canoga Avenue
Canogo Park, CA 91303
818-700-4880

Phillip Datte
SSC Laboratory, MS 1046
2550 Beckleymeade Avenue
Dallas, TX 75237
214-708-3450

Glenn Decker
Argonne National Laboratory
Bldg. 362
9700 S. Cass Avenue
Argonne, IL 60439
708-972-6635

Dong-Ping Deng
Brookhaven National Laboratory
Bldg. 911
Upton, NY 11973
516-282-2197 DENG@BNLGOV

Curt Dunnam
Cornell University
Wilson Laboratory
Ithaca, NY 14853
607-255-5749

Shumel Eylon
Lawrence Berkeley Laboratory
Bldg. 47/112
1 Cyclotron Road
Berkeley, CA 94720

Erich Feldl
CEBAF, MS 12A
12000 Jefferson Avenue
Newport News, VA 23606
804-249-7157 FELDL@CEBAF

Hsi Feng
CEBAF, MS 15A
12000 Jefferson Avenue
Newport News, VA 23606
804-249-7312 FENG@CEBAF

Kenneth Fertner
Advanced Technology Laboratories
1111 Street Road
Southampton, PA 18966
215-355-8111

Ralph Fiorito
Naval Surface Warfare Center
MS R42
Silver Springs, MD 20903
301-394-1908

Raymond Fuja
Argonne National Laboratory
Bldg. 371T
9700 S. Cass Avenue
Argonne, IL 60439
708-972-6442 FUJA@ANLAPS

Kenneth Fulett
Fermilab, MS 341
P. O. Box 500
Batavia, IL 60510
708-840-2507 FULLET@FNALAD

GMW Associates
P.O. Box 2578
Redwood City, CA 94064
415-368-4884

Floyd Gallegos
Los Alamos National Laboratory
MP-6, MS H812
Los Alamos, NM 87545
505-667-6848
GALLEGOS@LAMPF.LANL.GOV

David Gassner
Brookhaven National Laboratory
Upton, NY 11973

Hans Gerwers
Scientific International
P.O. Box 143
141 Snowden Lane
Princeton, NJ 08542
609-924-3011

William Ghiorso
Lawrence Berkeley Laboratory
Bldg. 47/112
1 Cyclotron Road
Berkeley, CA 94720
415-486-6372

Gianni Molinari
CERN
PS Division
1211 Geneva 23
Switzerland
VXCRN::MOLINA

Edward Gill
Brookhaven National Laboratory
Bldg. 911
Upton, NY 11973
516-282-7207

Doug Gilpatrick
Los Alamos National Laboratory
MS H808
Los Alamos, NM 87545
505-665-2904

Jim Griffin
Fermilab, MS 341
P.O. Box 500
Batavia, IL 60510
708-840-4828

Alan Hahn
Fermilab, MS 308
P.O. Box 500
Batavia, IL 60510
708-840-2987 AHAHN@FNAL

Thomas Hardek
Los Alamos National Laboratory
MS H838
P. O. Box 1663
Los Alamos, NM 87545
505-667-9132

Tom Hayes
Brookhaven National Laboratory
Bldg. 911A
Upton, NY 11973
516-282-7000

Jay Heefner
CEBAF, MS 15A
12000 Jefferson Avenue
Newport News, VA 23606
804-249-7271 HEEFNER@CEBAF

James Hinkson
Lawrence Berkeley Laboratory
MS 46-125
1 Cyclotron Road
Berkeley, CA 94720
415-486-4194

Tom Hofmeister
Ceramaseal
Box 260
New Lebanon, NY 12125
518-794-7800

James Hurd
SSC Laboratory
MS 1403 ADOD/LINAC
2550 Beckleymeade Avenue
Dallas, TX 75237
214-223-2789 HURD@SSCVX1

Keith Jobe
SLAC, Bin 66
2575 Sand Hill Road
Menlo Park, CA 94025
415-926-2084 RKJ@SLACSLC

Erik Johnson
Brookhaven National Laboratory
Bldg. 725C
Upton, NY 11973
516-282-4603

Ronald Johnson
SSC Laboratory, MS 1046
2550 Beckleymeade Avenue
Dallas, TX 75237
214-708-3450

Carol Johnstone
Fermilab, MS 341
P.O. Box 500
Batavia, IL 60510
708 840-3794 CJJ@FNAL

Jorway Corporation
27 Bond Street
Westbury, NY 11590
516-997-8120

Emanuel Kahana
Argonne National Laboratory
Bldg. 362
9700 S. Cass Avenue
Argonne, IL 60439
708-972-7383

Ivan Karabekov
Yerevan Physics Institute
Yerevan
Republic of Armenia

Quentin Kerns
Fermilab, MS 307
P.O. Box 500
Batavia, IL 60510
708-840-3964

Gene Klein
Advanced Technology Laboratories
1111 Street Road
Southampton, PA 18966
215-355-8111

Kevin Kleman
University of Wisconson
Synchrotron Radiation Center
3731 Schneider Drive
Stoughton, WI 53589
608-873-6651

Peter Kloeppel
CEBAF, MS 12A
12000 Jefferson Avenue
Newport News, VA 23606
804-249-7568 KLOEPPEL@CEBAF

Henry Lancaster
Lawrence Berkeley Laboratory
MS 46-125
1 Cyclotron Road
Berkeley, CA 94720
415-486-4261

LeCroy Corporation
110 West Madison
Oak Park, IL 60302
708-386-3628

Frank Lenkszus
Argonne National Laboratory
Bldg. 362
9700 S. Cass Avenue
Argonne, IL 60439
708-972-6972

John Mangino
SSC Laboratory, MS 1046
2550 Beckleymeade Avenue
Dallas, TX 75237

Donald Martin
SSC Laboratory, MS 1046
2550 Beckleymeade Avenue
Dallas, TX 75237
214-708-3066

Doug McCormick
SLAC
2575 Sand Hill Road
Menlo Park, CA 94075
415-926-2470

David McGinnis
Fermilab, MS 341
P.O. Box 500
Batavia, IL 60510
708-840-2789

C. B. McKee
Duke University
Department of Physics
Durham, NC 27706
919-681-8735
CBM@PHY.DUKE.EDU

L. Mestha
SSC Laboratory
2550 Beckleymeade Avenue
Dallas, TX 75237
214-708-3450

Mark Mills
SSC Laboratory, MS 1046
2550 Beckleymeade Avenue
Dallas, TX 75237
214-708-3450

Roman Nawrocky
Brookhaven National Laboratory
Bldg. 725C/NSLS
Upton, NY 11973
516-282-4449 NAWROCKY@BNL

David Neuffer
CEBAF, MS 12A
12000 Jefferson Avenue
Newport News, VA 23606
804-249-7613 NEUFFER@CEBAF

Stephen O'Day
Fermilab, MS 341
P.O. Box 500
Batavia, IL 60510
708-840-4827 ODAY@FNAL

Ralph Pasquinelli
Fermilab, MS 341
P. O. Box 500
Batavia, IL 60510
708-840-4724 PASQUIN@FNAL

Don Patterson
Argonne National Laboratory
Bldg. 362
9700 S. Cass Avenue
Argonne, IL 60439
708-972-6951

John Perry
CEBAF, MS 15A
12000 Jefferson Avenue
Newport News, VA 23606
804-249-7249 PERRY@CEBAF

David Peterson
Fermilab, MS 341
P.O. Box 500
Batavia, IL 60510
708-840-3073 PETERSON@FNAL

Mike Plum
Los Alamos National Laboratory
MP-5, MS H838
Los Alamos, NM 87545
505-667-7547 PLUM@LAMPF

William Rauch
Princeton Plasma Physics Laboratory
P.O. Box 451
Princeton, NJ 08543
609-243-2510

Kenneth Reece
Brookhaven National Laboratory
Bldg. 911B
Upton, NY 11973
516-282-4767

Luigi Rezzonico
Paul Scherrer Institute
CH-5232 Villigen PSI
Switzerland
41-56-99-33-77
REZZONICO@CAGEIR5A

Chris Rose
Los Alamos National Laboratory
MS H808
P.O. Box 1663
Los Alamos, NM 87505
505-665-0950

Marc Ross
SLAC, Bin 55
P.O. Box 4349
Stanford, CA 94309
415-726-3526 MCREC@SLACVM

Robert Rossmanith
CEBAF, MS 12A
12000 Jefferson Avenue
Newport News, VA 23606
804-249-7621
ROSSMANITH@CEBAF

Jeffrey Rothman
Brookhaven National Laboratory
NSLS/Bldg. 725C
Upton, NY 11973
516-282-4914

Joan Sage
SSC Laboratory, MS 1046
2550 Beckleymeade Avenue
Dallas, TX 75237

Bob Scala
Fermilab, MS 308
P.O. Box 500
Batavia, IL 60510
708-840-4110

Hermann Schmickler
CERN
Division SL-OP
CH 1211, Geneva 23
Switzerland
72-77-75-464
SCHMICKLER@VXCRNA

Rudiger Schmidt
CERN
CH1211, Geneva 23
Switzerland

Gerald Schobert
Princeton Plasma Physics Laboratory
P.O. Box 451
Princeton, NJ 08543
609-243-2815

Jim Sebek
Stanford Synchrotron Radiation Lab
Stanford, CA 94309
415-926-3164

William Sellyey
Argonne National Laboratory
Bldg. 362
9700 S. Cass Avenue
Argonne, IL 60439
708-972-2857

Scientific International
P.O. Box 143
Princeton, NJ 08542
609-924-3018

Robert Shafer
Los Alamos National Laboratory
MS H808
P.O. Box 1663
Los Alamos, NM 87545
505-667-5877

Tom Shea
Brookhaven National Laboratory
ADD/Bldg. 1005
Upton, NY 11973
516-282-2435
SHEA@BNLDAG.GOV

Robert Siemann
SLAC, Bin 26
P.O. Box 4349
Stanford, CA 94309
415-926-3892 SIEMANN@SLACVM

Gary Smith
Brookhaven National Laboratory
AGS/Bldg. 911C
Upton, NY 11973
516-282-3473

Todd Smith
Stanford University
High Energy Physics Laboratory
Stanford, CA 94305
415-723-1906
TISMITH@LELAND.STANFORD.
EDU

Walter Stark
Princeton Plasma Physics Laboratory
P.O. Box 451
Princeton, NJ 08543
609-243-2409

James Stiemel
Fermilab
P.O. Box 500
Batavia, IL 60510

Arnold Stillman
Brookhaven National Laboratory
AGS/Bldg. 911B
Upton, NY 11973
516-282-4944
STILLMAN@BNLDAG

Gregory Stover
Lawrence Berkeley Laboratory
MS 46-125
1 Cyclotron Road
Berkeley, CA 94720
415-486-6741

Tom Tallerico
Brookhaven National Laboratory
Bldg. 911B
Upton, NY 11973
516-282-4642

Tektronix
525 Butler Farm Road
Hampton, VA 23666
804-865-1588

William Van Asselt
Brookhaven National Laboratory
AGS/Bldg. 911B
Upton, NY 11973
516-282-7778

Olin Van Dyck
Los Alamos National Laboratory
MS H844
Los Alamos, NM 87545
505-667-7323 VANDYCK@LAMPF

Agoritsas Vassilis
CERN
PS Division
CH1211, Geneva 23
Switzerland
022--767-2590 AGORA@CERNVM

Lucien Vos
CERN
SL Division
CH1211, Geneva 23
Switzerland

Ian Walker
GMW Associates
P.O. Box 2578
Redwood City, CA 94064
415-368-4884

Xucheng Wang
Argonne National Laboratory
Bldg. 362
9700 S. Cass Avenue
Argonne, IL 60439
708-972-7511

Robert Webber
SSC Laboratory
2550 Beckleymeade Avenue
Dallas, TX 75237
214-708-3450

Frank D. Wells
Los Alamos National Laboratory
AT-3, MS H808
Los Alamos, NM 87545
505-665-0956

Neville Williams
Brookhaven National Laboratory
Upton, NY 11973

Richard Witkover
Brookhaven National Laboratory
AGS/Bldg. 911B
Upton, NY 11973
516-272-4607
WITKOVER@BNLDAG

Guan Hong Wu
Fermilab, MS 341
P.O. Box 500
Batavia, IL 60510
708-840-2026

Ying Wu
Duke University
FEL Group, Department of Physics
Durham, NC 27706
919-681-8735
WU@PHY.DUKE.EDU

Jim Zagel
Fermilab, MS 308
P.O. Box 500
Batavia, IL 60510
708-840-4076

Abbi Zolfaghari
MIT
P.O. Box 846
Middleton, MA 01949
617-245-6600 Abbi@MITBATES

Author Index

AIP Conference Proceedings

		L.C. Number	ISBN
No. 190	Radio-frequency Power in Plasmas (Irvine, CA, 1989)	89-45805	0-88318-397-8
No. 191	Advances in Laser Science–IV (Atlanta, GA, 1988)	89-85595	0-88318-391-9
No. 192	Vacuum Mechatronics (First International Workshop) (Santa Barbara, CA, 1989)	89-45905	0-88318-394-3
No. 193	Advanced Accelerator Concepts (Lake Arrowhead, CA, 1989)	89-45914	0-88318-393-5
No. 194	Quantum Fluids and Solids—1989 (Gainesville, FL, 1989)	89-81079	0-88318-395-1
No. 195	Dense *Z*-Pinches (Laguna Beach, CA, 1989)	89-46212	0-88318-396-X
No. 196	Heavy Quark Physics (Ithaca, NY, 1989)	89-81583	0-88318-644-6
No. 197	Drops and Bubbles (Monterey, CA, 1988)	89-46360	0-88318-392-7
No. 198	Astrophysics in Antarctica (Newark, DE, 1989)	89-46421	0-88318-398-6
No. 199	Surface Conditioning of Vacuum Systems (Los Angeles, CA, 1989)	89-82542	0-88318-756-6
No. 200	High T_c Superconducting Thin Films: Processing, Characterization, and Applications (Boston, MA, 1989)	90-80006	0-88318-759-0
No. 201	QED Stucture Functions (Ann Arbor, MI, 1989)	90-80229	0-88318-671-3
No. 202	NASA Workshop on Physics From a Lunar Base (Stanford, CA, 1989)	90-55073	0-88318-646-2
No. 203	Particle Astrophysics: The NASA Cosmic Ray Program for the 1990s and Beyond (Greenbelt, MD, 1989)	90-55077	0-88318-763-9
No. 204	Aspects of Electron–Molecule Scattering and Photoionization (New Haven, CT, 1989)	90-55175	0-88318-764-7
No. 205	The Physics of Electronic and Atomic Collisions (XVI International Conference) (New York, NY, 1989)	90-53183	0-88318-390-0
No. 206	Atomic Processes in Plasmas (Gaithersburg, MD, 1989)	90-55265	0-88318-769-8
No. 207	Astrophysics from the Moon (Annapolis, MD, 1990)	90-55582	0-88318-770-1
No. 208	Current Topics in Shock Waves (Bethlehem, PA, 1989)	90-55617	0-88318-776-0
No. 209	Computing for High Luminosity and High Intensity Facilities (Santa Fe, NM, 1990)	90-55634	0-88318-786-8
No. 210	Production and Neutralization of Negative Ions and Beams (Brookhaven, NY, 1990)	90-55316	0-88318-786-8
No. 211	High-Energy Astrophysics in the 21st Century (Taos, NM, 1989)	90-55644	0-88318-803-1

No. 212	Accelerator Instrumentation (Brookhaven, NY, 1989)	90-55838	0-88318-645-4
No. 213	Frontiers in Condensed Matter Theory (New York, NY, 1989)	90-6421	0-88318-771-X 0-88318-772-8 (pbk.)
No. 214	Beam Dynamics Issues of High-Luminosity Asymmetric Collider Rings (Berkeley, CA, 1990)	90-55857	0-88318-767-1
No. 215	X-Ray and Inner-Shell Processes (Knoxville, TN, 1990)	90-84700	0-88318-790-6
No. 216	Spectral Line Shapes, Vol. 6 (Austin, TX, 1990)	90-06278	0-88318-791-4
No. 217	Space Nuclear Power Systems (Albuquerque, NM, 1991)	90-56220	0-88318-838-4
No. 218	Positron Beams for Solids and Surfaces (London, Canada, 1990)	90-56407	0-88318-842-2
No. 219	Superconductivity and Its Applications (Buffalo, NY, 1990)	91-55020	0-88318-835-X
No. 220	High Energy Gamma-Ray Astronomy (Ann Arbor, MI, 1990)	91-70876	0-88318-812-0
No. 221	Particle Production Near Threshold (Nashville, IN, 1990)	91-55134	0-88318-829-5
No. 222	After the First Three Minutes (College Park, MD, 1990)	91-55214	0-88318-828-7
No. 223	Polarized Collider Workshop (University Park, PA, 1990)	91-71303	0-88318-826-0
No. 224	LAMPF Workshop on (π, K) Physics (Los Alamos, NM, 1990)	91-71304	0-88318-825-2
No. 225	Half Collision Resonance Phenomena in Molecules (Caracus, Venezuela, 1990)	91-55210	0-88318-840-6
No. 226	The Living Cell in Four Dimensions (Gif sur Yvette, France, 1990)	91-55209	0-88318-794-9
No. 227	Advanced Processing and Characterization Technologies (Clearwater, FL, 1991)	91-55194	0-88318-910-0
No. 228	Anomalous Nuclear Effects in Deuterium/Solid Systems (Provo, UT, 1990)	91-55245	0-88318-833-3
No. 229	Accelerator Instrumentation (Batavia, IL, 1990)	91-55347	0-88318-832-1
No. 230	Nonlinear Dynamics and Particle Acceleration (Tsukuba, Japan, 1990)	91-55348	0-88318-824-4
No. 231	Boron-Rich Solids (Albuquerque, NM, 1990)	91-53024	0-88318-793-4
No. 232	Gamma-Ray Line Astrophysics (Paris–Saclay, France, 1990)	91-55492	0-88318-875-9
No. 233	Atomic Physics 12 (Ann Arbor, MI, 1990)	91-55595	0-88318-811-2
No. 234	Amorphous Silicon Materials and Solar Cells (Denver, CO, 1991)	91-55575	0-88318-831-7

No. 235	Physics and Chemistry of MCT and Novel IR Detector Materials (San Francisco, CA, 1990)	91-55493	0-88318-931-3
No. 236	Vacuum Design of Synchrotron Light Sources (Argonne, IL, 1990)	91-55527	0-88318-873-2
No. 237	Kent M. Terwilliger Memorial Symposium (Ann Arbor, MI, 1989)	91-55576	0-88318-788-4
No. 238	Capture Gamma-Ray Spectroscopy (Pacific Grove, CA, 1990)	91-57923	0-88318-830-9
No. 239	Advances in Biomolecular Simulations (Obernai, France, 1991)	91-58106	0-88318-940-2
No. 240	Joint Soviet-American Workshop on the Physics of Semiconductor Lasers (Leningrad, USSR, 1991)	91-58537	0-88318-936-4
No. 241	Scanned Probe Microscopy (Santa Barbara, CA, 1991)	91-76758	0-88318-816-3
No. 242	Strong, Weak, and Electromagnetic Interactions in Nuclei, Atoms, and Astrophysics: A Workshop in Honor of Stewart D. Bloom's Retirement (Livermore, CA, 1991)	91-76876	0-88318-943-7
No. 243	Intersections Between Particle and Nuclear Physics (Tucson, AZ, 1991)	91-77580	0-88318-950-X
No. 244	Radio Frequency Power in Plasmas (Charleston, SC, 1991)	91-77853	0-88318-937-2
No. 245	Basic Space Science (Bangalore, India, 1991)	91-78379	0-88318-951-8
No. 246	Space Nuclear Power Systems (Albuquerque, NM, 1992)	91-58793	1-56396-027-3 1-56396-026-5 (pbk.)
No. 247	Global Warming: Physics and Facts (Washington, DC, 1991)	91-78423	0-88318-932-1
No. 248	Computer-Aided Statistical Physics (Taipei, Taiwan, 1991)	91-78378	0-88318-942-9
No. 249	The Physics of Particle Accelerators (Upton, NY, 1989, 1990)	XX-XXXXX	0-88318-789-2
No. 250	Towards a Unified Picture of Nuclear Dynamics (Nikko, Japan, 1991)	92-70143	0-88318-951-8
No. 251	Superconductivity and its Applications (Buffalo, NY, 1991)	92-52726	1-56396-016-8
No. 252	Accelerator Instrumentation (Newport News, VA, 1991)	92-70356	0-88318-934-8
No. 253	High-Brightness Beams for Advanced Accelerator Applications (College Park, MD, 1991)	92-52705	0-88318-947-X
No. 254	Testing the AGN Paradigm (College Park, MD, 1991)	92-52780	1-56396-009-5
No. 255	Advanced Beam Dynamics Workshop on Effects of Errors in Accelerators, Their Diagnosis and Corrections (Corpus Christi, TX, 1991)	XX-XXXXX	1-56396-006-0